Internet of Things

Technology, Communications and Computing

Series Editors

Giancarlo Fortino, Rende (CS), Italy

Antonio Liotta, Edinburgh Napier University, School of Computing, Edinburgh, UK

The series Internet of Things - Technologies, Communications and Computing publishes new developments and advances in the various areas of the different facets of the Internet of Things.

The intent is to cover technology (smart devices, wireless sensors, systems), communications (networks and protocols) and computing (theory, middleware and applications) of the Internet of Things, as embedded in the fields of engineering, computer science, life sciences, as well as the methodologies behind them. The series contains monographs, lecture notes and edited volumes in the Internet of Things research and development area, spanning the areas of wireless sensor networks, autonomic networking, network protocol, agent-based computing, artificial intelligence, self organizing systems, multi-sensor data fusion, smart objects, and hybrid intelligent systems.

** Indexing: *Internet of Things* is covered by Scopus and Ei-Compendex **

More information about this series at http://www.springer.com/series/11636

George Mastorakis ·
Constandinos X. Mavromoustakis ·
Jordi Mongay Batalla · Evangelos Pallis
Editors

Convergence of Artificial Intelligence and the Internet of Things

 Springer

Editors
George Mastorakis
Department of Management
Science and Technology
Hellenic Mediterranean University
Agios Nikolaos, Crete, Greece

Jordi Mongay Batalla
Department of Telecommunications
Warsaw University of Technology
Warsaw, Poland

Constandinos X. Mavromoustakis
Mobile Systems Laboratory (MoSys Lab)
Department of Computer Science
Research Centres (UNRF) – University
of Nicosia
Nicosia, Cyprus

Evangelos Pallis
Department of Electrical and Computer
Engineering
Hellenic Mediterranean University
Heraklion, Crete, Greece

ISSN 2199-1073   ISSN 2199-1081   (electronic)
Internet of Things
ISBN 978-3-030-44909-4   ISBN 978-3-030-44907-0   (eBook)
https://doi.org/10.1007/978-3-030-44907-0

This Springer imprint is published by the registered company Springer Nature Switzerland AG
The registered company address is: Gewerbestrasse 11, 6330 Cham, Switzerland

Introduction

The Internet of Things (IoT) paradigm has been introduced to characterize smart objects, supporting data exhange via the existing Web infrastructure, as well as via 5G mobile networking architectures. Artificial Intelligence (AI) plays a crucial role in such emerging networking environments, enabling for smarter services, applications, business processes, and social interaction among the IoT objects. In this respect, this book presents research approaches in converged IoT and Artificial Intelligence environments and emerging networking infrastructures. Major subjects of this book cover the analysis and the development of AI-powered mechanisms in future IoT applications and architectures.

Cloud-based IoT mechanisms [1] are utilized to enable the interconnection of a variety of smart objects, such as cars, cell phones, and sensors inside smart cities, smart businesses, and smart home environments [2]. IoT-based objects produce a vast volume of information frequently called as Big Data [3], which cannot be easily processed by traditional algorithms and schemes. However, future smart environments will enable efficient handling and processing of Big Data that is generated by the associated objects for an effective communication and co-operation among them. In any case, there is profoundly helpful data and many values in trying to analyze, as well as process it. IoT-based objects have an increased consideration from researchers in several scientific fields and disciplines. In this respect, a great potential exists for advanced technological solutions combined with Artificial Intelligence (AI) and IoT-enabled capabilities [4]. AI mechanisms applied in IoT environments can be exploited in smart homes to analyze human actions via motion or facial recognition sensors. AI plays a significant role in future IoT applications and infrastructures, by providing insights from collected data. This capability enables the identification of patterns and allows operational predictions with higher accuracy in small time periods. AI applications for smart devices also enable businesses to develop new products, reduce possible risks and increase efficiency during the production time, by predicting failures that are usually non-detectable by humans or simple devices [5].

The future of the implementation of AI-powered IoT infrastructures depends on the effective solutions to a number of technical challenges that such paradigms introduce [6]. These challenges include intelligent sensor capabilities improvement, smart Big Data analytics [7], automated remote data management, as well as open and secure composition of processes, which may be implemented into emerging AI-enabled IoT scenarios [8]. Some initiatives try to incorporate Artificial Intelligence schemes in IoT environments, but new frameworks have to be defined. In fact, the approach of IoT is to find the potential benefits of AI, in order to build extensive ecosystems, for increasing the number of automated services and their value. In this respect, this book addresses the major new technological developments in the field and reflects current research trends, as well as industry needs. It comprises a good balance between theoretical and practical issues, covering case studies, experience, and evaluation approaches as well as best practices, in utilizing AI applications in IoT networks. It also provides technical/scientific information about various aspects of AI technologies, ranging from basic concepts to research grade material, including future directions.

Research Solutions

Chapter "Fog Computing: Data Analytics for Time-Sensitive Applications" of this book elaborates on fog computing issues to intelligently reduce communications delays between users and cloud systems. The idea of fog computing allows users to interact with intermediate servers, while reaping the benefits of reliability and elasticity, which are inherent in cloud computing. Fog computing can leverage Internet of Things by providing a reliable service layer for time-sensitive applications and real-time analytics [9]. While the concept of fog computing is still evolving, it is pertinent to study the domain of fog computing and analyze its strengths and weaknesses [10]. Motivated by this need, chapter "Fog Computing: Data Analytics for Time-Sensitive Applications" describes the architecture of fog computing and explains its efficacy with respect to different smart applications. It highlights some of the key challenges associated with the evolving platforms along with future directions of research on this field.

Chapter "Medical Image Watermarking in Four Levels Decomposition of DWT Using Multiple Wavelets in IoT Emergence" of this book presents a process to exploit the characteristics that are offered by intelligent IoT systems in the healthcare sectoral issues [11]. This allows for more efficiency in the processing of a medical image. Medical images are exposed to the major risks through frequent attacks which may lead up to missing information by the physician in the diagnosis of the disease. Privacy remains the major challenge in the present-day world in IoT platforms for medical systems. In this regard, this chapter labors on a medical image based on digital watermarking that can be utilized to protect health sign. The proposed method performance is evaluated by utilizing MSE, PSNR, SSIM, and

NC, which are necessary to get the best result for performance metrics. This work is achieved in four levels Discrete Wavelet Transform (DWT). The proposed technique is highly robust against numerous sorts of attacks. The results refer that the proposed algorithm permits prevention at a higher level compared with other current structures and algorithms.

Chapter "Optimised Statistical Model Updates in Distributed Intelligence Environments" explores a sequential decision-making methodology of when to update statistical learning models in Intelligent Edge Computing devices given underlying changes in the contextual data distribution [12]. The proposed model update scheduling takes into consideration the optimal decision time for minimizing the network overhead while preserving the prediction accuracy of the models. This chapter reports on a comparison between the proposed approach with four other update delaying policies found in the literature, an evaluation of the performances using linear and support vector regression models over real contextual data streams and a discussion on the strengths and weaknesses of the proposed policy.

Chapter "Intelligent Vehicular Networking Protocols" introduces intelligent vehicular ad hoc networks, where nodes are either vehicles or fixed roadside units considered as VANET infrastructure [13]. This chapter discusses the five categories of VANET routing protocols with some of the most known examples in each. Each of the mentioned protocols has a brief overview of its design and implementation. Besides, authors mention some of the protocols' benefits, drawbacks, and enhancements. With the emergence of Internet of Things in the last decade, vehicles have a subcategory known as Internet of Vehicles (IoV) due to the special characteristics of vehicles and road topology. Therefore, the second section of this chapter discusses the classification of routing protocols in IoV which is the promising future in the vehicular network world.

Chapter "Towards Ubiquitous Privacy Decision Support: Machine Prediction of Privacy Decisions in IoT" presents a mechanism to predict privacy decisions of users in Internet of Things (IoT) environments, through data mining and machine learning techniques [14]. To construct predictive models, authors test several different machine learning models, combinations of features, and model training strategies on human behavioral data collected from an experience-sampling study. Experimental results show that a machine learning model called linear model and deep neural networks (LMDNN) outperforms conventional methods for predicting users' privacy decisions for various IoT services. Authors also present a feature vector, composed of both contextual parameters and privacy segment information with the best predictive performance. In addition, they propose a novel approach, which provides a common predictive model to a segment of users who share a similar notion of privacy. From a user perspective, authors' prediction mechanism takes contextual factors embedded in IoT services into account and only utilizes a small amount of information polled from the users. It is therefore less burdensome and privacy-invasive than the other mechanisms.

Chapter "Energy-Efficient Design of Data Center Spaces in the Era of IoT Exploiting the Concept of Digital Twins" represents the power management of server units and server room spaces to ease the installation of a Data Center, inferring cost reduction, and environmental protection for its operation. This type of research focuses either on the restriction of power consumption of the devices of a data center, or on the minimization of energy exchange between the data center space as a whole and its environment. In particular, authors attempt to estimate the effect that can infer on the energy performance of a data room, various interventions on the structural envelop of the building that hosts the data center. The performance of the data center is evaluated in this work by measuring its PUE index. Authors' modeling methodology includes the selection of a proper simulation tool for the creation of the digital twin, that is, of a digital clone, of a real data center, and then building its thermal model via energy measurements. In this work, authors exploit the potential of EnergyPlus software for modeling and simulating the behavior of a data center space. The energy measurements are conducted for a long period, using a low cost, smart IoT infrastructure, in the data center space of an educational building in Heraklion, Greece.

Chapter "In-Network Machine Learning Predictive Analytics: A Swarm Intelligence Approach" addresses the problem of collaborative predictive modeling via in-network processing of contextual information captured in Internet of Things environments. In-network predictive modeling allows the computing and sensing devices to disseminate only their local predictive Machine Learning (ML) models instead of their local contextual data [15]. The data center, which can be an Edge Gateway or the Cloud, aggregates these local ML predictive models to predict future outcomes. Given that communication between devices in IoT environments and a centralized data center is energy consuming and communication bandwidth demanding, the local ML predictive models in the proposed in-network processing are trained using Swarm Intelligence for disseminating only their parameters within the network. Authors further investigate whether dissemination overhead of local ML predictive models can be reduced by sending only relevant ML models to the data center. This is achieved since each IoT node adopts the Particle Swarm Optimisation algorithm to locally train ML models and then collaboratively with their network neighbors one representative IoT node fuses the local ML models. Authors provide comprehensive experiments over Random and Small World network models using linear and non-linear regression ML models to demonstrate the impact on the predictive accuracy and the benefit of communication-aware in-network predictive modeling in IoT environments.

Machine learning is a rapidly evolving paradigm that has the potential to bring intelligence and automation to low-powered devices. In parallel, there have been considerable improvements made in the ambient backscatter communications due to high research interest over the last few years [16]. The combination of these two technologies is inevitable which would eventually pave the way for intelligent Internet of Things (IoT) in 6G wireless networks. There are several use cases for machine learning-enabled ambient backscatter networks that range from healthcare network, industrial automation, and smart farming. Besides this, it would also be helpful in enabling services like ultra-Reliable and Low-Latency Communications (uRLLC), massive Machine-Type Communications (mMTC), and enhanced Mobile Broadband (eMBB). Also, machine learning techniques can help backscatter communications to overcome its limiting factors. The information-driven machine learning does not require the need of a tractable scientific model as the models can be prepared to deal with channel imperfections. On-going examinations have likewise demonstrated that machine learning methodologies can be used to protect backscatter devices to enhance their ability to handle security and privacy vulnerabilities. These previously mentioned propositions alongside the ease-of-use of machine learning techniques inspire the authors of chapter "Machine Learning Techniques for Wireless-Powered Ambient Backscatter Communications: Enabling Intelligent IoT Networks in 6G Era" to investigate the feasibility of machine learning-based methodologies for backscatter communications. To do such, they start the chapter by talking about the basics and different favors of machine learning. This includes supervised learning, unsupervised learning, and reinforcement learning.

Deep learning models are taking place at many artificial intelligence tasks [17]. These models are achieving better results but need more computing power and memory. Therefore, training and inference of deep learning models are made at cloud centers with high-performance platforms. In many applications, it is more beneficial or required to have the inference at the edge near the source of data or action requests avoiding the need to transmit the data to a cloud service and wait for the answer. In many scenarios, transmission of data to the cloud is not reliable or even impossible, or has a high latency with uncertainty about the round-trip delay of the communication, which is not acceptable for applications sensitive to latency with real-time decisions. Other factors like security and privacy of data force the data to stay in the edge. With all these disadvantages, inference is migrating partial or totally to the edge. The problem is that deep learning models are quite hungry in terms of computation, memory, and energy which are not available in today's edge computing devices. Therefore, artificial intelligence devices are being deployed by different companies with different markets in mind targeting edge computing. Chapter "Processing Systems for Deep Learning Inference on Edge Devices" describes the actual state of algorithms and models for deep learning and analyzes the state of the art of computing devices and platforms to deploy deep learning on edge. Authors describe the existing computing devices for deep learning and analyze them in terms of different metrics, like performance, power, and flexibility. Different technologies are to be considered including GPU, CPU, FPGA, and ASIC. Authors explain the trends in computing devices for deep learning on edge, what has been researched, and what should we expect from future devices.

As one of the promising radio access techniques in two-way relaying network, authors of chapter "Power Domain Based Multiple Access for IoT Deployment: Two-Way Transmission Mode and Performance Analysis" consider Power Domain-based Multiple Access (PDMA) in an effective way to deploy in 5G communication networks. In this chapter, a cooperative PDMA two-way scheme with two-hop transmission is proposed to enhance the outage performance under consideration on how exact Successive Interference Cancellation (SIC) performs at each receiver. In the proposed scheme, performance gap of two NOMA users is examined, and power allocation factors are main impairment in such performance evaluation. In order to reveal the benefits of the proposed scheme, authors choose full-duplex at relay to improve bandwidth efficiency. As important result, its achieved outage probability is mathematical analyzed with imperfect SIC taken into account. Authors' examination shows that the proposed scheme can significantly outperform existing schemes in terms of achieving an acceptable outage probability. Then, authors go on to describe some of the potential uses of machine learning in ambient backscatter communications. They also provide a detailed analysis of reinforcement learning for wireless-powered ambient backscatter devices and give some insightful results along with relevant discussion. In the end, they present some concluding remarks and highlight some future research directions.

Using statistical learning theory and machine learning techniques surrounding the principles of Rival Penalised Competitive Learning (RPCL), Chapter "Big Data Thinning: Knowledge Discovery from Relevant Data" proposes a novel approach aiming to aid Big Data Thinning, i.e., analyzing the potential data sub-spaces and not the entire extensive data space. Data scientists, data analysts, IoT applications, and Edge-centric services are in need for predictive modeling and analytics [18]. This is achieved by learning from past issued analytics queries and exploiting the analytics query access patterns over the large distributed data-sets revealing the most interested and important sub-spaces for further exploratory analysis. By analyzing user queries and respectively mapping them into relatively small-scale predictive local regression models, we can yield higher predictive accuracy. This is done by thinning the data space and freeing it of irrelevant and non-popular data sub-spaces; thus, making use of less training data instances. Experimental results and statistical analysis support the research idea proposed in this work.

Blockchain is one of the new tools which is still experiencing a serious lack of understanding and awareness in general public. Besides, to say that blockchain is a versatile technology with many applications would be an understatement. The several features of blockchain technology have led to an immense amount of research interest in the blockchain systems, especially from the perspective of enabling the Internet of Things (IoT) [19]. It is expected that its adoption will be gradual, starting with researchers and leading up to startups and companies who will discover its potential for change. Then the general public who will demand changes would use it as an everyday technology. Finally, organizations that had until then resisted change would adopt it at a large scale, thus, paving the way for globalization of blockchain. In this regard, chapter "Optimizing Blockchain Networks with Artificial Intelligence: Towards Efficient and Reliable IoT Applications" elaborates on studying that the blockchain is not a single object, a single trend, or a particular feature, but a composition and collection of several autonomous as well as manually operated entities.

Internet of Things is a concept that proposes the inclusion of physical devices as a new form of communication, connecting them with various information systems. Nowadays, IoT cannot be reduced to smart homes as due to the recent technological advances, this concept has evolved from small to large-scale environments. There was also a need to adopt IoT in several business sectors, such as manufacturing, logistics, or transportation in order to converge information technologies and technological operations. Due to this convergence, it was possible to arrive at a new IoT paradigm, called the Industrial Internet of Things (IIOT). However, in IIoT, it is also necessary to analyze and interact with a real system through a virtual production, which refers to the use of Augmented Reality (AR) as a method to achieve this interaction [20]. AR is the overlay of digital content in the real world, and even play an important role in the life cycle of a product, from its design to its support, thus allowing greater flexibility. The adoption of interconnected systems and the use of IoT have motivated the use of Artificial Intelligence because much of the data coming from several sources is unstructured. Several AI algorithms have been used for decades aiming at "making sense" of unstructured data and transforming it into relevant information. Therefore, converging IoT, AR, and AI makes systems become increasingly autonomous and problem-solving in many scenarios. Chapter "Industrial and Artificial Internet of Things with Augmented Reality" elaborates on such convergence issues.

It is challenging to detect burn severity due to the long-time period needed to capture the ecosystem characteristics. Whether there is a warning from the IoT devices, multitemporal remote sensing data is being received via satellites to contribute to multitemporal observations before, during and after a bushfire, and enhance the difference on detection accuracy. In chapter "IoT Detection Techniques for Modeling Post-Fire Landscape Alteration Using Multitemporal Spectral Indices", authors strive to design an infrastructure to fire detection, to perform a qualitative assessment of the condition as quickly as possible in order to avoid a major disaster. Studying the multitemporal spectral indicators such as Normalized Difference Vegetation Index (NDVI), Enhanced Vegetation Index (EVI), Normalized Burn Ratio (NBR), Soil-Adjusted Vegetation Index (SAVI), Normalized Difference Moisture (Water) Index (NDMI or NDWI), and Normalized Wildfire Ash Index (NWAI), authors draw reliable conclusions about the seriousness of an area. The scope of this chapter is to examine the correlation between multitemporal spectral indices and field-observed conditions and provide a practical method for immediate fire detection to assess its burn severity in advance.

Furthermore, a quantified mapping model is presented to illustrate the spatial distribution of fire severity across the burnt area. The study focuses on the recent bushfire that took place in Mati Athens in Greece on 24th of July 2018 which had as a result not only material and environmental disaster but also was responsible for the loss of human lives.

Hardware/Software (hw/sw) systems changed the human way of living. Internet of Things and Artificial Intelligence are intended and expected to change it more [21]. Chapter "Internet of Things and Artificial Intelligence—A Wining Partnership?" elaborates on relevant challenges associated with the development of a "society" of intelligent smart objects. Humans and smart objects are expected to interact. Such issues are addressed in this chapter from the engineering and educational points of view. In fact, when dealing with "decision-making systems", not only design and test should guarantee the correct and safe operation, but also the soundness of the decisions such smart objects take, during their lifetime. The concept of Design for Accountability (DfA) is, thus, proposed and some initial guidelines are outlined.

Chapter "AI Architectures for Very Smart Sensors" describes modern neural network designs and discusses their advantages and disadvantages. The state-of-the-art neural networks are usually too much computationally difficult which limits their use in mobile and IoT applications. However, they can be modified with special design techniques which would make them suitable for mobile or IoT applications with limited computational power. These techniques for designing more efficient neural networks are described in great detail. Using them opens a way to create extremely efficient neural networks for mobile or even IoT applications. Such neural networks make the applications very intelligent which paves the way for very smart sensors [22].

Conclusion

This book is intended for researchers, practitioners in the field, engineers, and scientists involved in the design and development of protocols and AI applications for IoT-related devices. It can also be used as the recommended textbook for undergraduate or graduate courses. The intended audience includes college students, researchers, scientists, engineers, and technicians in the field of IoT technologies and issues related to the associated AI-enabled applications. As the book covers a wide range of applications and scenarios, where IoT technologies can be applied, the material covered is readable and a solid base for penetration into the comprehensive reference material on advanced communication concepts in the related research field. It is also a reference for selection by the audience with different but close to the field backgrounds. Finally, the book has a broad appeal to electrical, electronic, computer, software, and telecommunications engineers. It adopts an interdisciplinary approach and its form reflects both theoretical and practical approaches, in order to be targeted in multiple audiences.

References

1. Biswas, A.R., Giaffreda, R.: IoT and Cloud convergence: opportunities and challenges. IEEE World Forum on Internet of Things (WF-IoT). IEEE (2014)
2. Fortino, G., Trunfio, P. (eds.): Internet of Things Based on Smart Objects: Technology, Middleware and Applications. Springer Science & Business Media (2014)
3. Mohammadi, M., et al.: Deep learning for IoT big data and streaming analytics: a survey. IEEE Commun. Surv. Tutorials 20(4), 2923–2960 (2018)
4. Serrano, M., Nguyen Dang, H., Mau Quoc Nguyen, H.: Recent advances on artificial intelligence and internet of things convergence for human-centric applications: internet of things science. In: Proceedings of the 8th International Conference on the Internet of Things (2018)
5. Al-Turjman, F. (ed.): Artificial Intelligence in IoT. Springer (2019)

6. Fortino, G., et al.: Agent-oriented cooperative smart objects: from IoT system design to implementation. IEEE Trans. Syst. Cybern.: Syst. **48**(11), 1939–1956 (2017)
7. Kibria, M.G., et al.: Big data analytics, machine learning, and artificial intelligence in next-generation wireless networks. IEEE Access **6**, 32328–32338 (2018)
8. Da Xu, L., He, W., Li, S.: Internet of things in industries: a survey. IEEE Trans. Ind. Inform. **10**(4), 2233–2243 (2014)
9. Dastjerdi, A.V., Buyya, R.: Fog computing: helping the Internet of Things realize its potential. Computer **49**(8), 112–116 (2016)
10. Bellavista, P., Corradi, A., Zanni A.: Integrating mobile internet of things and cloud computing towards scalability: lessons learned from existing fog computing architectures and solutions. Int. J. Cloud Comput. **6**(4), 393–406 (2017)
11. Ravì, D., et al.: Deep learning for health informatics. IEEE J. Biomed. Health Inform. **21**(1), 4–21 (2016)
12. Liu, Y., et al.: Intelligent edge computing for IoT-based energy management in smart cities. IEEE Netw. **33**(2), 111–117 (2019)
13. Sakiz, F., Sen, S.: A survey of attacks and detection mechanisms on intelligent transportation systems: VANETs and IoV. Ad Hoc Netw. **61**, 33–50 (2017)
14. Jagannath, J., et al.: Machine learning for wireless communications in the Internet of Things: a comprehensive survey. Ad Hoc Netw. **93**, 101913 (2019)
15. Siryani, J., Tanju B., Eveleigh T.J.: A machine learning decision-support system improves the internet of things' smart meter operations. IEEE Int. Things J. **4**(4), 1056–1066 (2017)
16. Robb, E., et al.: Classifying SuperDARN backscatter using machine learning algorithms. AGU Fall Meeting Abstracts (2018)
17. He, L., Ota, K., Dong, M.: Learning IoT in edge: deep learning for the Internet of Things with edge computing. IEEE Netw. **32**(1), 96–101 (2018)
18. Mahdavinejad, M.S., et al.: Machine learning for Internet of Things data analysis: a survey. Digital Commun. Netw. **4**(3), 161–175 (2018)
19. Fernández-Caramés, T.M., Fraga-Lamas, P.: A review on the use of blockchain for the Internet of Things. IEEE Access **6**, 32979–33001 (2018)
20. Norouzi, N., et al.: A systematic review of the convergence of augmented reality, intelligent virtual agents, and the internet of things. Artif. Intell. IoT. Springer, Cham. 1–24 (2019)
21. Xiao, M., Guo, M.: Research on key technologies of Internet of Things and artificial intelligence. In Recent Trends in Intelligent Computing, Communication and Devices, pp. 531–537. Springer, Singapore (2020)
22. Islam, T., Mukhopadhyay S.C., Suryadevara N.K.: Smart sensors and internet of things: a postgraduate paper. IEEE Sens. J. **17**(3), 577–584 (2016)

Contents

Fog Computing: Data Analytics for Time-Sensitive Applications

Jawwad A. Shamsi, Muhammad Hanif, and Sherali Zeadally

Abstract Fog computing has been initiated to reduce communications delays between users and cloud systems. The idea of Fog computing allows users to interact with intermediate servers, while reaping the benefits of reliability and elasticity, which are inherent in cloud computing. Fog computing can leverage Internet of Things (IoT) by providing a reliable service layer for time-sensitive applications and real-time analytics. While the concept of fog computing is still evolving, it is pertinent to study the domain of fog computing and analyze its strengths and weaknesses. Motivated by this need, this chapter describes the architecture of fog computing and explain its efficacy with respect to different applications. The chapter highlights some of the key challenges associated with this evolving platform along with future directions of research.

1 Introduction

Cloud computing has been a standard platform for many distributed applications. The inherent availability of cloud systems integrated with their capability of elasticity has provided a de-facto platform for many applications which requires resiliency and load balancing. In cloud systems, there are several requirements that need to be resolved. First and foremost, as cloud systems are centralized, they are inappropriate for delay-sensitive applications. Such applications have strict delay requirements. But due to the distant location of a cloud server, computing a timely response becomes a challenge. Additionally, many cloud applications (such as mobile cloud

J. A. Shamsi (✉) · M. Hanif
National University of Computer and Emerging Sciences, Karachi, Pakistan
e-mail: jawwad.shamsi@nu.edu.pk

M. Hanif
e-mail: hanif.soomro@nu.edu.pk

S. Zeadally
University of Kentucky, Lexington, USA
e-mail: szeadally@uky.edu

© Springer Nature Switzerland AG 2020
G. Mastorakis et al. (eds.), *Convergence of Artificial Intelligence and the Internet of Things*, Internet of Things,
https://doi.org/10.1007/978-3-030-44907-0_1

1

computing and real-time data analytics) require location-awareness for their optimal operations. However, considering the centralized location of cloud systems, meeting this requirement remains a significant challenge.

Fog computing has emerged as an enhancement to cloud computing. Using fog services, the goals of delay and location-awareness can be fulfilled. In fog computing, time-sensitive services are provided through edge nodes, whereas delay-tolerant applications are provided through centralized cloud computing resources [32]. Considering the wide-scale usability of machine learning, emergence of smart city and Internet of Things (IoT), fog computing has been gaining a lot of attention. Fog computing can leverage IoTs by providing a platform for data analytics, data integration, and location-awareness. It can also provide a parallel platform for real-time computation utilizing machine learning and deep learning.

In this context, it is important to understand the architecture of fog computing and assess its strengths and weaknesses [19, 30]. This chapter explains architecture of fog computing, analyzes related challenges, and explores relevant opportunities particularly in the context of machine learning.

2   Fog Computing Applications

In the literature, four scenarios have been described where fog computing can be useful:

1. Time sensitive applications, where latency is critical.
2. Analytical applications where machine learning can be applied for data analytics.
3. Geographically dispersed applications where network connectivity is irregular.
4. Network constrained environment, where network bandwidth is critical.

Fog computing was introduced to provide services to IoT, where most end devices have limited processing, memory, and storage resources [12]. The scarcity of resources in the IoT infrastructure could be better managed by using fog computing. Some important scenarios could benefit from fog computing are discussed below:

1. **Health care** Health care service provisioning generates a large amount of local data which can be processed locally using fog computing and could minimize delays and energy consumption. Heart beat rate, pulse rate, and similar data of patients are collected through the use of sensors. The collected data is sent to fog nodes and servers for further analysis and processing by specialist doctors available in the vicinity. Machine learning algorithms can be applied to detect alerts in patient monitoring systems. In such a scenario, classification algorithms can either be trained on the cloud or they can be trained on fog nodes. However, testing for real-time alert generation is needed to be implemented on fog nodes for timely detection. Figure 1a illustrates one such scenario for a fog-based health care system. Whenever patients are discharged from the hospital, the patients data is sent to the cloud for permanent storage and further analysis as needed [6].

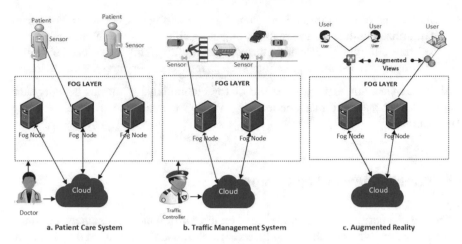

Fig. 1 Use of Fog computing for various applications

2. **Smart Traffic Light System** A traffic light system measures the speed and distance of each vehicle and sense the presence of ambulance or pedestrians crossing the road. Consequently, it can generate alert messages to nearby vehicle to slow down by traffic light signals. This scenario is shown in Figure 1b. For this real-time application, an instant action is needed. Utilizing cloud computing resources may delay the response, which may cause accidents. In this case, fog computing is more suitable as it minimizes latency and can efficiently analyze data in real-time. This smart system may include a few more data collection devices to collect the speed, volume, and weight of the vehicles and notify authorities regarding overweight or speed automatically with the vehicle identification number [8]. For such an application, video recognition such as number plate reading and detection of ambulance or pedestrian emergency response alert needs to be implemented at fog nodes [25]. Different steps of analytics such as feature extraction as well as model training and testing can be implemented at fog nodes for timely processing [34]. To enhance computational capacity, nodes at the fog layer can be interconnected with high speed links for cluster-based processing.

3. **Augmented Reality** It provides a view of a digitally generated environment which is similar to the real world. Data is collected and processed in real time to provide a visual, real-world environment, whose components are digitally enhanced with sound, video or graphical data. It requires lot of data for collection, processing, and transfer. One of the major characteristics of augmented reality applications is their real-time response, which makes fog computing suitable for providing an instantaneous response through its edge processing [39]. Consequently, the augmented environment could be efficiently managed by fog computing as shown in Fig. 1c. Fog computing supports real time response, by providing immediate resource availability to augmented reality applications [4]. Fog nodes can also enhance the augmented reality experience by creating an environment (including

images and objects) through machine learning and neural networks. Fog nodes can utilize intelligence to create an object-specific pixel-wise alpha map which can fuse different objects and images together [26]. This fusion can be applied to numerous real-time applications such as remote surgery and tele-medicine.

Above applications elaborate the use of fog computing for various benefits such as delay resistance and machine learning. We now describe the architecture of fog computing.

3   Architecture of Fog Computing

The fog computing architecture is based on three different layers namely, the smart layer, the fog layer, and the cloud layer. Figure 2 illustrates the layered architecture along with possibilities of configuration at the fog layer.

3.1   Smart Layer

This layer consists of smart devices which are used for sensing and measurements. These are generally small devices such as IoT, Rasberry Pi, or smart phones, with limited storage and networking capabilities. These devices may communicate with fog nodes either through wired or through a wireless communication infrastructure [6, 11]. Various communication protocols such as 6Lowpan, ZigBee, WiMAX are used for communication among end devices [16, 39].

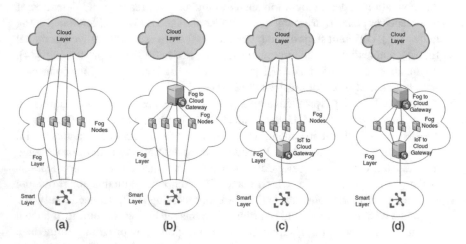

Fig. 2 Fog architecture

3.2 Fog Layer

The purpose of the fog layer is to facilitate computation, storage, and communication infrastructure by providing intermediary services between the cloud layer and the smart layer. A fog layer consists of fog nodes and gateways. Fog nodes aggregate data from various connected smart devices through a Fog instance. In this way, significant amount of data is processed without incurring considerable delays. A Fog instance is a virtualized instance, which helps fog nodes to manage end users requests for services on a dynamic basis. During computation, if more resources are required, then the virtualized instance may spawn itself into other instances. If the workload of a fog node increases, then the request can be distributed among multiple fog nodes. On the other end, fog gateways are lightweight clouds which forward the data received from the smart layer and distribute it among the fog nodes. Gateways can be useful in resource and task scheduling as well. Significant features such as location awareness, support for mobile users [23], data analytics, and low latency of fog computing are mainly dependent on gateways [6]. A gateway can also exist between fog nodes and the cloud layer in order to aggregate data and conserve bandwidth. It can also schedule offload data processing to cloud servers. Various services such as computation, evaluation, virtualization [24], storage, and protocol implementation can be supported on gateway.

Figure 2 illustrates four sample scenarios for arrangement of nodes in the fog layer. Figure 2a shows a few fog nodes for intermediary services. Figure 2b includes a fog to cloud gateway as well. Besides conserving network bandwidth, a gateway may also serve as a coordinator for fog nodes. Figure 2c shows a gateway between fog nodes and the smart layer. Such a gateway would be useful for low-power devices in the smart layer. Figure 2d integrates gateways on both sides of fog nodes. This can be useful in reducing network traffic on both sides of the fog layer.

3.3 Cloud Services

The Cloud layer includes systems with a lot of resources. A cloud supports a global architecture where all gateways and nodes can send data to. It needs to support computation-intensive operations and long-term storage. A cloud can be connected to multiple fog nodes at a time and mimics the notion of unlimited computational power and storage.

4 Benefits of Fog Computing

By implementing the fog layer, applications can benefit in many ways. For instance, the fog layer can incorporate machine learning to enhance security. Threat detection and malware analysis can be executed at the fog layer to enable the timely detection of security attacks. Fog computing also reduces distance for information exchange thereby reducing chances for eavesdropping and man-in-the-middle attacks. Similarly, fog can provide improved location-awareness, timely alerts on data analytics, enhanced support for mobility, a higher degree of resilience against faults, improved availability in case of disasters, and better opportunities for context-aware computing. Brief explanation of major benefits of fog computing are discussed below:

1. **Security** Fog computing can enhance security by distributing resources such that in case of a security attack on a few fog nodes, the remaining ones may still continue services. Opportunities for network attacks such as eavesdropping and man-in-the-middle attack may also reduce due to shorter distance between smart devices and fog nodes [29]. Fog computing, machine learning and data analytics can be applied to detect security threats in real-time. For instance, DDoS attacks can be detected by executing pattern analysis and DDoS detection modules at fog nodes [15].

2. **Mobility** Mobile devices are generally battery operated with limited processing and storage resources. Fog computing extends useful support for mobile devices for data collection, integration, aggregation, and access. For mobile computing platforms, data analytics can be applied at fog nodes to limit network bandwidth. For instance, a mobile cloud system implementing video recognition can execute feature extraction as well as training and testing of video streams [9]. This will eventually conserve network bandwidth and will also reduce processing time.

3. **Location awareness**
   Fog nodes can detect location of clients in order to provide location-specific services. Location can be obtained through various methods such as Geo-IP location and nearby server approximation. Location-specific services may include finding nearby services and context-aware search [14]. Location awareness is particularly useful in scenarios where context-aware services are delivered. For this purpose, location-aware rule-based machine learning algorithms can also be exploited. For instance, for smart phone users, location can be recognized by using k-nearest neighbor and decision tree algorithms. The recognized location can be used to predict destination using hidden Markov models [13]. Location-aware data can be useful for machine learning algorithms. This mechanism is significant for high mobility scenarios [20].

4. **Low Latency**
   Fog computing can greatly reduce the delay by introducing intermediary servers for storage and processing. In this way, delay-sensitive applications are processed at the nearest available physical fog node thereby reducing communication delays and yielding faster response times. Consequently, more processes could be completed within the required timeline [22]. Fog nodes can be used to select appropriate nearby servers which can lead to low latency. The selection of servers can be implemented by using soft K-Means clustering [7]. Soft clustering allows a fog node to be included in multiple clusters which eventually leads to incorporation of dynamic policies for network routing and load balancing.

5. **Low Network Bandwidth and High Scalability** Network traffic is reduced in fog computing because many computations are performed near the end nodes. This conserves network bandwidth because not all the information needs to be sent to the distant cloud. Although cloud servers have enormous resources to process various data requests, they may suffer from constrained network resources if many requests are sent to a cloud within a short amount of time. In contrast, fog computing is highly scalable as edge devices could be easily added.

6. **Load Balancing and Improved Availability**
   Efficiency and availability can also be increased by implementing load balancing among fog nodes. Fog computing enables continuous availability of services with the help of its resources, which are shareable to manage accessibility. To support high resources availability both vertical and horizontal scaling may be performed.

7. **Parallelization**
   Fog computing can also provide a distributed and parallel platform for data analytics. This could be highly beneficial for big data analytics such as machine learning and deep learning. In such a setup, fog nodes are needed to be interconnected through high speed networking equipment such as Infiniband [21] or Fibre Channel over Ethernet (FCoE) [38]. A high speed network is useful in achieving high speed data transfer—a key requirement for big data analytics.

8. **Economy**
   The cost of fog computing includes networking cost, deployment cost, and execution cost. Fog computing can reduce the overall cost of the system. Data aggregation at the fog layer is expected to conserve network traffic thereby lowering the networking cost of the system. Fog computing is also likely to reduce the resource requirements at the cloud layer. As fog nodes are commodity machines, the cost of deployment of fog nodes is likely to be low. For end users, fog computing provides a similar pay per use mechanism as offered by cloud computing, and users are charged based on their usage requirements [22]. Fog computing is also economical if analytics is performed at the fog layer thereby providing timely results and saving network bandwidth [2].

Fog computing provides a useful solution to encounter the high latency problem associated with cloud. Fog nodes can provide rapid response for time-sensitive applications. In the next section, we discuss various challenges related to fog computing.

5 Challenges of Fog Computing

There are many challenges which could affect the progress of fog computing. In this section, some of the challenges that are needed to be addressed are explained:

1. **Data Analytics Model** Use of fog computing for data analytics opens up new opportunities and challenges related to efficient and capable models for computation. Considering that fog nodes may have limited resources and that data

streaming rate at the fog layer could vary over time, these requirements and challenges are further exacerbated.. In a multi-tenant environment where resources are shared by different applications, we need an efficient model for data analytics which remains a challenge. In a fog layer, fog nodes needed to be connected via high speed links in order to facilitate high speed data analytics. In addition, advanced algorithms that would enhance the computational capabilities of fog nodes such that multiple fog nodes can coherently execute algorithms in parallel, are needed to be developed.

2. **Security** Security threats in fog computing pose real concerns as various attacks can affect the functionality of fog computing. For instance, through the man-in-the middle attack, an adversary can eavesdrop messages and inject fabricated messages in the system. Similarly, a Denial of Service (DoS) attack can allow an attacker to disrupt the normal operation. The aftermath of all these attacks can be lethal. An insider attack can also be launched through which authorized users may intentionally or unintentionally launch an attack on fog nodes or on the interconnected networks. Rogue fog nodes are also a major security concern. In this case, an attacker penetrates the network and takes control of the fog node or the complete data center and can respond to users requests with incorrect data. In all these attacks, reliability, and availability of the fog computing architecture can be affected leading to interruption of services [29]. Attack prevention mechanisms such as firewalls and Intrusion Detection Systems (IDS) can be incorporated to prevent these attacks. However, ensuring fool proof security to protect critical infrastructure remains a major challenge for the users and providers of fog computing [5, 10].

3. **Privacy** In fog computing, privacy can be managed by hiding the users data, location, and usage patterns. As data acquired by the fog architecture is mostly generated by sensors and end devices, privacy preservation mechanisms are needed to protect the confidentiality of the sensed data. Efficient techniques are also required for obfuscation of users confidential data, while providing necessary services for fog computing.

4. **Mobility** Fog computing can be used to support mobility of users. In this context, various services such as providing user specific data, context-aware services, and supporting seamless transitions are needed. As the size of the network and the number of users increases, meeting these requirements become a real challenge. Implementing real-time, uninterruptable transfer of data is also a challenge. During mobility, data is also more vulnerable to attacks, so network and system level security mechanisms are needed to prevent such data from being compromised.

5. **Transient Data and Load Balancing**
   End user devices may send data to one or more fog nodes for processing. It is difficult to evaluate the requirements for all applications a priori and optimize the load on all available fog nodes. It is also a challenging task to predict how long data will be stored in fog nodes before sending it to the permanent storage at cloud servers.

6. **Computational Model and Job Scheduling** In fog computing, in case of shortage of local resources, a user can request for fog services. At the fog layer, the total load is divided among fog nodes by the fog gateway. If no fog node is available for processing, then the fog gateway forwards user requests to the cloud layer. The resources available in the fog and cloud layers can be scheduled for user jobs with the help of either virtual machines or containers. Different algorithms such as First Come First Serve (FCFS), Round Robin, Min-Min, and Min-Max algorithms can be used to schedule jobs for processing at fog layers. Due to limited resources at fog nodes, meeting strict Service Level Agreement (SLA) may not always be possible and current implementations rely on best effort criteria for resource allocation [3, 28]. Improved resource allocation techniques and computational models are needed for higher efficiency [18]. Depending on the actual workload, the fog layer could also be used to provide data-intensive as well as CPU-intensive tasks. Multiple fog nodes can be used to create a cluster. Parallel computing frameworks such as Spark, Hadoop, MPI can be explored. Research opportunities exist in providing improved coordination among fog nodes for cluster computing and designing efficient data storage to reduce data transfers between fog nodes.

7. **Data Consistency** Fog computing has been based on an eventual consistency model. That is, eventually all replicas will be consistent. This flexible model ensures high availability. However, as the usage of fog computing increases in the future, stricter consistency models are needed. This is specifically true for building solutions for delay-sensitive applications. Providing strong and reliable consistency in a mobile environment remains a real challenge [36].

8. **Energy Requirements** Fog computing provides computational and storage services near end users. In contrast to the central cloud system, fog computing requires more devices. Consequently, it may consume more energy. New mechanisms and techniques are needed to save energy in fog computing [18].

9. **Communication and Aggregation Model** Fog layer has been designed to facilitate instant computation against the distant cloud. Aggregated data can then be shared with the cloud for verification and backup. Research opportunities exist in designing appropriate models for data aggregation and sending communication updates from the fog layer to the cloud.

10. **Hardware Requirements** As discussed earlier, the fog layer could be used for various tasks such as analytics, storage, gaming, decision support system and so on. The widespread use of fog computing motivates us to review and assess different hardware requirements for such systems. For instance, in the storage model, fog nodes can be connected as a storage cluster such as Storage Area Network (SAN). High speed network connectivity between nodes of the fog layer is critical. Remote Direct Memory Access (RDMA) based systems such as InfiniBand can be adopted to facilitate high speed data transfer. Similarly, for data analytics, highly capable Graphics Processing Unit (GPU) clusters can be deployed. In general, there are many research opportunities related to determining and integrating improved hardware systems while maintaining low cost.

Table 1 Fog computing applications: benefits and challenges

Fog application	Objectives and benefits	Challenges
Mobiles games	Reduced network traffic from edge to Cloud, Reduce computation time, Load balancing	Context awareness, mobility management, utility computing, elasticity
Military application	Reduced network traffic from edge to Cloud Reduce computation time Enhanced security Low cost	Security, load balancing, communication and aggregation
Face recognition	Reduced network traffic Data aggregation	Data analytics, efficient energy consumption
Smart city Traffic, air pollution, garbage collection	Reduced network Traffic Low cost	Data consistency, security and privacy, data analytics, load balancing
Monitoring of critical infrastructure such as power plant	Reduce network traffic Time-sensitive decisions Enhanced security	Security, data analytics, data aggregation and load balancing, hardware
Remote surgery, patient monitoring system	Irregular network connectivity and Time-sensitive decision	Security, timely communication

11. **Elasticity** As fog is an emerging platform, it is significant to explore elasticity at the fog layer. This would enable better resource management. Container-based orchestration may be preferred over VM-based orchestration due to its smaller resource footprint [17]. Fog nodes can experience a variation in resource usage due to changes in network traffic and the number of users; elasticity can play improved role in such a scenario.

12. **Utility Computing** So far, the fog model has been implemented for private clouds. However, benefits of fog computing are also needed to be explored for public clouds involving utility computing. In Table 1, few examples are explained where fog computing can help meet application needs and improve the overall application performance.

6 Conclusion and Discussions

Fog computing is an emerging platform to support IoT communication with the cloud. The platform has brought various benefits including support of delay-sensitive applications, mobility management, bandwidth conservation, and appropriate use of resources. Fog computing has also opened up several open issues related to security, efficient usage, communication, and data analytics. Research opportunities exist in determining appropriate solutions for these challenges and in developing suitable applications which can benefit from this emerging platform.

Many applications and emerging paradigms can leverage fog computing. For instance, Software Defined Networks (SDN) can be integrated with fog computing to improve network flexibility and adaptability from IoT layer to the cloud layer. Fog computing can also benefit from improved security mechanisms such as cost effective encryption techniques, efficient intrusion detection systems, and incorporating context-aware security mechanisms. Similarly, advances such as power optimization techniques [31], data analytics [33], deep learning, and network protocols could be very useful in improving fogs system performance and minimizing operational cost, as well as supporting timely computation of results, and effective utilization of bandwidth [27]. Fog computing can be best utilized through wide range of applications which can provide improved services to users. It will require consistent and thorough efforts from the research community to unleash the true potential of fog computing [27, 37].

References

1. Aazam, M., Huh, E.N.: Fog computing and smart gateway based communication for cloud of things. In: 2014 International Conference on Future Internet of Things and Cloud (FiCloud), pp. 464–470. IEEE (2014)
2. Aazam, M., Zeadally, S., Harras, K.A.: Offloading in fog computing for IoT: review, enabling technologies, and research opportunities. Future Gener. Comput. Syst. (2018)
3. Agarwal, S., Yadav, S., Yadav, A.K.: An efficient architecture and algorithm for resource provisioning in fog computing. Int. J. Inf. Eng. Electron. Bus. 8(1), 48 (2016)
4. Alhaija, H.A., Mustikovela, S.K., Mescheder, L., Geiger, A., Rother, C.: Augmented reality meets deep learning for car instance segmentation in urban scenes. In: British Machine Vision Conference, vol. 1, p. 2 (2017)
5. Ali, M.: Green cloud on the horizon. In: IEEE International Conference on Cloud Computing, pp. 451–459. Springer (2009)
6. Andriopoulou, F., Dagiuklas, T., Orphanoudakis, T.: Integrating IoT and fog computing for healthcare service delivery. In: Components and Services for IoT Platforms, pp. 213–232. Springer (2017)
7. Balevi, E., Gitlin, R.D.: Unsupervised machine learning in 5g networks for low latency communications. In: 2017 IEEE 36th International Performance Computing and Communications Conference (IPCCC), pp. 1–2. IEEE (2017)
8. Bonomi, F., Milito, R., Natarajan, P., Zhu, J.: Fog computing: a platform for internet of things and analytics. In: Big Data and Internet of Things: A Roadmap for Smart Environments, pp. 169–186. Springer (2014)
9. Chen, N., Chen, Y., You, Y., Ling, H., Liang, P., Zimmermann, R.: Dynamic urban surveillance video stream processing using fog computing. In: 2016 IEEE Second International Conference on Multimedia Big Data (BigMM), pp. 105–112. IEEE (2016)
10. Chen, Y., Abraham, A., Yang, B.: Hybrid flexible neural-tree-based intrusion detection systems. Int. J. Intell. Syst. 22(4), 337–352 (2007)
11. Chiang, M., Ha, S., Chih-Lin, I., Risso, F., Zhang, T.: Clarifying fog computing and networking: 10 questions and answers. IEEE Commun. Mag. 55(4), 18–20 (2017)
12. Chiang, M., Zhang, T.: Fog and IoT: an overview of research opportunities. IEEE Internet Things J. 3(6), 854–864 (2016)
13. Cho, S.-B.: Exploiting machine learning techniques for location recognition and prediction with smartphone logs. Neurocomputing 176, 98–106 (2016)

14. Dastjerdi, A.V., Gupta, H., Calheiros, R.N., Ghosh, S.K., Buyya, R.: Fog computing: principles, architectures, and applications. In: Internet of Things, pp. 61–75. Elsevier (2016)
15. Diro, A.A., Chilamkurti, N.: Distributed attack detection scheme using deep learning approach for internet of things. Future Gener. Comput. Syst. **82**, 761–768 (2018)
16. Dsouza, C., Ahn, G.J., Taguinod, M.: Policy-driven security management for fog computing: preliminary framework and a case study. In: 2014 IEEE 15th International Conference on Information Reuse and Integration (IRI), pp 16–23. IEEE (2014)
17. Hoque, S., de Brito, M.S., Willner, A., Keil, O., Magedanz, T.: Towards container orchestration in fog computing infrastructures. In: 2017 IEEE 41st Annual Computer Software and Applications Conference (COMPSAC), vol. 2, pp. 294–299. IEEE (2017)
18. Kaur, K., Dhand, T., Kumar, N., Zeadally, S.: Container-as-a-service at the edge: trade-off between energy efficiency and service availability at fog nano data centers. IEEE Wirel. Commun. **24**(3), 48–56 (2017)
19. Lee, K., Kim, D., Ha, D., Rajput, U., Oh, H.: On security and privacy issues of fog computing supported internet of things environment. In: 2015 6th International Conference on the Network of the Future (NOF), pp. 1–3. IEEE (2015)
20. Luan, T.H., Gao, L., Li, Z., Xiang, Y., Wei, G., Sun, L.: Fog computing: focusing on mobile users at the edge. arXiv preprint arXiv:1502.01815 (2015)
21. MacArthur, P., Liu, Q., Russell, R.D., Mizero, F., Veeraraghavan, M., Dennis, J.M.: An integrated tutorial on infiniband, verbs, and MPI. IEEE Commun. Surv. Tutor. **19**(4), 2894–2926 (2017)
22. Mahmud, R., Kotagiri, R., Buyya, R.: Fog computing: a taxonomy, survey and future directions. In: Internet of Everything, pp. 103–130. Springer (2018)
23. Markakis, E., Mastorakis, G., Mavromoustakis, C.X., Pallis, E.: Cloud and Fog Computing in 5G Mobile Networks: Emerging Advances and Applications. Institution of Engineering and Technology (2017)
24. Markakis, E.K., Karras, K., Zotos, N., Sideris, A., Moysiadis, T., Corsaro, A., Alexiou, G., Skianis, C., Mastorakis, G., Mavromoustakis, C.X., et al.: Exegesis: extreme edge resource harvesting for a virtualized fog environment. IEEE Commun. Mag. **55**(7), 173–179 (2017)
25. Nikoloudakis, Y., Panagiotakis, S., Markakis, E., Pallis, E., Mastorakis, G., Mavromoustakis, C.X., Dobre, C.: A fog-based emergency system for smart enhanced living environments. IEEE Cloud Comput. (6), 54–62 (2016)
26. Pauly, O., Diotte, B., Fallavollita, P., Weidert, S., Euler, E., Navab, N.: Machine learning-based augmented reality for improved surgical scene understanding. Comput. Med. Imag. Graph. **41**, 55–60 (2015)
27. Perera, C., Qin, Y., Estrella, J.C., Reiff-Marganiec, S., Vasilakos, A.V.: Fog computing for sustainable smart cities: a survey. ACM Comput. Surv. (CSUR) **50**(3), 32 (2017)
28. Pham, X.Q., Huh, E.N.: Towards task scheduling in a cloud-fog computing system. In: 2016 18th Asia-Pacific Network Operations and Management Symposium (APNOMS), pp. 1–4. IEEE (2016)
29. Roman, R., Lopez, J., Mambo, M.: Mobile edge computing, fog et al.: A survey and analysis of security threats and challenges. Future Gener. Comput. Syst. **78**, 680–698 (2018)
30. Salonikias, S., Mavridis, I., Gritzalis, D.: Access control issues in utilizing fog computing for transport infrastructure. In: International Conference on Critical Information Infrastructures Security, pp. 15–26. Springer (2015)
31. Sheikh, F., Fazal, H., Taqvi, F., Shamsi, J.: Power-aware server selection in nano data center. In: 2015 IEEE 40th Local Computer Networks Conference Workshops (LCN Workshops), pp. 776–782. IEEE (2015)
32. Shi, W., Cao, J., Zhang, Q., Li, Y., Lanyu, X.: Edge computing: vision and challenges. IEEE Internet Things J. **3**(5), 637–646 (2016)
33. Tang, B., Chen, Z., Hefferman, G., Pei, S., Wei, T., He, H., Yang, Q.: Incorporating intelligence in fog computing for big data analysis in smart cities. IEEE Trans. Ind. Inf. **13**(5), 2140–2150 (2017)

34. Tang, B., Chen, Z., Hefferman, G., Wei, T., He, H., Yang, Q.: A hierarchical distributed fog computing architecture for big data analysis in smart cities. In: Proceedings of the ASE BigData & SocialInformatics, p. 28. ACM (2015)
35. Teerapittayanon, S., McDanel, B., Kung, H.T.: Distributed deep neural networks over the cloud, the edge and end devices. In: 2017 IEEE 37th International Conference on Distributed Computing Systems (ICDCS, pp. 328–339. IEEE (2017)
36. Vaquero, L.M., Rodero-Merino, L.: Finding your way in the fog: towards a comprehensive definition of fog computing. ACM SIGCOMM Comput. Commun. Rev. **44**(5), 27–32 (2014)
37. Varghese, B., Wang, N., Nikolopoulos, D.S., Buyya, R.: Feasibility of fog computing. arXiv preprint arXiv:1701.05451 (2017)
38. Williams, J.B.: Fibre channel over ethernet, 8 July 2014. US Patent 8,774,215 (2014)
39. Yi, S., Li, C., Li, Q.: A survey of fog computing: concepts, applications and issues. In: Proceedings of the 2015 Workshop on Mobile Big Data, pp. 37–42. ACM (2015)

Medical Image Watermarking in Four Levels Decomposition of DWT Using Multiple Wavelets in IoT Emergence

Tamara K. Al-Shayea, Constandinos X. Mavromoustakis,
Jordi Mongay Batalla, George Mastorakis, Evangelos Pallis,
Evangelos K. Markakis, Spyros Panagiotakis, and Imran Khan

Abstract Medical images will be an inseparable part for evaluating medical conditions of a person in real-time. This process will become efficient by exploiting the characteristics that are offered by IoT in the healthcare sectoral issues. This will allow more efficiency in the processing of a medical image. However, the medical images are exposed to the major risks through frequent attacks which may lead up to misinformation the physician in the diagnosis of the disease. Piracy eradication remains

T. K. Al-Shayea (✉) · C. X. Mavromoustakis
Mobile Systems Laboratory (MoSys Lab), Department of Computer Science, University of
Nicosia, Nicosia, Cyprus
e-mail: alshayeh.t@unic.ac.cy

C. X. Mavromoustakis
e-mail: mavromoustakis.c@unic.ac.cy

J. M. Batalla
National Institute of Telecommunications and Warsaw University of Technology, Szachowa Str. 1
and Nowowiejska Str. 15/19, Warsaw, Poland
e-mail: jordim@tele.pw.edu.pl

G. Mastorakis
Department of Management Science and Technology, Hellenic Mediterranean University, Agios
Nikolaos, 72100 Crete, Greece
e-mail: gmastorakis@hmu.gr

E. Pallis · E. K. Markakis · S. Panagiotakis
Department of Electrical and Computer Engineering, Hellenic Mediterranean University,
Heraklion, 71500 Crete, Greece
e-mail: pallis@pasiphae.eu

E. K. Markakis
e-mail: markakis@pasiphae.eu

S. Panagiotakis
e-mail: spanag@hmu.gr

I. Khan
Department of Electrical Engineering, University of Engineering and Technology, P.O.B. 814,
Peshawar, Pakistan
e-mail: ikn.eup121@gmail.com

the major challenge in the present-day world in IoT platform. Subsequently, the medical watermarked image is significant technique of ensuring the clinical information that exists in the medical images. In this regard, this paper labors on a medical image based on digital watermarking can be utilized to protect health sign embedding in a medical image within an invisible status. The proposed method performance is evaluated by utilizing MSE, PSNR, SSIM, and NC, which are necessary to get the best result for performance metrics. This work is achieved in four levels Discrete Wavelet Transform (DWT). Each level is utilized different wavelet family. These wavelets family are composed of biorthogonal wavelet, reverse biorthogonal wavelet, discrete meyer wavelet, symlet wavelet, and coiflets wavelet transform. The proposed technique is highly robust against numerous sorts of attacks. The results refer that this proposed algorithm permits prevention at a higher level compared with other current structures and algorithms.

1   Introduction

The medical images have faced up considerable problems that are increased in the internet environment in IoT. In consequence of the gigantic attacks which are susceptive to affect the information for the medical image. Therefore, it is necessary to highlight the information of medical image and the content of medical image should be retained by providing a robust medical image watermarking.

The medical images are transferred, stored, retrieved through networks. In fact, image watermarking has been mostly recognized as a pertinent technique for enhancing data security, integrity and authenticity. In order to prohibit any possibility of access, modification or destruction. Private medical data embedded in the image should be imperceptible and can only be retrieved by authorized persons [1]. The image quality should be protected to avoid the wrong diagnosis. Wherefore, significant attention is wanted before embedding watermarking information in medical images [2]. Watermarking algorithms need to be solved that consider important difficulties. The valid watermarking algorithms should satisfy the requirement of robustness against different unintentional or malicious attacks as well as watermarking schemes are needed to supply trustworthy evidence for saving rightful ownership [3]. Image watermarking is one of the varied technicalities that keep digital image against various attacks by embedding a signature into the host image [4]. Li et al. [5] presented an algorithm of robust various watermarks by using DWT and DFT. Moniruzzaman et al. [6] proposed discrete wavelet transform (DWT) domain and chaotic system based medical image watermarking scheme for embedding patient information in a medical image to authenticate, and also to track it the origin of the image. Mavromoustakis et al. [7] proposed a new offload-aware recommendation algorithm in the IoT environment, towards permitting the efficient monitoring of high energy consumption applications which operate in the mobile devices. AL-Shayea et al. [8] presented a novel watermarking algorithm based on the biorthogonal family (biorthogonal 2.2, biorthogonal 3.5 and biorthogonal 5.5) wavelet transform.

Skourletopoulos et al. [9] presented a novel game theory technique for effective storage allocation in IoT systems. AL-Shayea et al. [10] proposed a hybrid measurement technique for digital image watermarking that uses medical images (X-ray, MRA, and CT) in IoT infrastructure. As privacy is dependably a major issue over each framework yet in IoT, it is the most crucial area towards embedding information or data that will be utilized in the future inter-connected networks [11–13]. The Internet of Things (IoT) paradigm has made a vital part of all significant services provision, it plays a decisive role in today's ubiquitous systems by utilizing machine-to-machine (M2M) communication [14–17]. The use of the Internet tends to grow, as the number of devices and the end-users' applicability scenarios increase. The Internet of Things (IoT) paradigm is the future for the coming decades, which helps in constructing the smart world where everything is associated with one system [18, 19].

An algorithm to embed a healthcare marker inside a medical digital image without obvious changing in the appearance or evident distortion in the host image is proposed in this paper. The proposed algorithm uses four levels discrete wavelet decomposition. Each level uses different wavelets family transform to submit the watermarked image. The proposed algorithm uses discrete meyer, reverse biorthogonal, biorthogonal, coiflets, and symlet wavelet transform independently for each level. After that, it is extracted the medical watermark image from the watermarked image without loss or distortion of any of its features. In the end, attacking experiments are accomplished on the host image, by utilizing several attacking methods. Experimental results are implemented using four statistical parameters. The statistical parameters that were used to measure the performance in this labor are Mean Square Error (MSE), Peak Signal-to-Noise Ratio (PSNR), Structural Similarity Index (SSIM) and Normalized Cross-Correlation (NC). Statistical parameters result show that the proposed algorithm has elevated performance characteristics.

The rest of this paper is organized as follows: Sect. 2 gives digital image watermarking algorithm and multiple wavelets family transform. Section 3 describes the proposed watermarking algorithm, while Sect. 4 provides the experimental results and the evaluation of the proposed algorithm. Section 5 concludes this work.

2 Digital Image Watermarking Algorithms

The Discrete Wavelet Transform (DWT) has a huge number of applications in many areas. It has become a powerful technique in watermarking [20]. Wavelet basis is a doubly-indexed family of L2(R) functions. The mother wavelet function is:

$$\psi_{j,k}(t) = \frac{1}{\sqrt{2^j}} \psi \left(\frac{t - k2^j}{2^j} \right) \tag{1}$$

where j is the scale parameter and k is the shift parameter both of them are integers [21].

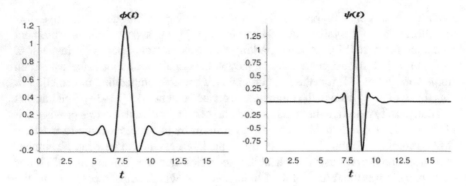

Fig. 1 Bior6.8. **a** Scaling function φ. **b**. Wavelet function ψ

A sufficient condition for the reconstruction of any signal x of finite energy by the formula:

$$x(t) = \sum_{j \in z} \sum_{k \in z} (x, \psi_{j,k}) . \psi_{j,k}(t) \tag{2}$$

That is the function $\{\psi_{j,k} : j, k \in z\}$ form an orthonormal basis of L2(R).

Diverse families of wavelets that have proven to be especially beneficial are included in this section [22].

2.1 Biorthogonal Wavelet

This family of wavelet shows the property of linear phase, which is required for signal and image reconstruction [22]. Figure 1 shows wavelet, and scaling function for biorthogonal 6.8(bior6.8) wavelet transform [23].

2.2 Reverse Biorthogonal Wavelet

Reverse biorthogonal wavelet family is gained by the biorthogonal wavelet coupled. These wavelets families are guided by biorthogonal spline wavelets, so the symmetrical condition and reconstruction can be assured [22] (Figs. 2 and 3).

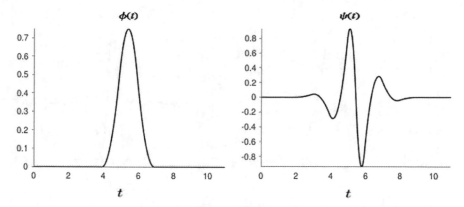

Fig. 2 Rbio3.5. **a** Scaling function φ. **b** Wavelet function ψ

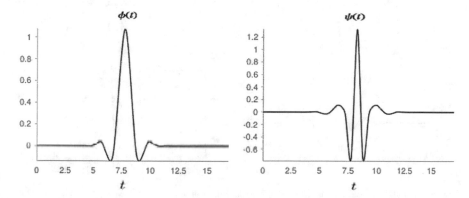

Fig. 3 Rbio6.8. **a** Scaling function φ. **b** Wavelet function ψ

2.3 Symlet Wavelet

The symlets are almost symmetrical wavelets presented by Daubechies as amendments to the Daubechies family. The properties of the two wavelet families are identical [22]. Figure 4 shows symlets5 (sym5) wavelet transform wavelet, and scaling function for wavelet transform while Fig. 5 shows symlets8 (sym8) wavelet transform [23].

2.4 Coiflets Wavelet

This wavelet is existing under the name of Coiflets, but it is in fact constructed by I. Daubechies at the demand of Coifman [24, 25]. Coiflet wavelet transform has a

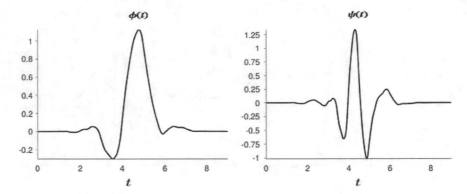

Fig. 4 Symlets5. **a** Scaling function φ. **b** Wavelet function ψ

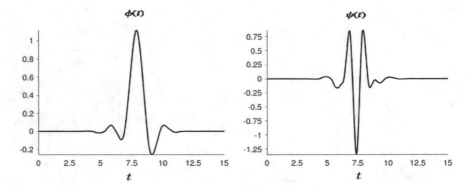

Fig. 5 Symlets8. **a** Scaling function φ. **b** Wavelet function ψ

short name of Coif, and for the order N is written with CoifN. The length of the Coiflet wavelet filter is 6N [26]. Figure 6 shows wavelet, and scaling function for coiflets (coif4) wavelet transform while Fig. 7 shows wavelet, and scaling function for coiflets (coif5) wavelet transform [23].

2.5 *Discrete Meyer Wavelet*

Discrete Meyer wavelet is band finite and has an infinite number of backings but has quicker decomposing than sync wavelet and is endlessly times differentiable, discrete Meyer wavelet is orthogonal and has symmetric scaling wavelet function [27]. Figure 8 shows wavelet, and scaling function for discrete Meyer (dmey) wavelet transform [23].

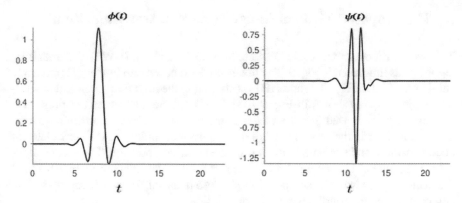

Fig. 6 Coifelts4. **a** Scaling function φ. **b** Wavelet function ψ

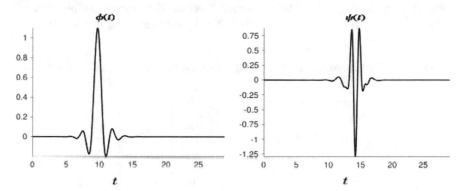

Fig. 7 Coifelts5. **a** Scaling function φ. **b** Wavelet function ψ

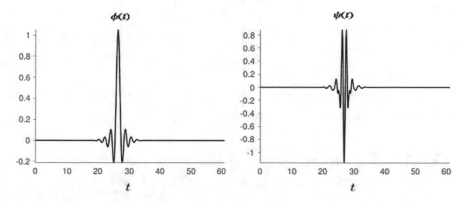

Fig. 8 Dmey. **a** Scaling function φ. **b** Wavelet function ψ

3   The Proposed Medical Image Watermarking Algorithm

This work proposes four levels Discrete Wavelet Transform (DWT) watermarking algorithm as illustrated in Fig. 9. The image is decomposed into four equal frequency sub-bands (LL1, LH1, HL1, and HH1) according to the high frequency and low frequency in DWT as shown in Fig. 9 [28]. Two multiple algorithms are presented. The first algorithm used discrete meyer in the first level of decomposition, reverse biorthogonal 6.8 in the second level of decomposition, coiflets 5 in the third level of decomposition, and symlet 8 in the fourth level of decomposition. The second algorithm used reverse biorthogonal 3.5 in the first level of decomposition, biorthogonal 6.8 in the second level of decomposition, symlet 5 in the third level of decomposition, and coiflets 4 in the fourth level of decomposition.

The block diagram of the embedding process is shown in Fig. 10 while Fig. 11 shows the block diagram of the extracting process. The embedding algorithm is illustrated in Algorithm 1, while the extracting algorithm is illustrated in Algorithm 2.

Algorithm 1 Pseudocode for Watermark Embedding Algorithm

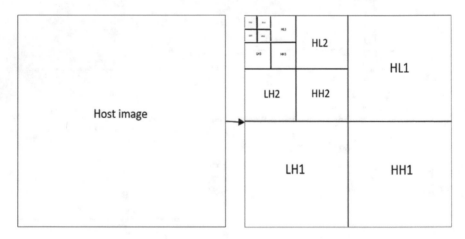

Fig. 9   Four level discrete wavelet decomposition

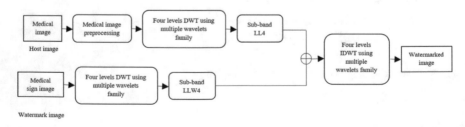

Fig. 10   Block diagram of the embedding process

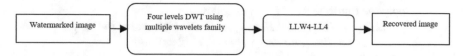

Fig. 11 Block diagram of the extraction process

1. **Inputs**: Host image, watermarks
2. **Output**: Watermarked image
3. for each host image H do
4.   Read the image.
5.   Convert the image to double.
6.   Apply the first level of DWT as follows:
   [LL1,LH1,HL1,HH1] = dwt2(H, one of the wavelet
       Families (i.e.: dmey, rbio6.8, coiflets5 and
   sym8));
7.   Apply the second level of DWT as follows:
   [LL2,LH2,HL2,HH2] = dwt2(LL1, one of the wavelet
   families (i.e.: dmey, rbio6.8, coiflets5 and
   sym8));
8.   Apply the third level of DWT as follows:
   [LL3,LH3,HL3,HH3] = dwt2(LL2, one of the wavelet
   families (i.e.: dmey, rbio6.8, coiflets5 and
   sym8));
9.   Apply the fourth level of DWT as follows:
   [LL4,LH4,HL4,HH4] = dwt2(LL3, one of the wavelet
   families (i.e.: dmey, rbio6.8, coiflets5 and
   sym8));
10. end for
11. for each watermarks image W do
12.   Read the image.
13.   Convert the image to double.
14.   Apply the first level of DWT as follows:
   [LLw1,LHw1,HLw1,HHw1] = dwt2(W, one of the
   wavelet families (i.e.: dmey, rbio6.8, coiflets5 and
   sym8));
15.   Apply the second level of DWT as follows:
   [LLw2,LHw2,HLw2,HHw2] = dwt2(LLw1, one of the
   wavelet families (i.e.: dmey, rbio6.8, coiflets5 and
   sym8));
16.   Apply the third level of DWT as follows:
   [LLw3,LHw3,HLw3,HHw3] = dwt2(LLw2, one of the
   wavelet families (i.e.: dmey, rbio6.8, coiflets5 and
   sym8));

17.   Apply the fourth level of DWT as follows:
          [LLw4,LHw4,HLw4,HHw4] = dwt2(LLw3, one of the
          wavelet families (i.e.: dmey, rbio6.8, coiflets5 and
          sym8));
18.   end for
19.   *Watermarked image = LL4 + α ∗ LLw4*
          where α is the parameter for embedding strength coefficient
          that controls the embedding strength. (α = 0.01).
20.   for each watermarked image do
21.   Apply the first level of inverse DWT as follows: *Watermarkedimage_level1 =*
          idwt2(Watermarkedimage,LH4,HL4,HH4, one of the wavelet families
          (i.e.: dmey, rbio6.8, coiflets5 and sym8));
22.   Apply the second level of inverse DWT as follows: *Watermarkedimage_level2*
          = idwt2(Watermarkedimage_level1,LH3,HL3,HH3, one of the wavelet families
          (i.e.: dmey, rbio6.8, coiflets5 and sym8));
23.   Apply the third level of inverse DWT as follows: *Watermarkedimage_level3 =*
          idwt2(Watermarkedimage_level2,LH2,HL2,HH2, one of the wavelet families
          (i.e.: dmey, rbio6.8, coiflets5 and sym8));
24.   Apply the fourth level of inverse DWT as follows: *Watermarkedimage_final =*
          idwt2(Watermarkedimage_level3,LH1,HL1,HH1, one of the wavelet families
          (i.e.: dmey, rbio6.8, coiflets5 and sym8));
25.   end for

Algorithm 2 Pseudocode for Watermark Extraction Algorithm

1.   **Input**: Watermarked image
2.   **Output**: Extracted watermarks
3.   for each watermarked image do
4.   Read the watermarked image.
5.   Apply the first level of DWT as follows:
             [a b c d] = dwt2(Watermarked image, one of the wavelet
             families (i.e.: dmey, rbio6.8, coiflets5 and
             sym8));
6.   Apply the second level of DWT as follows:
             [aa bb cc dd] = dwt2(a, one of the wavelet
             Families (i.e.: dmey, rbio6.8, coiflets5 and
             sym8));
7.   Apply the third level of DWT as follows:
             [aaa bbb ccc ddd] = dwt2(aa, one of the wavelet
             families (i.e.: dmey, rbio6.8, coiflets5 and
             sym8));
8.   Apply the fourth level of DWT as follows:
             [aaaa bbbb cccc dddd] = dwt2(aaa, one of the wavelet
             families (i.e.: dmey, rbio6.8, coiflets5 and
             sym8));

9. *End for*
10. *Do the following to extract the image:*
11. *Extracted image = aaaa-LL4;*

4 Experimental Results and Evaluation

During this section, an evaluation analysis is performed. The proposed algorithm is implemented in a MATLAB-based software.

The host medical image and the watermarked image that has been utilized is a 512 × 512 grayscale image. Three sorts of medical images (X-ray image, MRA image, CT image) are utilized as a host image. Figure 12 shows the medical host images with various types while Fig. 13 shows the watermark image. The watermarked image is shown in Fig. 14, while the extracted image is shown in Fig. 15.

Peak Signal-to-Noise Ratio (PSNR) is implemented to measure the performance of the proposed algorithm by finding the difference between the host image and the

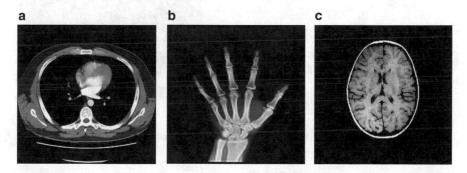

Fig. 12 **a** X-ray image. **b** Brain image. **c** Chest image

Fig. 13 Watermark image

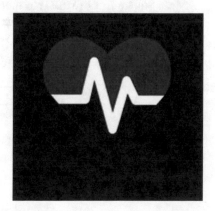

Fig. 14 Watermarked image

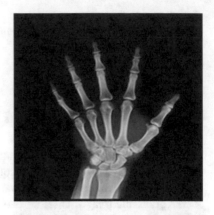

Fig. 15 Extracted image

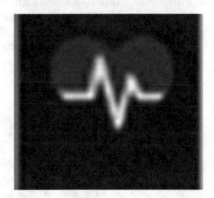

watermarked image. The PSNR is measured in dB value.

$$PSNR = 10 \times log_{10}\left(\frac{255^2}{MSE}\right) \tag{3}$$

Mean Square Error (MSE) clarified the difference between the original pixels in a host image and the watermarked pixels in the watermarked image, which demonstrates the quality of the distortion value of the watermarked image. The MSE is measured in real value.

$$MSE = \frac{1}{M \times N} \sum_{x=0}^{M-1} \sum_{y=0}^{N-1} (H(x, y) - W(x, y))^2 \tag{4}$$

where H, is the original host image and W is a watermarked image, respectively. M is the size of the host image and M is row size and N is the column size of the host image, respectively.

The Normalized Cross-Correlation (NC) is executed to measure the quantitative similarity between the host image and the watermarked image. NC is defined as:

$$NC = \frac{\sum_i \sum_j [W(i,j).W'(i,j)]}{\sum_i \sum_j [W(i,j)]^2} \tag{5}$$

where W refers the embedded watermarking and W' refers to extracted watermarking. We are seeking to increase the NC value that indicates more similarity between the host image and the watermarked image.

Structural Similarity Index (SSIM) is implemented for measuring image quality. SSIM is based on the computation of three terms, called the luminance term, the contrast term, and the structural term. The total index is a multiplicative mixture of the three terms [22].

$$SSIM(x, y) = [I(x, y)]^\alpha \cdot [C(x, y)]^\beta \cdot [S(x, y)]^\gamma \tag{6}$$

where

$$I(x, y) = \frac{2\mu_x \mu_y + C_1}{\mu_x^2 + \mu_y^2 + C_1}, \ C(x, y) = \frac{2\sigma_x \sigma_y + C_2}{\sigma_x^2 \sigma_y^2 + C_2}, \ S(x, y) = \frac{\sigma_{xy} + C_3}{\sigma_x \sigma_y + C_3}$$

where μx, μy, σx, σy, and σxy are the local means, standard deviations, and cross-covariance for images x, y. If $\alpha = \beta = \gamma = 1$ (the default for Exponents), and $C_3 = C_{2/2}$ (default selection of C_3) the index simplifies to:

$$SSIM(x, y) = \frac{(2\mu_x \mu_y + C_1)(2\sigma_{xy} + C_2)}{(\mu_x^2 + \mu_y^2 + C_1)(\sigma_x^2 + \sigma_y^2 + C_2)} \tag{7}$$

We are looking to minimize the MSE and maximize the PSNR, SSIM, and NC. The statistical parameters as the most frequent watermarked image quality metric that utilized the preceding research works. Towards this study, four statistical parameters are applied for determining the levels of strength and weakness of watermarking algorithms. So as implementing PSNR, MSE, SSIM, and NC measures to assess the robustness and imperceptibility of the medical images. Several processing attacks had been applied to distort the watermarked image to check the robustness of the medical image. In accord with that, the statistical parameters are used to evaluate the robustness of the medical image after various attacks to the watermarked medical images.

Table 1 shows the MSE, PSNR, SSIM, and NC with no attacks to evaluate the performance of the proposed algorithm. Three different medical images types are utilized as a host image (X-ray image, MRA image, and CT image). The proposed multiple wavelet family transform retains the image quality after processing very well as shown in Table 1.

Table 1 MSE, PSNR, SSIM, and NC with no attacks

Wavelet	Image	MSE	PSNR	SSIM	NC
Multiple wavelet1	X-ray	0.000005	121.8424	0.9983	1.0000
Multiple wavelet1	MRA	0.000005	121.3890	0.9971	1.0000
Multiple wavelet1	CT	0.000005	121.6397	0.9863	1.0000
Multiple wavelet2	X-ray	0.000005	121.8314	0.9983	1.0000
Multiple wavelet2	MRA	0.000005	121.3782	0.9971	1.0000
Multiple wavelet2	CT	0.000005	121.6293	0.9860	1.0000

Tables 2 and 3 shows the MSE, PSNR, SSIM, and NC with various attacks for the proposed multiple wavelets family transform. By executing attacks on the watermarked image and calculating the MSE, PSNR, SSIM, and NC, the robustness of the proposed watermark algorithm is assessed. The results are promised as represented in Tables 2 and 3.

Table 2 MSE, PSNR, SSIM and NC of different attacks for multiple wavelet1 (dmey, rbio6.8, coif5, sym8)

Type of attack	Image	MSE	PSNR	SSIM	NC
Noise attack (salt and pepper)	X-ray	0.0043	55.4792	0.7653	0.9647
Median attack		0.00009	94.0730	0.9893	0.9992
Mean attack		0.0002	86.0760	0.9836	0.9983
Gaussian noise attack		0.0007	72.3646	0.5941	0.9934
Adjust image attack		0.0004	79.7763	0.9569	0.9993
Rotation attack		0.0201	40.0094	0.7055	0.8274
Noise attack (salt and pepper)	MRA	0.0043	55.0214	0.7812	0.9743
Median attack		0.0002	84.8482	0.9836	0.9987
Mean attack		0.0005	75.9807	0.9761	0.9968
Gaussian noise attack		0.0006	74.0265	0.4847	0.9967
Adjust image attack		0.0002	82.6569	0.9900	0.9995
Rotation attack		0.0237	37.8977	0.7314	0.8534
Noise attack (salt and pepper)	CT	0.0044	54.9152	0.7650	0.9735
Median attack		0.0002	85.8437	0.9680	0.9988
Mean attack		0.0007	73.2751	0.9505	0.9957
Gaussian noise attack		0.0007	73.2951	0.5443	0.9961
Adjust image attack		0.0004	78.5835	0.9693	0.9989
Rotation attack		0.0303	35.6945	0.6097	0.8141

Table 3 MSE, PSNR, SSIM and NC of different attacks for multiple wavelet2 (rbio3.5, bior6.8, sym5, coif4)

Type of attack	Image	MSE	PSNR	SSIM	NC
Noise attack (salt and pepper)	X-ray	0.0042	55.6635	0.7702	0.9653
Median attack		0.00009	94.0748	0.9893	0.9992
Mean attack		0.0002	86.0777	0.9836	0.9983
Gaussian noise attack		0.0007	72.3436	0.5934	0.9933
Adjust image attack		0.0004	79.7785	0.9568	0.9993
Rotation attack		0.0201	40.0092	0.7056	0.8274
Noise attack (salt and pepper)	MRA	0.0045	54.5967	0.7732	0.9732
Median attack		0.0002	84.8483	0.9836	0.9987
Mean attack		0.0005	75.9802	0.9760	0.9968
Gaussian noise attack		0.0006	74.0290	0.4852	0.9967
Adjust image attack		0.0002	82.6568	0.9899	0.9995
Rotation attack		0.0237	37.8991	0.7314	0.8534
Noise attack (salt and pepper)	CT	0.0045	54.7719	0.7642	0.9731
Median attack		0.0002	85.8419	0.9678	0.9988
Mean attack		0.0007	73.2745	0.9503	0.9957
Gaussian noise attack		0.0007	73.2720	0.5434	0.9961
Adjust image attack		0.0004	78.5828	0.9689	0.9989
Rotation attack		0.0303	35.6951	0.6094	0.8141

5 Conclusion

This paper proposes two medical watermark algorithms that have utilized the different wavelet transform families in four levels discrete wavelet decomposition. The first algorithm applied dmey in the first level of decomposition, rbio 6.8 in the second level of decomposition, coiflet 5 in the third level of decomposition, and symlet 8 in the fourth level of decomposition while the second algorithm applied rbio 3.5 in the first level of decomposition, bior 6.8 in the second level of decomposition, symlet 5 in the third level of decomposition, and coiflets 4 in the fourth level of decomposition. This work is substantially robust against varying types of attacks deterring piracy and keep authenticity in IoT. Three kinds of medical images have been used to achieve the proposed watermarking algorithms. To evaluate the robustness of the proposed watermarking algorithms, various attacks are performed. PSNR, MSE, SSIM, and NC are used to evaluate the performance of the proposed algorithms. The experimental results give us pleasing values comparing with other results gained from other algorithms that had been proposed before. In future work, more families of wavelets are used and hybridization between wavelet families.

References

1. Mestiri, A., Kricha, A., Sakly, A., Mtibaa, A.: Watermarking for integrity, authentication and security of Medical Imaging. In: IEEE 14th International Multi-Conference on Systems, Signals and Devices (SSD) (2017)
2. Kunhu, A., Al-Ahmad, H., Taher, F.: Medical images protection and authentication using hybrid DWT-DCT and SHA256-MD5 hash functions. In: 24th IEEE International Conference on Electronics, Circuits and Systems (ICECS) (2017)
3. Mashalkar, S.D., Shirgan, S.S.: Design of watermarking scheme in medical image authentication using DWT and SVD technique. In: IEEE 2017 International Conference on Computing Methodologies and Communication (ICCMC) (2017)
4. Benyoussef, M., Mabtoul, S., El marraki, M., Aboutajdine, D.: Medical image watermarking for copyright protection based on visual cryptography. In: 2014 International Conference on Multimedia Computing and Systems (ICMCS) (2014)
5. Li, J., Han, X., Dong, C., Chen, Y.W.: Robust multiple watermarks for medical image based on DWT and DFT. In: IEEE 6th International Conference on Computer Sciences and Convergence Information Technology (ICCIT) (2011)
6. Md. Moniruzzaman, Md. Hawlader, A., Md. Hossain, F.: Wavelet based watermarking approach of hiding patient information in medical image for medical image authentication. In: IEEE 17th International Conference on Computer and Information Technology (ICCIT) (2014)
7. Mavromoustakis, C.X., Mastorakis, G., Batalla, J.M.: A mobile edge computing model enabling efficient computation offload-aware energy conservation. IEEE Access 7, 102295–102303 (2019)
8. Alshayea, T.K., Mavromoustakis, C.X., Batalla, J.M., Mastorakis, G., Mukherjee, M., Chatzimisios, P.: Efficiency-aware watermarking using different wavelet families for the internet of things. In: Accepted in IEEE International Conference on Communications (ICC) 2019 in track Communication QoS, Reliability and Modeling Symposium—Communication QoS, Reliability and Modeling, 20–24 May 2019, Shanghai, China
9. Skourletopoulos, G., Mavromoustakis, C.X., Mastorakis, G., Sahalos, J.N., Batalla, J.M., Dobre, C.: Cost-benefit analysis game for efficient storage allocation in cloud-centric internet of things systems: a game theoretic perspective. In: 2017 IFIP/IEEE Symposium on Integrated Network and Service Management (IM), Lisbon, Portugal, 8–12 May 2017
10. Al-Shayea, T.K., Mavromoustakis, C.X., Batalla, J.M., Mastorakis, G.: A hybridized methodology of different wavelet transformations targeting medical images in IoT infrastructure. Measurement 148 (2019)
11. Kryftis, Y., Mastorakis, G., Mavromoustakis, C.X., Batalla, J.M., Rodrigues, J.J.P.C., Dobre, C.: Resource usage prediction models for optimal multimedia content provision. IEEE Syst. J. 11, 2852–2863 (2016)
12. Bourdena, A., Mavromoustakis, C.X., Mastorakis, G., Rodrigues, J.J.P.C., Dobre, C.: Using socio-spatial context in mobile cloud process offloading for energy conservation in wireless devices. IEEE Trans. Cloud Comput. 7, 392–402 (2019)
13. Mavromoustakis, C.X., Mastorakis, G., Pallis, E., Mysirlidis, C., Dagiuklas, T., Politis, I., Dobre, C., Papanikolaou, K.: On the perceived quality evaluation of opportunistic Mobile P2P Scalable Video streaming. In: IEEE 2015 International Wireless Communications and Mobile Computing Conference (IWCMC), Dubrovnik, Croatia, 24–28 Aug 2015
14. Mavromoustakis, C.X., Batalla, J.M., Mastorakis, G., Markakis, E., Pallis, E.: Socially-oriented edge computing for energy-awareness in IoT architectures. IEEE Commun. 56, 139–145 (2018)
15. Mavromoustakis, C.X., Mastorakis, G., Batalla, J.M., Chatzimisios, P.: Social-oriented mobile cloud offload processing with delay constraints for efficient energy conservation. In: IEEE ICC 2017 Next Generation Networking and Internet Symposium (main track), 21–25 May 2017, Paris, France
16. Markakis, E.K., Politis, I., Lykourgiotis, A., Rebahi, Y., Mastorakis, G., Mavromoustakis, C.X., Pallis, E.: Efficient next generation emergency communications over multi-access edge computing. IEEE Commun. Mag. 55, 92–97 (2017)

17. Xu, X., Li, D., Sun, M., Yang, S., Yu, S., Manogaran, G., Mastorakis, G., Mavromoustakis, C.X.: Research on key technologies of Smart campus teaching platform based on 5G network. IEEE Access **7**, 20664–20675 (2019)
18. Markakis, E.K., Karras, K., Zotos, N., Sideris, A., Moysiadis, T., Corsaro, A., Alexiou, G., Skianis, C., Mastorakis, G., Mavromoustakis, C.X., Pallis, E.: EXEGESIS: extreme edge resource harvesting for a virtualized fog environment. IEEE Commun. Mag. **55**, 173–179 (2017)
19. Batalla, J.M., Krawiec, P., Mavromoustakis, C.X., Mastorakis, G., Chilamkurti, N., Négru, D., Queyreix, J.B., Borcoci, E.: Efficient media streaming with collaborative terminals for the smart city environment. IEEE Commun. Mag. **55**, 98–104 (2017)
20. Al-Shayea, T.K., Mavromoustakis, C.X., Batalla, J.M., Mastorakis, G., Markakis, E.K., Pallis, E.: On the efficiency evaluation of a novel scheme based on Daubechies wavelet for watermarking in 5G. In: IEEE 23rd International Workshop on Computer Aided Modeling and Design of Communication Links and Networks (CAMAD) (2018)
21. Mallat, S.: A Wavelet Tour of Signal Processing The Sparse Way. Elsevier Inc., (2009)
22. https://www.mathworks.com/help/images/ref/ssim.html
23. http://wavelets.pybytes.com/wavelet/db5/. Last accessed 4 April 2019
24. Daubechies: Orthonormal bases of compactly supported wavelets. J. SIAM Math. Anal. **24**(2), 499–519 (1993)
25. Beylkin, G., Coifman, R., Rokhlin, V.: Fast wavelet transforms and numerical algorithms. J. Commun. Pure Appl. Math. **44**, 141–183 (1991)
26. Zaeni, A., Kasnalestari, T., Khayam, U.: Application of wavelet transformation symlet type and coiflet type for partial discharge signals denoising. In: 5th International Conference on Electric Vehicular Technology (ICEVT) (2018)
27. Chui, C.: Wavelet Analysis and its Applications. Wavelets: A Tutorial in Theory and Applications. Academic Press, San Diego, CA (1992)
28. Gao, R., Yan, R.: Discrete Wavelet Transform. Wavelets Theory and Applications for Manufacturing. Springer (2011)

Optimised Statistical Model Updates in Distributed Intelligence Environments

Ekaterina Aleksandrova and Christos Anagnotopoulos

Abstract This paper explores a sequential decision making methodology of *when to update* statistical learning models in Intelligent Edge Computing devices given underlying changes in the contextual data distribution. The proposed model update scheduling takes into consideration the optimal decision time for minimizing the network overhead while preserving the prediction accuracy of the models. The paper reports on a comparison between the proposed approach with four other update delaying policies found in the literature, an evaluation of the performances using linear and support vector regression models over real contextual data streams and a discussion on the strengths and weaknesses of the proposed policy.

1 Introduction

The Internet of Things (IoT) environments have been gaining popularity since the end of the 20th century. They have allowed the transformation of small computing environments into large scale ecosystems, whose cores (cloud data centers) require a massive computational power to process all the received contextual data along with a substantial network overhead in order to receive all the raw data.

However, as this has proven to be inefficient, a computing paradigm has been provided in the form of Edge Computing (EC), whose rationale is to *push most of the computations to the edge of the network*, e.g. sensors, mobile devices, etc. This allows to take advantage of the computational and sensing capabilities of modern devices and deliver the locally processed contextual data to the cloud in the form of partially or entirely extracted knowledge [3]. The edge-centric rationale also contributes as a security measure as raw data is processed locally on the sensing device

E. Aleksandrova (✉) · C. Anagnotopoulos
Pervasive & Distributed Intelligence Laboratory, University of Glasgow,
Glasgow G12 8QQ, UK
e-mail: 2133352a@student.gla.ac.uk

C. Anagnotopoulos
e-mail: christos.anagnostopoulos@glasgow.ac.uk

© Springer Nature Switzerland AG 2020
G. Mastorakis et al. (eds.), *Convergence of Artificial Intelligence and the Internet of Things*, Internet of Things,
https://doi.org/10.1007/978-3-030-44907-0_3

and as a method for decreasing network traffic, as the delivered statistical learning model representations of the data have significantly smaller size compared to the raw contextual sensed data. The EC methodology is expected to significantly reduce the required computational power. However, if the acquired knowledge is still transferred to the Cloud on each iteration, then the considerable communication overhead remains along with other complications explained further ahead in this paper.

1.1 Problem Statement

The main problem which this work aims is to estimate an *optimal waiting time criteria* based on made sequential observations on multivariate contextual data. The proposed scheme should be able to decide when a model representation of the sensed data is to be sent from an edge node to an edge gateway in order to minimise the communication overhead and, in parallel, to preserve or increase the accuracy of the generated statistical models. This can be expressed as the following research question:

How can the delivery waiting time and model accuracy at the edge gateway be maximised while minimising the communication overhead and computational complexity at the sensing edge node level?

1.2 Paper Organisation

The paper is organised as follows: Sect. 2 presents related work on the current state of the art on this research topic. Section 3 introduces the proposed methodology followed for this project including the rationale behind the proposed algorithm built on the principles of the Optimal Stopping Theory. Section 4 reports on the performance assessment and comparison of the proposed method with other sequential decision making policies and algorithms using real data sets and provides a summarised analysis. Section 5 includes the concluding remarks and ideas on future development of this work.

2 Related Work and Background

Given a sample path in an IoT system consisting of multiple sensing nodes, a local edge gateway and a central data node as en endpoint, the multi-hop transmission can increase energy consumption when the gateway is busy and the data has to be passed to the central node [16]. When data processing is not performed at the edge level, if N multivariate data points are sensed and sent through the network, that results in N number of transmissions. However, if the data X is locally processed at the edge, the result can be a linear model representation of the data e.g., $y = \alpha \cdot X + \beta$, where only the parameters α, β are being passed through the network resulting in a reduced energy consumption [19].

Wireless Sensor Networks (WSNs) in IoT environments used in remote locations (e.g. rain forest sensors [21], surface and mine monitoring [2], water pollution detectors [1], etc.) are usually being powered by a battery. Therefore, the energy consumption of the sensing devices should be minimised as much as possible to reduce the amount of human interaction required to replace the exhausted energy source. Such interaction with the device may distort the environment and introduce incorrect data. If the sensors are placed at locations with difficult access, that could also increase the financial cost or even cause danger to the person responsible for the replacement.

Other similar sensor systems rely on renewable energy, most widely used is solar energy, which could be slower to harvest in environments such as rain forests [21], therefore, having a power efficient device is required in order to have a well-functioning system. Another concern with communicating processed data on every sensing and reporting iteration is that it is possible that the data contains bias, e.g. missing or corrupted data points [4]. When passed to the gateway it can potentially be used to make inaccurate predictions of the sensed environment. This can be avoided by *delaying* the model update before making sure that the bias is not in fact a novelty in the data.

A way of minimising the energy cost of an edge device even more is by reducing the number of the data transmissions. This could be performed by introducing a certain delay in the delivery of the up-to-date model [3–5], at the cost of allowing a reasonable error in the data. Based on this delay-based rationale, the approach adopted in this work is as follows: *a data model is generated at the edge node and then sent to the edge gateway, on the next iteration. In the case that the node does not detect a significant change between the initially modelled data and the currently sensed dataset, no communication is made between the edge node and the edge gateway. Making the decision when to send a new updated model to the edge gateway before the accuracy has degraded beyond a reasonable threshold should be looked from a slightly different angle.*

2.1   Optimised Sequential Decision Making

The Optimal Stopping Theory (OST) attempts to solve exactly the problem stated above, i.e., choosing a *best* time instance to take a given action: stop observing changes on the derived model and send the model to the edge gateway. The action is based on sequentially observed random variables from a distribution in order to maximise an expected reward or minimise an expected cost. Based on this abstraction of the considered problem, several OST-based variations are adopted as a basis to problems in Computing Science notably: the called *House-Selling* problem and the *Quickest Change Detection* problem [6].

The House-Selling problem explores the attempt to stop and sell a house at the highest offer by taking into account a cost, such as living cost or real estate agency commission, for each rejected offer. This problem relates to the approach that waiting can be penalised, the way Tian et al. use it in their "cost-aware" update policy algorithm [20]. In our context, the decision on whether to update an outdated statistical learning model with the current one is based on how much it was lost (how big the "regret" is) since the update with the better model was not performed in the past. Once a prediction quality tolerance is exceeded, an update is inevitable, which also prevents additional retraining computations on the edge node.

The Change Point Detection problem deals with the exploration of the context distribution and stopping when a change is detected but penalising when the algorithm has signalled a false alarm [6]. Research has been mainly focusing on two major approaches based on assumptions on the underlying data distribution. Shiryaev [17] provides an optimal solution to the problem using a Bayesian approach. While, on the other hand, there is a non-Bayesian approach using the well-known CuSum algorithm provided by Page [15]. The latter was proven to be optimal by Moustakides [14] based on Lorden's formulation of the problem [11].

In a recent publication, Lau and Tay [18] introduce the idea of having a "critical change" and a "nuisance change" in a distribution, where the critical change is of highest importance and should be detected, as opposed to the nuisance change which should be ignored by the algorithm. They propose two algorithms using a Bayesian and a non-Bayesian approach, which then are compared to the performance for both types of change against the naive 2-stage procedures by Page [15] and Shiryaev [17]. For their non-Bayesian approach, the authors compare their proposed algorithm with the naive 2-stage CuSum. The naive approach uses two separate thresholds, β_c for the critical change and β_n for the nuisance change. First they are applied on the cumulative sum of the log-likelihood ratio of the normal distribution and then on the distributions after the nuisance and the critical change. According to the authors, their non-Bayesian procedure, which is based on the Generalised Likelihood Ratio test, performs more accurately than the naive 2-stage optimal CuSum algorithm.

2.2 Contribution

The contributions of this paper are as follows:

1. An analytical optimisation model derived from the optimal stopping rule aimed at minimising the communication overhead at EC systems.
2. A comparison of the derived rule to an established change detection rule and an intuitive base method.
3. A method for hyperparameter optimisation when applying the optimal postponing rule.

3 Methodology

The main focus of this work is to present and experiment with the proposed *time-optimised model update postponing* policy/algorithm for reducing the communication rate but preserving the quality of the context. We assume that the distributed/EC environment contains at least sensor node and an edge gateway device.

Initially, the sensing node is responsible for gathering environmental/multivariate contextual data and generating a statistical/Machine Learning (ML) model from the first w contextual data observations. The ML model is then temporarily stored on the sensing device but also is communicated with the edge gateway. On each next observation, the piece of data is received and appended to the currently stored sensor data set of size w, while the datum with the oldest timestamp is discarded. A new model is generated or incrementally updated representing the current updated dataset. By fitting the dataset on the old and the new ML models, we obtain the two Mean Squared Errors (MSE):

- MSE e from the ML model the edge gateway, which has been previously received;
- MSE e' from the ML model based on the most up-to-date sensed data at the edge node.

Given that the predictability of the latest ML model of the edge node satisfies a certain criteria, the decision is made to communicate the current ML model to the edge gateway. This also requires that the temporarily stored outdated ML model at the edge device is also updated to correctly represent the disseminated local knowledge in the eco-system.

3.1 *Optimal Postponing Policy*

The Optimal Postponing (OP) policy for ML model update is based on the change in the distribution using the cumulative sum of the absolute error difference between the two MSEs e and e' at time instance/observation $t > 0$. Consider the following definitions:

$$Z_t = \Delta e_t = |e'_t - e_t| \tag{1}$$

$$S_{t+1} = \Delta e_{t+1} + \Sigma_{k=0}^t \Delta e_k \tag{2}$$

$$S_t = S_{t-1} + Z_t \tag{3}$$

A prediction quality tolerance $\Theta > 0$ is defined to determine the acceptable error sum, which will be used to determine if the edge node should proceed with a ML model update to the edge gateway or not. Under this context, we define the following random variable V_t as the reward of an update decision:

$$V_t = \begin{cases} t, & \text{if } S_t \leq \Theta, \\ -B, & \text{if } S_t > \Theta. \end{cases} \tag{4}$$

The penalty factor $B > 0$ denotes that the current cumulative error exceed the error tolerance Θ, thus, the edge node could have sent the updated ML model to the edge gateway. On the other hand, to be communication efficient, the edge node desires to postpone the ML model update, thus, increasing the value of V_t. However, as long as the edge node delays the model update, then the cumulative error S_t approaches Θ. The problem is for the cumulative error S_t to reach as close to the tolerance Θ as possible, but without exceeding this. Hence, we then formalise the OST problem:

Problem 1 The edge node should find the best time instance t^* such that the expected reward for delaying a ML model update is maximized, i.e., the optimal stopping time t^* obtains the following essential supremum

$$\text{ess} \sup_t \mathbb{E}[V_t].$$

We then provide the solution of the optimal stopping time estimate as stated in the Theorem 1.

Theorem 1 *The optimal stopping time t^* that maximizes the essential supremum* $\text{ess} \sup_t \mathbb{E}[V_t]$ *of the reward function on the edge node is the first time instance $t > 0$ such that:*

$$F_Z(\Theta - S_t) \leq \frac{t + B}{t + 1 + B},$$

where $F_Z(z)$ is the cumulative distribution function (CDF) of the error differences Z_t of the old and the updated ML models e_t and e_t', respectively.

Proof From the reward function in (4) we can derive the expected current reward $\mathbb{E}[V_t]$.

$$\begin{aligned} \mathbb{E}[V_t] &= t \cdot P(S_t \leq \Theta) + (-B \cdot P(S_t > \Theta)) \\ &= t \cdot P(S_t \leq \Theta) - B \cdot (1 - P(S_t \leq \Theta)) \\ &= (t + B) \cdot P(S_t \leq \Theta) - B, \end{aligned} \tag{5}$$

Let now the filtration $\mathbb{F}_t = \{S_1, S_2, \ldots, S_t\} \cup \{Z_1, Z_2, \ldots, Z_t\}$ be the realisation of all random variables up to t. This is the whole information the edge node has accumulated up to time instance t. Then, the conditional expectation of the reward V_{t+1} given filtration $\mathbb{F}_t$ is then expressed as a martingale:

$$\mathbb{E}[V_{t+1}|\mathbb{F}_t] = (t + 1 + B) \cdot P(S_{t+1} \leq \Theta|\mathbb{F}_t) - B \tag{6}$$

However, it is known from Eq. (3) that:

$$S_{t+1} = S_t + Z_{t+1} \tag{7}$$

Therefore, the probability of the sum at time instance $t + 1$ being less then or equal to Θ given filtration $\mathbb{F}$ equals the cumulative distribution function of Z taking a value less than or equal to $\Theta - S_t$

$$P(S_{t+1} \leq \Theta | \mathbb{F}_t) = P(S_t + Z_{t+1} \leq \Theta | \mathbb{F}_t) \tag{8}$$
$$= P(Z_{t+1} \leq \Theta - S_t | \mathbb{F}_t) \tag{9}$$
$$= F_Z(\Theta - S_t) \tag{10}$$

Hence, we obtain that:

$$\mathbb{E}[V_{t+1} | \mathbb{F}_t] = (t + 1 + B) \cdot P(S_{t+1} \leq \Theta | \mathbb{F}_t) - B \tag{11}$$
$$= (t + 1 + B) \cdot P(S_t + Z_{t+1} \leq \Theta | \mathbb{F}_t) - B \tag{12}$$
$$= (t + 1 + B) \cdot P(Z_{t+1} \leq \Theta - S_{t+1} | \mathbb{F}_t) - B \tag{13}$$
$$= (t + 1 + B) \cdot F_Z(\Theta - S_t) - B \tag{14}$$

We aim to postpone sending the ML model if we know that the next iteration will increase the reward value. That is why we stop at the first instance t where the current reward is more than the expected future reward at time instance $t + 1$ or $V_t \geq \mathbb{E}[V_{t+1} | \mathbb{F}_t]$. Based on this, we obtain that:

$$V_t \geq \mathbb{E}[V_{t+1} | \mathbb{F}_t] \tag{15}$$
$$t \geq (t + 1 + B) \cdot F_Z(\Theta - S_t) - B \tag{16}$$
$$F_Z(\Theta - S_t) \leq \frac{t + B}{t + 1 + B}, \tag{17}$$

which completes the proof of Theorem 1.

Algorithm 3.1 illustrates the local process in the edge node.

Algorithm 3.1: Time-optimised Policy (OP)

$rewardDist \leftarrow$ obtainRewardDist()   ▷ Appendix 2
receive(X_0, y_0, X_1, y_1); $initModel \leftarrow$ **fit**(X_0, y_0); $model \leftarrow$ **fit**(X_1, y_1)
$e \leftarrow$ **error**$(initModel, X_1, y_1)$
$e' \leftarrow$ **error**$(model, X_1, y_1)$
$Z \leftarrow | e - e' |$
$S \leftarrow Z$; $t \leftarrow 1$
while $true$ **do**
    receive(X, y); $newModel \leftarrow$ **fit**(X, y)
    $e \leftarrow$ **error**$(model, X, y)$
    $e' \leftarrow$ **error**$(newModel, X, y)$

$Z \leftarrow |e - e'|$
if cdf($rewardDist, (\Theta - S)) \leq \frac{(t+B)}{(t+1+B)}$ **then**
    $model \leftarrow newModel$ ▷ update & communicate model
    $S \leftarrow Z; t \leftarrow 1$
else
    $S \leftarrow S + Z; t \leftarrow t + 1$
end if
end while

3.2 Policies Under Comparison

In order to evaluate the performance of the presented algorithm 4 other policies are implemented, which rely either on certain statistics of the data or on pure randomness: the Median-based Policy, the Random Policy, the Accuracy Policy, and the optimal CuSum Policy [7].

3.2.1 Median-Based Policy

The Median Policy has an initially learned median from the previously seen data. It is calculated from the absolute error difference values given that the initial model is never updated. The rationale behind this is to explore the worst case scenario of the error difference. This assumes that the closer the timestamps of two models are, the lower the error difference between them is and on the contrary, the further away the timestamps of two models are, the higher their error difference is.

Using a fraction α of the median value the policy determines whether the current absolute error difference is a tolerable amount or if it indicates that the newly sensed data is significantly outdated and needs to be communicated with the edge gateway.

This policy is expected to reduce communication by being more tolerant towards initial small abrupt changes but detects continuously increasing values.

Algorithm 3.2.1: Median-based Policy

procedure OBTAINMEDIAN
    receive(X, y); $model \leftarrow$ **fit**(X, y)
    for $t \leftarrow 1$ **to** T **do**
        receive(X, y); $newModel \leftarrow$ **fit**(X, y)
        $e \leftarrow$ **error**$(model, X, y)$
        $e' \leftarrow$ **error**$(newModel, X, y)$
        $E \leftarrow E \cup \{|e - e'|\}$
    end for
    $median \leftarrow$ **median**(E)
end procedure

```
receive(X, y); model ← fit(X, y)
while true do
    receive(X, y); newModel ← fit(X, y)
    e ← error(model, X, y)
    e' ← error(newModel, X, y)
    if | e − e' |> α · median then
        model ← newModel                    ▷ update & communicate model
    end if
    if t  mod T = 0 then
        updateMedian()
    end if
end while
```

3.2.2   Random-Based Policy

The random policy is intended to be sending updates of the latest model to the edge gateway with the same probability as the OST policy.

The aim of the policy is to show that even if the number of updates is approximated to the optimal value, the accuracy of the models at the edge gateway would still suffer a decrease.

Algorithm 3.2.2: Random-based Policy

```
receive(X, y); model ← fit(X, y)
while true do
    receive(X, y); newModel ← fit(X, y)
    rand ← generate({0, 1}, probability)
    if rand = 1 then
        model ← newModel                    ▷ update & communicate model
    end if
end while
```

3.2.3   Accuracy-Based Policy

The accuracy policy works by comparing the MSE of the old model and the most up-to-date model. Once it detects a decrease in the accuracy the latest model is sent to the edge gateway. The policy aims to present a naive algorithm for reducing communication but preserving accuracy as high as possible.

Algorithm 3.2.3: Accuracy-based Policy

```
receive(X, y); model ← fit(X, y)
while true do
    receive(X, y); newModel ← fit(X, y)
    e ← error(model, X, y)
    e' ← error(newModel, X, y)
    if e > e' then
        model ← newModel                    ▷ update & communicate model
    end if
end while
```

3.2.4   CuSum Policy

The algorithm implemented here relies on the direct form of the CuSum algorithm explained by Granjon [7].

For the execution of the algorithm, two assumptions of the distributions are made representing the two hypothesis, where the algorithm aims to decide which hypothesis represents the current sample of data. Firstly, the $\mathcal{H}_0$ no change hypothesis refers to the "good distribution" which is assumed to be the distribution of the absolute error difference when the edge gateway model is communicated on every iteration. On the contrary, the "bad distribution" represented by the changed distribution hypothesis $\mathcal{H}_1$ is assumed to be the distribution of the absolute error difference when the model is communicated only on the first iteration and then never again.

The assumption is made that the absolute error differences $Z_0, Z_1, \ldots, Z_n$ are continuous independent random variables. In this case a gamma distribution is a good choice to represent the data (Fig. 1), relying on the two flexible parameters for scale and shape of the probability density function of the distribution.

When a new absolute error difference is calculated it is then fitted in the probability density function for each of the two distributions. The logarithmic ratio of the probability of the absolute error belonging to either the good or the bad distribution determines whether $\mathcal{H}_0$ or $\mathcal{H}_1$ is rejected. Once the cumulative sum of the logarithmic ratios passes an initially determined threshold Φ the model is communicated to the edge gateway.

Algorithm 3.2.4: *CuSum* Policy

```
badDist ← obtainBadDist(); goodDist ← obtainGoodDist()        ▷ Appendix 1
receive(X_0, y_0, X_1, y_1); initModel ← fit(X_0, y_0); model ← fit(X_1, y_1)
e ← error(model, X_1, y_1)
e' ← error(initModel, X_1, y_1)
Z ←| e − e' |
P_0 ← pdf(goodDist, Z); P_1 ← pdf(badDist, Z)
logRatio ← log(P_1/P_0)
```

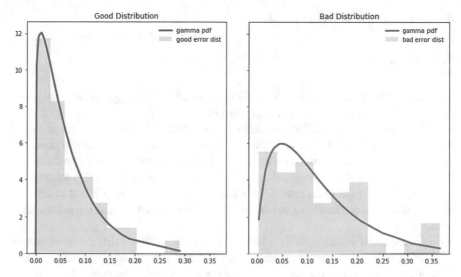

Fig. 1 Gamma distribution approximation on GNFUV data using Linear Regression

$S \leftarrow logRatio; logVector \leftarrow \{S\}$
while $True$ **do**
  receive$(X, y); newModel \leftarrow$ fit(X, y)
  $e \leftarrow$ **error**$(model, X, y)$
  $e' \leftarrow$ **error**$(newModel, X, y)$
  $Z \leftarrow | e - e' |$
  $P_0 \leftarrow$ **pdf**$(goodDist, Z),$ $P_1 \leftarrow$ **pdf**$(badDist, Z)$
  $logRatio \leftarrow$ **log**(P_1/P_0)
  $S \leftarrow S + logRatio$
  $decisionValue \leftarrow S - \min(logVector)$
  $logVector \leftarrow logVector \cup \{S\}$
  if $decisionValue > \Phi$ **then**
    $model \leftarrow newModel$ $\triangleright$ update & communicate model
    $S \leftarrow logRatio;$
    $logVector \leftarrow \{S\}$
  end if
end while

4   Performance Evaluation

4.1   Data Sets

Two time-series datasets will be used in order to apply the proposed approach and perform the previously discussed experiments.

• **GNFUV Unmanned Surface Vehicles Sensor Data Data Set** [8]—a timestamped dataset produced by 4 Unmanned Surface Vehicle sensing devices which observed and recorded the multivariate contextual environmental data using temperature and humidity sensors. The collected data will be associated with the Linear Regression task performed in the experiments for each policy. The generated linear regression models are aimed at predicting the environmental humidity based on the sensed temperature.

• **Gas sensors for home activity monitoring Data Set** [9]—a timestamped dataset containing readings from temperature, humidity and 8 metal-oxide(MOX) gas sensors of a contained environment whether or not a stimuli is presented. The collected data will be used in the Support Vector Regression(SVR) (with an RBF kernel) task in order to assess the 5 policies. The generated support vector models are aimed at predicting the level of MOX gas based on the vector of the detected humidity and temperature.

4.2   Experimentation with Linear Regression Models

The experiment using linear regression models was performed using the initial 100 datapoints of the dataset for any pre-processing analysis depending on the targeted policy and the rest of the dataset was used to simulate an online approach using each of the 5 policies.

Three of the policies require parameterisation which needs to be mentioned in advance along with the global parameter on the window size, which is set to be 25 for all simulations.

The Median policy (policy M) was performed using an α fraction of the median of the first 100 datapoints, in this case α was set to 0.5.

The CuSum policy (policy C) relies on a threshold parameter $\Phi = 2$, up to which the cumulative sum of the logarithmic ratio is allowed to increase.

The OP policy requires two parameter values, one for the error sum threshold and one for the penalty value. The boxplot analysis was performed by specifying a threshold Θ for the OP policy and running the simulation with penalty values ranging from 2 to 15.

In Fig. 2 for the USV sensor "pi4" is shown that the highest mean on update delay is achieved using a threshold $\Theta = 3$ and a penalty value $B = 8$. These values are then used in the complete simulation of the policy comparison. The absolute error rate and the communication rate for each policy are depicted in Figs. 3 and 4 respectively.

Fig. 2 Waiting time boxplot for $\Theta = 3$ and B values in the range 2–15 for Linear Regression; waiting time has the highest mean with $B = 8$

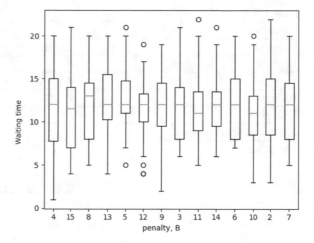

Fig. 3 Absolute error difference for USV sensor system, s = pi4, w = 25, using Linear Regression

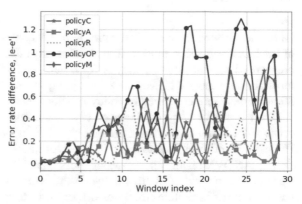

Fig. 4 Communication for USV sensor system, s = pi4, w = 25, using Linear Regression

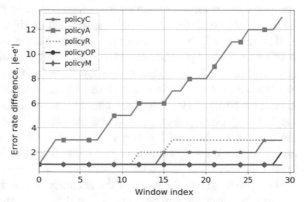

Table 1 Tukey's HSD test results

Waiting time	
Sensor name	s.s.diff. (higher mean for policyOP)
USV pi2	✓
USV pi3	✓
USV pi4	✓
USV pi5	✗
Absolute error difference	
Sensor name	No s.s.diff.
USV pi2	✓
USV pi3	✗ (higher mean for policyOP)
USV pi4	✓
USV pi5	✓

As can be seen from Fig. 3, the plot is very inconsistent due to the low accuracy of the linear regression models. The absolute error difference for the optimal policy preserves a similar rate to policy A for 6 iterations which is aimed at keeping the most accurate model. However, as seen in Fig. 4, the optimal policy waits close to 30 iterations more compared to policy A before sending an update.

The number of iterations between two model updates are referred to as waiting times and they are further investigated by performing an ANOVA test and a Tukey's Honestly Significant Difference (HSD) test for multiple mean differences.

The ANOVA results show p-values less than 5% which point to a statistically significant difference (s.s.diff.) between the policy means for each sensor on its waiting time and absolute error rate. We perform a follow-up Tukey HSD test, which compares the means for each couple of policies and the results are summarise in Table 1. The results for sensor pi4 is illustrated in Figs. 5 and 6.

The plot in Fig. 5 shows that optimal policy has a statistically significant higher waiting time compared to policy M, policy C and policy A and no significant difference with the Random policy (policy R) as intended. This shows the advantage in communication reduction introduced by the optimal policy where the update is introduced further in the future compared to the other policies. Moreover, we perform the Tukey's HSD test on the absolute error difference of each policy obtained during the simulation along with policy E. Policy E communicates the up-to-date models on each iteration and is intended as a lower bound which shows the absolute error rate of the models at its lowest.

The results in Fig. 6 show that the accuracy of the models does not suffer a significant deterioration compared to policy E and the other communication reduction policies but preserves the highest postponing time. Moreover, the plot shows that randomly updating the models leads to a significant increase in the absolute error, therefore, the optimal policy does not just correctly calculate the optimal number of updates but also the optimal time between the updates.

Fig. 5 Policy comparison on waiting time using using Linear Regression

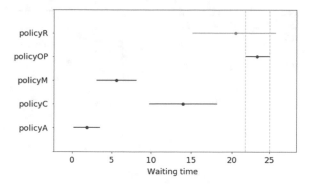

Fig. 6 Policy comparison on absolute error rate time using Linear Regression

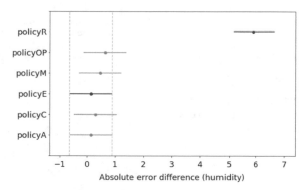

The results for sensor "pi5" show that the optimal policy does not reduce communication, however, that reflects in a low absolute error difference compared to the accuracy-based (policy A) policy. The experiment on sensor "pi3" has caused a significantly higher waiting time, however, it has also resulted in a low absolute error rate. That may be as a result from not performing hyperparameter tuning on the prediction tolerance parameter and instead setting a static value. The statically set threshold value for all sensors has proven to keep the absolute error low for three of the sensors.

4.3 Experimentation with Support Vector Regression Models

After some analysis of the data it was found that the initial 100 datapoints of the HT sensor dataset contain a single drift in the observed pattern. However, the rest of the data used in the simulation of the online algorithm has no noticeable changes, which results in no communication between the edge node and the gateway. This called for placing an artificial change in the data as shown in Fig. 7: one incremental long term change and one abrupt short term change affecting only the MOX sensor data and preserving the initial datastream for the humidity and temperature. This is aimed to simulate a new pattern in the data which will cause a change in the communicated models.

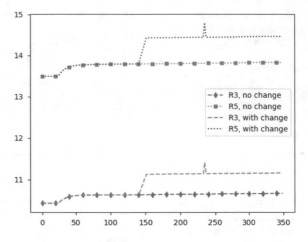

Fig. 7 Artificial change for the data from sensors R3 and R5

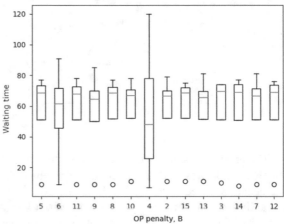

Fig. 8 Waiting time boxplot for $\Theta = 1$ and B values in the range 2–15 for Support Vector Regression; waiting time has the highest mean with $B = 3$

The incremental long term change occurs from iteration 50 of the start of the simulation, where the MOX sensor values are increased with 10% of the mean value in the next 12 iterations and after that the change is kept persistent throughout the rest of the simulation. This drift in the data is intended to serve as a critical change [18] which should be detected by the algorithm and acted upon as soon as possible.

The second abrupt short term change occurs in iteration 108. The change in the MOX sensor values is increased with 2.5% of the mean value and then decreased to the genuine datastream in the next 4 iterations. This drift in the data is intended to serve as a nuisance change [18] which should be of least importance to the algorithm.

The global parameters for this experiment were set the same way as in the experimentation using the linear regression models: window size 25 and initial preprocessing data containing 100 datapoints. The Median policy uses the α with a set value of 0.5 and the cumulative sum threshold for the CuSum policy is set to $\Phi = 0.5$.

Fig. 9 Absolute error difference for HT sensor system, s = R5, w = 25, using Support Vector Regression with RBF Kernel given an artificial change at 40th window iteration

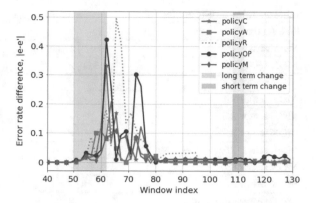

Fig. 10 Communication for HT sensor system, s = R5, w = 25, using Support Vector Regression with RBF Kernel

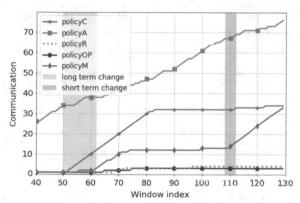

For the optimal policy, the cumulative sum threshold was set to $\Theta = 1$ as it allowed the needed sensitivity for the detection of the critical change. The boxplot in Fig. 8 shows the waiting times when performing the simulation for MOX sensor R5 using penalty values in the range 2–15. The highest mean waiting time is achieved using penalty $B = 3$. The simulation results using this setting is depicted in Figs. 9 and 10.

In Fig. 9, the absolute error line is distorted as expected right after the critical change and less affected after the nuisance change. The error appears to increase the most for the CuSum policy and the optimal policy and less for the Median policy.

However, if we focus our attention on the communication rate of the policies in Fig. 10, despite the allowed error by the CuSum policy, the critical change appears to cause a communication overhead mid and post change, where even the Median policy manages to decrease the absolute error at the cost of less communication. On the contrary, the optimal policy allows a higher error rate but at the cost of just two sent messages as a result of the critical change, which happens shortly after the change and the nuisance change does not cause an update even 20 iterations after the change.

The waiting times for each policy is investigated further again by performing an ANOVA test and a Tukey's HSD test for multiple mean differences.

Table 2 Tukey's HSD test results

Waiting time	
Sensor name	s.s.diff. (higher mean for policyOP)
MOX sensor R1	✓
MOX sensor R2	✓
MOX sensor R3	✓
MOX sensor R4	✓
MOX sensor R5	✓
MOX sensor R6	✓
MOX sensor R7	✓
MOX sensor R8	✓
Absolute error difference	
Sensor name	No s.s.diff.
MOX sensor R1	✓
MOX sensor R2	✓
MOX sensor R3	✓
MOX sensor R4	✗ (higher mean for policyOP)
MOX sensor R5	✓
MOX sensor R6	✓
MOX sensor R7	✓
MOX sensor R8	✓

Fig. 11 Policy comparison on waiting time using Support Vector Regression with RBF Kernel; optimal policy (policyOP) is the target policy and is coloured in blue and any policy coloured in red has a significant statistical difference from the target policy

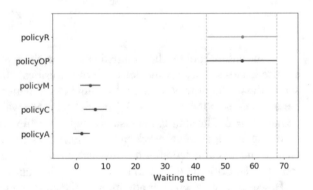

The ANOVA test produces p-values less than 5% which show that there is a statistically significant difference between the means of the waiting time and the absolute error for all sensors. Since the hypothesis that the policy means were the same was rejected, we performed the Tukey's HSD test with results summarised for all sensors in Table 2 and just for sensor "R5" plotted in Figs. 11 and 12.

The plots for sensor "R5" show a significantly higher waiting time for the optimal policy but no statistically significant difference with the base accuracy of policy E in

Fig. 12 Policy comparison on absolute error rate time using Support Vector Regression with RBF Kernel; optimal policy (policyOP) is the target policy and is coloured in blue and any policy coloured in red has a significant statistical difference from the target policy

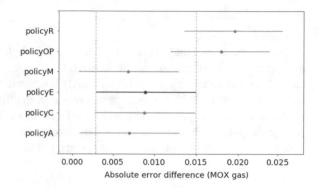

Fig. 12. Overall, that seems to be the trend for all eight sensors except sensor "R4". There we observed as expected a significantly higher waiting time for the optimal policy, however, that had resulted in causing a deterioration in the prediction accuracy of the models.

4.4 Evaluation Summary

The results of this experiment can be concluded in the summarised Table 3 which shows the trade-offs for each policy in different scenarios. For example, when dealing with models which have high quality predictions over data of a small window, if we are willing to sacrifice some of the waiting time then the CuSum policy can deliver the wanted prediction accuracy results. However, if we aim for both low error rate and less communication and we can allow tuning a prediction quality tolerance and penalty parameter then the Optimal Policy would have higher advantage over the other policies.

On the contrary, if the generated models are expected to produce low quality predictions, then the optimal policy the most appropriate option. It will reduce the communication between the sensing nodes and the edge gateway but will preserve the initial quality of the contextual data. However, if the aim is to avoid tuning parameters despite higher communication rate, then the accuracy-based policy will complete the task with lower prediction error models.

Table 3 Summarised analysis

Policy	High quality predictions	Low quality predictions
Policy C	Low error/high communication	High communication
Policy M	High communication	High communication
Policy OP	✓	✓
Policy A	High communication	Low error/high communication

5 Conclusions

This paper presented and evaluated a decision technique for optimal model update postponing at the edge. The results showed that the algorithm preserves the accuracy of the generated models and significantly reduces the communication overhead for models with both higher and lower prediction quality. Further work should be involved in deploying the algorithm in an IoT environment and evaluating and comparing the policies based on computational cost. This will show a more detailed image of any existing trade-offs of the optimal policy.

Acknowledgements This research is funded by the EU H2020 GNFUV Project RAWFIE–OC2–EXP–SCI (Grant No. 645220), under the EC FIRE+ initiative.

Appendix 1

CUSUM Policy

Function to Obtain Bad Distribution

function OBTAINBADDIST
    receive(X, y); $model \leftarrow$ **fit**(X, y)
    $\mathbf{z} \leftarrow \emptyset$
    for $t \leftarrow 1$ **to** T **do**
        receive(X, y); $newModel \leftarrow$ **fit**(X, y)
        $e \leftarrow$ **error**$(model, X, y)$; $e' \leftarrow$ **error**$(newModel, X, y)$
        $Z \leftarrow |\, e - e'\, |$
        $\mathbf{z} \leftarrow \mathbf{z} \cup \{Z\}$
    end for
    $\mu \leftarrow$ **mean**$(\mathbf{z})$; $\sigma^2 \leftarrow$ **var**$(\mathbf{z})$
    $shape \leftarrow \mu^2/\sigma^2$
    $scale \leftarrow \sigma^2/\mu$
return Gamma$(shape, scale)$
end function

Function to Obtain Good Distribution

function OBTAINGOODDIST
    receive(X, y); $model \leftarrow$ **fit**(X, y)
    $\mathbf{z} \leftarrow \emptyset$
    for $t \leftarrow 1$ **to** T **do**
        receive(X, y); $newModel \leftarrow$ **fit**(X, y)
        $e \leftarrow$ **error**$(model, X, y)$; $e' \leftarrow$ **error**$(newModel, X, y)$

$$Z \leftarrow | e - e' |$$
$$\mathbf{z} \leftarrow \mathbf{z} \cup \{Z\}$$
$$model \leftarrow newModel$$
end for
$$\mu \leftarrow \mathbf{mean}(\mathbf{z}); \sigma^2 \leftarrow \mathbf{var}(\mathbf{z})$$
$$shape \leftarrow \mu^2/\sigma^2$$
$$scale \leftarrow \sigma^2/\mu$$
return Gamma(*shape*, *scale*)
end function

Appendix 2

Time-Optimised Policy

Obtain Reward Distribution

function OBTAINREWARDDIST
   receive(X, y); $model \leftarrow$ **fit**(X, y)
   $S \leftarrow 0; \mathbf{z} \leftarrow \emptyset$
   for $t \leftarrow 1$ **to** T **do**
      receive(X, y); $newModel \leftarrow$ **fit**(X, y)
      $e \leftarrow$ **error**($model$, X, y); $e' \leftarrow$ **error**($newModel$, X, y)
      $Z \leftarrow | e - e' |; S \leftarrow S + Z$
      $\mathbf{z} \leftarrow \mathbf{z} \cup \{Z\}$
      if $S > \Theta$ **then**
         $S \leftarrow Z$
      end if
   end for
   $\mu \leftarrow$ **mean**($\mathbf{z}$); $\sigma^2 \leftarrow$ **var**($\mathbf{z}$)
   $shape \leftarrow \mu^2/\sigma^2$
   $scale \leftarrow \sigma^2/\mu$
return Gamma(*shape*, *scale*)
end function

Appendix 3

Results for Linear Regression

See Figs. 13 and 14.

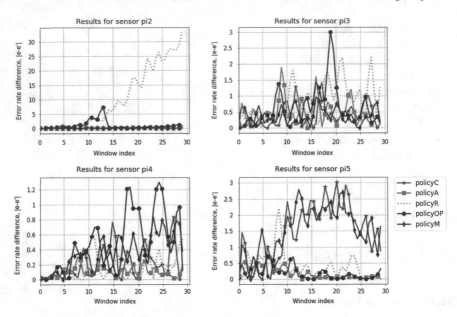

Fig. 13 Absolute error difference rate for sensors pi2 to pi5

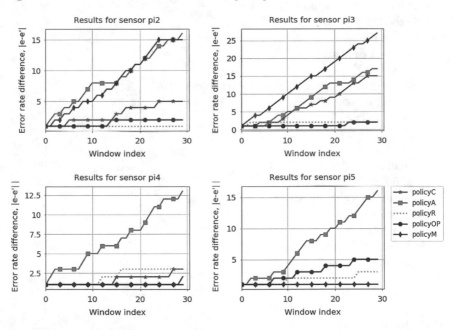

Fig. 14 Communication rate for sensors pi2 to pi5

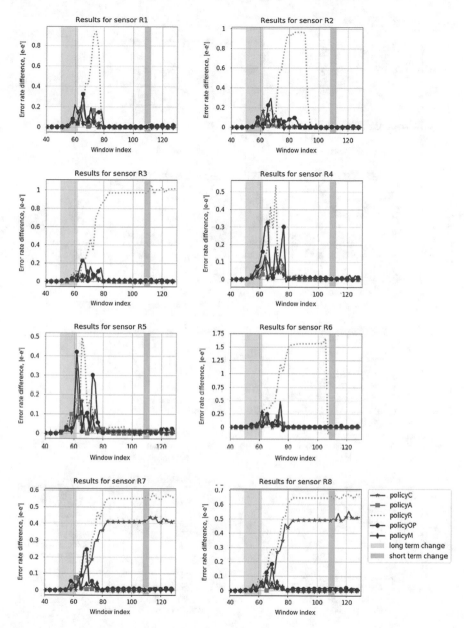

Fig. 15 Absolute error difference rate for sensors R1 to R8

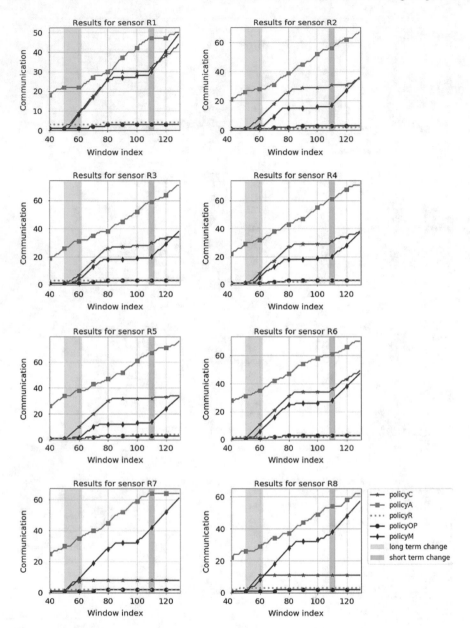

Fig. 16 Communication rate for sensors R1 to R8

Appendix 4

Support Vector Regression

See Figs. 15 and 16.

References

1. Adu-Manu, K.S., Tapparello, C., Heinzelman, W., Katsriku, F.A., Abdulai, J.D.: Water quality monitoring using wireless sensor networks: current trends and future research directions. ACM Trans. Sen. Netw. **13**(1), 4:1–4:41 (2017). https://doi.org/10.1145/3005719
2. Akkaş, M.A.: Using wireless underground sensor networks for mine and miner safety. Wireless Netw. **24**(1), 17–26 (2018). https://doi.org/10.1007/s11276-016-1313-0
3. Anagnostopoulos, C.: Time-optimized contextual information forwarding in mobile sensor networks. J. Parallel Distrib. Comput. **74**(5), 2317–2332 (2014). https://doi.org/10.1016/j.jpdc.2014.01.008
4. Anagnostopoulos, C.: Quality-optimized predictive analytics. Appl. Intell. **45**, (2016). https://doi.org/10.1007/s10489-016-0807-x
5. Anagnostopoulos, C., Kolomvatsos, K.: A delay-resilient and quality-aware mechanism over incomplete contextual data streams. Inf. Sci. **355–356**, 90–109 (2016). https://doi.org/10.1016/j.ins.2016.03.020. http://www.sciencedirect.com/science/article/pii/S0020025516301670
6. Ferguson, T.S.: Optimal stopping and applications. http://www.math.ucla.edu/~tom/Stopping/sr1.pdf
7. Granjon, P.: The cusum algorithm: a small review (2013)
8. Harth, N., Anagnostopoulos, C.: Edge-centric efficient regression analytics (2018). http://eprints.gla.ac.uk/160937/
9. Huerta, R., Mosqueiro, T., Fonollosa, J., Rulkov, N.F., Rodriguez-Lujan, I.: Online decorrelation of humidity and temperature in chemical sensors for continuous monitoring. Chemom. Intell. Lab. Syst. **157**, 169–176 (2016). https://doi.org/10.1016/j.chemolab.2016.07.004. http://www.sciencedirect.com/science/article/pii/S0169743916301666
10. Liu, Y., Yang, C., Jiang, L., Xie, S., Zhang, Y.: Intelligent edge computing for iot-based energy management in smart cities. IEEE Netw. **33**(2), 111–117 (2019). https://doi.org/10.1109/MNET.2019.1800254
11. Lorden, G.: Procedures for reacting to a change in distribution. Ann. Math. Stat. **42**, (1971). https://doi.org/10.1214/aoms/1177693055
12. Markakis, E.K., Karras, K., Zotos, N., Sideris, A., Moysiadis, T., Corsaro, A., Alexiou, G., Skianis, C., Mastorakis, G., Mavromoustakis, C.X., Pallis, E.: Exegesis: extreme edge resource harvesting for a virtualized fog environment. IEEE Commun. Mag. **55**(7), 173–179 (2017). https://doi.org/10.1109/MCOM.2017.1600730
13. Mavromoustakis, C., Batalla, J., Mastorakis, G., Markakis, E., Pallis, E.: Socially oriented edge computing for energy awareness in iot architectures. IEEE Commun. Mag. **56**(7), 139–145 (2018). https://doi.org/10.1109/MCOM.2018.1700600
14. Moustakides, G.: Optimal stopping times for detecting changes in distributions. Ann. Stat. **14**, (1986). https://doi.org/10.1214/aos/1176350164
15. Page, E.S.: Continuous inspection schemes. Biometrika **41**(1–2), 100–115 (1954). https://doi.org/10.1093/biomet/41.1-2.100
16. Shi, W., Cao, J., Zhang, Q., Li, Y., Xu, L.: Edge computing: vision and challenges. IEEE Internet Things J. **3**, 1–1 (2016). https://doi.org/10.1109/JIOT.2016.2579198

17. Shiryaev, A.: On optimum methods in quickest detection problems. Theory Probab. Appl. **8**(1), 22–46 (1963). https://doi.org/10.1137/1108002
18. Siang Lau, T., Tay, W.P.: Quickest change detection under a nuisance change, pp. 6643–6647 (2018). https://doi.org/10.1109/ICASSP.2018.8462436
19. Tan, Z., Liu, Y., Zhang, Z.: Performance requirements on energy efficiency in WSNS. **3**, (2011). https://doi.org/10.1109/ICCRD.2011.5764269
20. Tian, H., Yu, M., Wang, W.: Continuum: a platform for cost-aware, low-latency continual learning. In: Proceedings of the ACM Symposium on Cloud Computing—SoCC 18 (2018). https://doi.org/10.1145/3267809.3267817
21. Wark, T., Hu, W., Corke, P., Hodge, J., Keto, A., Mackey, B., Foley, G., Sikka, P., Bruenig, M.: Springbrook: challenges in developing a long-term, rainforest wireless sensor network. pp. 599 – 604 (2009). https://doi.org/10.1109/ISSNIP.2008.4762055

Intelligent Vehicular Networking Protocols

Grace Khayat, Constandinos X. Mavromoustakis, George Mastorakis,
Hoda Maalouf, Jordi Mongay Batalla, Evangelos Pallis,
and Evangelos K. Markakis

Abstract This chapter will introduce Vehicular Ad Hoc Network; abbreviated as VANET; which is a variation of Mobile Ad Hoc Network, MANET. In VANET the nodes are either vehicles or fixed roadside units considered as VANET infrastructure. This chapter will discuss the five categories of VANET routing protocols with some of the most known examples in each. Each of the mentioned protocols will have a brief overview of its design and implementation. Besides, we will mention some of the protocols' benefits, drawbacks, and enhancements. With the emergence of the Internet of Things (IoT) in the last decade, vehicles have a subcategory known

G. Khayat (✉) · C. X. Mavromoustakis
Department of Computer Science, University of Nicosia, Nicosia, Cyprus
e-mail: khayat.g@live.unic.ac.cy

C. X. Mavromoustakis
e-mail: mavromoustakis.c@unic.ac.cy

G. Mastorakis
Department of Management Science and Technology, Hellenic Mediterranean University, Agios Nikolaos, Crete, Greece
e-mail: gmastorakis@hmu.gr

H. Maalouf
Department of Computer Science, Notre Dame University – Louaize, Zouk Mosbeh, Lebanon
e-mail: hmaalouf@ndu.edu.lb

J. M. Batalla
National Institute of Telecommunications and Warsaw University of Technology, Nowowiejska Str. 15/19, Warsaw, Poland
e-mail: jordim@tele.pw.edu.pl

E. Pallis · E. K. Markakis
Department of Electrical and Computer Engineering, Hellenic Mediterranean University, Estavromenos, Heraklion, Crete, Greece
e-mail: pallis@pasiphae.eu

E. K. Markakis
e-mail: markakis@pasiphae.eu

© Springer Nature Switzerland AG 2020
G. Mastorakis et al. (eds.), *Convergence of Artificial Intelligence and the Internet of Things*, Internet of Things,
https://doi.org/10.1007/978-3-030-44907-0_4

as Internet of Vehicles (IoV) due to the special characteristics of vehicles and road topology. Therefore, the second section of this chapter will discuss the classification of routing protocols in IoV which is the promising future in the vehicular network world.

1   Introduction

Intelligent Transportation Systems (ITS) is one of the emerging applications developed to exchange relevant information to increase safety by avoiding collisions, distributing road and map information. Vehicular Ad Hoc Network (VANET) is a core subject of ITS. VANETs are a sub-category of Ad hoc networks where vehicles are mobile nodes sharing different information such as safety, traffic, and traveler's information. In VANET, the contributors in packet delivery are vehicles and access points which are fixed and connected to the internet. VANET are characterized from other ad hoc networks by high mobility, self-organization, distributed communication, road pattern restrictions, and unbounded network. VANET has different applications classified into the following classes: safety, non-safety, commercial, convenience, and productive applications.

In VANET, vehicles communicate among themselves directly; Vehicle to Vehicle communication (V2V); or communicate with fixed equipment at the side of the road constituting Vehicle to Infrastructure communication (V2I) as shown in Fig. 1. V2V is communication between moving vehicles that act as source, destination or router.

Fig. 1 VANET
communication

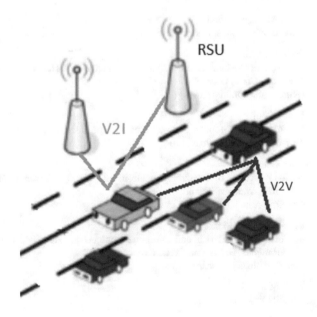

In V2V intermediate vehicles transport packets to the destination node. V2V has two types of communication: one hop, direct vehicle to vehicle communication, and multi-hop communication where the vehicle relies on other nodes to retransmit. V2V communication has link breakdown due to high nodes' mobility yet it is efficient and cost-effective. Neighbor vehicles exchange packets using short-range wireless technologies like Wi-Fi and WAVE. On the other hand, Wi-Fi hotspots or long/wide range wireless technologies are used in V2I communication where vehicles communicate with fixed infrastructural units, Roadside Units (RSUs).

VANET's vehicles are supposed to have wireless transceiver and data sharing functionalities which allow them to create a dynamic network while being driven on the roads. Packets are routed via neighboring vehicles to destinations that are not within direct communication range. VANET is characterized by high mobility which results in fast topology changes, unbounded network size, varying vehicles' density, and velocity. For example, low vehicles' count will lead to poor connectivity and performance degradation. All these characteristics make routing in VANET very challenging.

Some applications in VANET present road traffic conditions, safety-related information, road condition warning, and weather information thus requiring the information to be transferred without any delay. Therefore, some of the challenges in VANET are quality of service with minimum delay and retransmission, scalable and robust routing algorithm, and network security.

VANET applications effectiveness mainly depend on packet routing with at least one intermediate node. Routing involves two activities which are determining the optimal path and transferring the information as packets. Packet transfer may take place in a single hop or multiple hops. Single hop transmission occurs when the destination node is within the transmission range of the sender node thus being a neighbor node. On the other hand, multi-hop transmission occurs when the destination is out of the source node's range. In this case, the neighbor node forwards the packets to their neighbors and so on till the destination is reached.

In the last several years, the Internet of Things (IoT), cloud computing, and Big Data have emerged. IoT is a dynamic global network infrastructure with self configuring capabilities based on standard and interoperable communication protocols. Nowadays vehicles are equipped with a powerful processing unit and large storage devices thus vehicles can be used either in short term applications or in small scale services. Therefore, VANET evolved into the Internet of Vehicles (IoV) through connecting those numerous intelligent vehicles. IoV has special characteristics due to vehicle mobility, road topology, available energy, etc. Routing protocols in IoV is a very challenging field for researchers in the coming years.

2   Routing Protocols in VANET

The goal of routing protocols is to specify the best path for nodes' communication. Several characteristics such as location, scope, speed, security, amount of information, etc. must be taken into consideration when implementing a protocol. Routing protocols look for the optimal path to ensure a reliable packet transmission to the destination. One of the vital needs of these protocols is the least communication time and network resources. VANET is a promising technology with a major challenge for finding a protocol that adapts to a highly dynamic environment with frequent links disconnections. Wireless ad hoc routing algorithms are compared based on several network performance metrics, such as an end to end delay, switching delay and backoff delay [1]. Many difficulties face VANET routing design such as security, privacy, connectivity, and quality of services. The mobility of a wireless device is limited and can lead to network partitioning if energy is critical. This issue is solved by efficient energy conservation (EC) scheme which is mainly exploited either at a link layer or a network layer or a combination of both [2].

The following metrics must be used in VANET routing protocols [3]:

- Neighbor discovery
- Destination identification
- Data transmission
- Network partition
- Movements prediction.

Many routing protocols have been refined for VANET. Those protocols are classified into several groups based on different aspects such as protocol's characteristics, used techniques, routing information, quality of services, network structures, routing algorithms, and so on.

VANET protocols are classified according to their routing characteristics and techniques into five classes as below:

- Topology based routing protocols
- Position based routing protocols
- Broadcast based routing protocols
- Geocast based routing protocols
- Cluster based routing protocols.

VANET routing protocols can further be grouped into three classes based on the network structures:

- Hierarchical routing
- Flat routing
- Position-base routing.

Furthermore, VANET routing protocols are grouped based on routing strategies into two classes:

- Proactive
- Reactive.

Finally, VANET routing protocols are grouped into two categories according to packet forwarding information:

- Geographic-based
- Topology-based.

This chapter will discuss in detail the 5 routing protocols according to their characteristics and techniques.

2.1 Topology Based Routing Protocols

Topology-based routing protocols transfer data packets using link's information [4]. During routing, these protocols need node topology information. They maintain routing tables for storing the link information; source to destination information. Based on the routing tables packets are forwarded. The overhead in this protocol is due to bandwidth and time involved in routing the packets. In VANET, these protocols perform slower than other routing protocols. Based on the information collection method, these protocols are further divided into 3 classes. Proactive, Reactive and Hybrid.

2.1.1 Proactive Routing Protocols

Proactive routing protocols; also called table-driven protocols; build routing tables before the application is made and store all network routing information in it. Every node maintains a routing table with each record in it pointing to the next hop in the destination's path. The need for route discovery is eliminated as the destination route is available in the routing tables. To keep track of topology changes, the nodes periodically exchange packets with updated topology information, and routing tables should be broadcasted to the neighbors either by periodic or triggered updates. The updated topology results in minimum waiting time in packet forwarding. To choose the routes, proactive protocols mainly use shortest path algorithms which increases the routing table size.

Below is a flowchart showing the protocols to be discussed in the following section (Fig. 2).

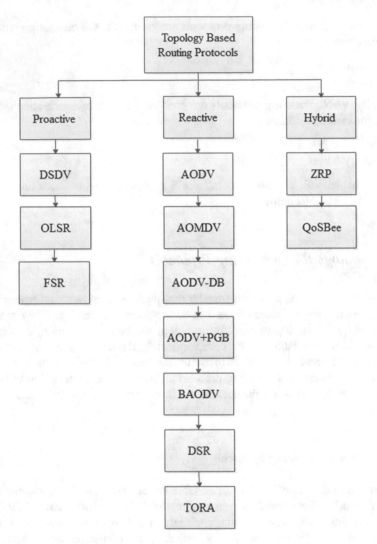

Fig. 2 Topology based routing protocols

Destination Sequence Distance Vector Routing (DSDV)

DSDV protocol is one of the oldest ad hoc routing protocols implementing distance-vector strategy and using shortest path algorithm. It makes available loop-free routes using a single source to destination route. This route knows all about available nodes and the total needed hops to reach the destination. In the routing table, the destination node sets the record's sequence number. To maintain routes reliability, routing tables are periodically broadcasted to neighbor nodes using two types of packets: full dump packets, containing information about every participating vehicle in the VANET, and incremental packets [5]. This protocol excludes frequent traffic updates, reduces control message overhead, keeps every node's optimal path thus reducing routing table size. However, in the large network DSDV increases the overhead due to unnecessary updates. DSDV doesn't provide multi routes to the destination. Route efficiency is decreased since DSDV has no control over the network congestion. Because of these limitations, Randomized DSDV protocol (R-DSDV) is proposed to support congestion control but producing more overhead. Nodes' randomized decision allows each node to decide whether to forward or discard the packet.

DSDV guarantees loop-free paths, reduces infinity problem counts, reduces routing table size by saving the best path to every destination. Whereas DSDV does not support multipath routing. In some cases, bandwidth waste might take place due to unnecessary advertisement of routing information.

Optimized Link State Routing Protocol (OLSR)

OLSR protocol is a proactive point-to-point routing protocol. It keeps a routing table containing information about all possible routes based on the traditional link-state algorithm. Whenever a change in the network topology occurs each node sends an update message to some selective nodes which rebroadcast it to their selective nodes and so forth. Unselected nodes are not allowed to retransmit the update messages. Note that the selected node gets updated only once. It uses a multipoint relaying (MPR) for route setup or route maintenance thus minimizing the number of active relays. OLSR can built-in different operating systems due to its friendly architecture. Also, OLSR works well in dynamic topology and is suitable for applications that require minimum latency in data transmission. However, due to the frequent control packets, the network topology update may cause network congestion. OLSR ignores the node's transmission range, bandwidth, directional antenna, etc. Therefore, Hierarchical Optimized Link State Routing (HOLSR) protocol was proposed to decrease routing overhead in large networks. HOLSR maximizes routing performance through defining network hierarchy architecture with multiple networks. Also, QoS-OLSR (QOLSR) was proposed to provide a path so the available bandwidth of each node on the path is not less than the required bandwidth. These protocols provide improvements in QoS but with higher complexity and packet overhead.

Fisheye State Routing (FSR)

Fisheye State Routing (FSR) is a protocol that maintains a nodes' topology table and updates the network information to other nodes thus reducing the size of update message. Periodically, each node updates its table based on the information received from its neighbors. Entries that are far from the node are broadcasted at a lower frequency as compared to entries that are near. In FSR, whenever the topology changes frequently the route to a remote destination is invalid. Besides, the route can't be discovered if the destination drives outside the source node's scope.

Summary of Topology Based Proactive Protocols

The table below shows a summary of the above discussed proactive protocols showing their pros, cons, and enhancements (Table 1).

Table 1 Summary of topology based proactive protocols

Proactive protocol	Strategy	Pros	Cons	Enhancement
Destination Sequence Distance Vector (DSDV)	– Uses the shortest path algorithm	– Reduces control message overhead – Reduces routing table size	– Excludes frequent traffic changes – No multiroute – No congestion control	Randomized DSDV (R-DSDV): – Support congestion – Increase overhead
Optimized Link State (OLSR)	– Uses link-state algorithm – Uses multipoint relaying for route setup and maintenance	– Minimizes number of active relays – Friendly operating system – Very good in dynamic topology	– May have congestion – Ignores the node's transmission range, bandwidth, and antenna direction	Hierarchical OLSR (HOLSR): – Maximize routing performance QoS-OLSR(QOLSR): – Better QoS – Higher complexity and packet overhead
Fisheye State Routing (FSR)	– Maintains a nodes' topology table and updates the network information	– Reduces size of update message	– Route to remote destination might be invalid	

2.1.2 Reactive Routing Protocols or On-Demand Routing Protocols

These protocols have no information on the entire network topology which is constantly changing. In these protocols, each node establishes the path based on demand which reduces the overhead. Reactive protocols discover the route by flooding the Route Request and Route Reply messages. Once the destination node receives the query packet, the route information reply is reverted to the source in a route reply message. Any node on the route leading to the destination will send back a unicast route acknowledgment message after receiving the request message. The bandwidth will be used for data transmission once the route is defined [6]. For large and highly mobile ad hoc networks with frequently changing topology, these routing protocols are the best choice. On the other hand, these protocols have huge control packets due to route discovery thus having higher latency.

Ad Hoc On-Demand Distance Vector (AODV)

AODV, proposed for mobile ad hoc network, reduces messages flooding thus having low network overhead. AODV only keeps entries for recent active routes which reduces routing tables and memory size requirements. The next hop is only saved rather than the complete route. AODV uses destination sequence numbers to dynamically update route conditions and eliminate loops. It is featured for large-scale and highly dynamic networks, but it increases delays. A new route discovery might be necessary after a route failure. The new route discovery may produce extra delays which decrease the data transmission rate and increase the network overhead. The redundant broadcasts will consume extra bandwidth which increases as the network get congested. Also, the lost packet will cause a lot of collisions.

Ad Hoc On-Demand Multipath Distance Vector Routing (AOMDV)

AOMDV protocol, an extension of AODV, is a multi-path on-demand protocol with better performance than single path protocol. Control messages used in AODV and AOMDV are identical. AOMDV has additional fields that reduce the overhead for multiple paths discovery. AOMDV routing table saves all available paths from which the source chooses one, thus any of the existing redundancy paths can be used. In multi-path protocols, the new route discovery takes place when all known paths are failing. This protocol provides lower overhead since it decreases route discovery transmission rate.

Ad Hoc On-Demand Distance Vector with Broadcasting Data Packets (AODV_DB)

AODV_BD broadcasts the data packet instead of the route request packet for discovering and maintaining a route. The data packet's header has the reverse path. The intermediate nodes save the reverse path in the header and rebroadcast the data packet. The destination node sends a route reply when the data packet is received. AODV_BD reduces the transmission route setup time. If there is no route to destination AODV_BD increases network overhead due to data packet flooding.

Ad Hoc On-Demand Distance Vector Preferred Group Broadcasting (AODV + PGB)

The AODV protocol is enhanced by the Preferred Group Broadcasting (PGB) algorithm which reduces control message overhead and offers routes availability. AODV + PGB predicts the routing path possibility in a defined critical domain thus maintaining a substituting routing path [7]. The route discovery might be longer if the rebroadcasting node for the route request is not the nearest node to the destination. If there is no node having a rebroadcast permit the broadcast will be stopped. Packet duplication may take place whenever any two nodes rebroadcast the same packet simultaneously.

The Bus Ad Hoc on-Demand Distance Vector (BAODV)

BAODV was designed to allow drivers to avoid congested roads. It selects a route having the minimum vehicle number thus minimizing end-to-end packet delay. The

AODV protocol is not efficient in VANET due to its limitations. BAODV tracks the vehicles' characteristics and behaviors. BAODV added additional factors such as type and behavior of the vehicle to the lower hop count of AODV to get better routing performance. Also, BAODV added the number of vehicles in the route to RREP and RREQ messages. BAODV drawback is that it didn't take into consideration the frequent changes in topology.

Dynamic Source Routing Protocol (DSR)

DSR provides a highly reactive routing protocol with very low overhead for frequently changing networks. DSR is a multi-hop protocol decreasing periodic messages exchanged. DSR has route discovery and maintenance procedures. The route discovery is followed whenever a source requires an unavailable route. A route request message will be broadcasted, and all nodes receiving it will rebroadcast as well. The intermediate node does not rebroadcast the route request message if it is the destination or if it has a route to the destination. On the other hand, it sends a route reply message back to the source. The received route is saved in the source routing table for future use. An error message will inform the source node in case the route failed thus the source node deletes this route from its routing table. In the case of failure route, an alternative route is selected and replaces the deleted one. If there is no alternative route, the sender node will start a route discovery process. DSR is perfect for a low mobility network since it can use alternative routes before initiating a route discovery process.

Temporally Ordered Routing Algorithm (TORA)

TORA is a distributed routing protocol aiming at reducing the communication overhead caused by frequent network changes. The shortest path algorithm is not implemented in this protocol. TORA builds a directed graph with the source node being the tree root. In TORA, packets flow from higher to lower nodes. When a packet is broadcasted, neighbor nodes will broadcast a route reply if it has a downward link to the destination, otherwise, it drops the packet. TORA guarantees multi path loop-Tolerant Network Protocols free routing since packets always flow downward. In TORA every node has a route reducing the control messages broadcast. In highly dynamic networks, routing overhead may be caused while maintaining routes to all nodes.

Summary of Reactive Routing Protocols

The table below shows a summary of the above discussed reactive protocols showing their pros and cons (Table 2).

Table 2 Summary of reactive protocols

Protocol	Strategy	Pros	Cons
Ad Hoc On-Demand Distance Vector (AODV)	– Only keeps entries for recent active routes – Uses destination sequence numbers – Featured for large-scale and highly dynamic networks	– Reduces messages flooding – Low network overhead – Reduces routing tables and memory size requirement	– Increases delays – Consumes extra bandwidth – Network may get congested – Many collisions might occur
Ad Hoc On-Demand Multipath Distance Vector Routing (AOMDV)	– Saves all available paths	– Reduces the overhead for multiple paths discovery – Decreases route discovery transmission rate	– A route discovery procedure is needed after each link failure
Ad Hoc On-Demand Distance Vector with Broadcasting Data Packets (AODV_DB)	– Broadcasts the data packet instead of route request packet	– Reduces the transmission route setup time	– Might increases network overhead due to data packet flooding
Ad Hoc On-demand Distance Vector Preferred Group Broadcasting (AODV + PGB)	– Repairs the routing path by substituting the lost relay node	– Reduces control message overhead	– Route discovery might be longer – Packet duplication may take place
Bus Ad hoc On-demand Distance Vector (BAODV)	– Tracks the vehicles' characteristics and behaviors	– Minimum end-to-end packet delay	– Didn't accommodate for frequent topology changes
Dynamic Source Routing Protocol (DSR)	– Route discovery whenever a source needs an unavailable route	– Low overhead for frequently changing networks – Perfect for low mobility network	– Increased routing packet size due to adding node's information in the header
Temporally Ordered Routing Algorithm (TORA)	– Builds a directed graph	– Guarantees multi path loop free routing	– In highly dynamic networks, routing overhead may increase

2.1.3   Hybrid Routing Protocols

Hybrid protocols, a merge of proactive and reactive protocols, decrease both the proactive control overhead and route discovery delay of the reactive routing. The hybrid protocols split the network into several zones which provide better route discovery, reliability, and maintenance. Every node splits the network into an inside and outside regions. Inside the region, nodes use a proactive routing mechanism to maintain routes and uses a reactive route discovery mechanism to reach the nodes of the outside region.

Zone Routing Protocol (ZRP)

ZRP, the first hybrid routing protocol, divides the network into zones based on several factors such as the power of transmission, signal strength, and speed. ZRP uses independent protocols for the outside and inside zones. For the outside zone, the ZRP reactively discover a route where route request packets including a unique sequence number, source address and destination address are sent to border nodes. Border nodes checks the destination within its inside zone. In case the destination is found, a route reply on the reverse path is sent otherwise the border node adds its address to the route request packet and forwards it to its border nodes. Whenever the source receives a reply, the path stored will be used for data transmission. ZRP is considered a proactive protocol for large zones and performs a reactive protocol for small ones.

Quality of Service Multipath VANET Routing Protocol (QoSBee)

QoSBee inspired by bee's swarm is a protocol that resembles the food source searching technique. Like bees this protocol is self-configured. QoSBee has two types of packets: scout and forager. Scout packet is used for route request with the target to find the destination. Once the destination is found scout packet returns to the source node. Forager packet is used for data transmission.

2.2   Position-Based Routing Protocols or Geographic Routing Protocols

Position based routing protocols are the most promising protocols since they support the vehicle's geographical position to offer routing. These protocols are highly effective for highly mobile and dynamic networks. In urban scenarios with obstacles, these protocols do not perform well. The node's transmission decision is essentially based on the destination's and neighbors' positions. Geographic routing requires resources like GPS (Global Positioning System) or periodic beacon messages. Vehicular nodes must know their own and neighboring nodes' positions for optimal path calculation with no previous route discovery. The source node attaches to the packet a header with the destination location information. Data packets transmission decision is based on a greedy approach where the node forwards the packet to the geographically closest neighbors to the destination. This process goes on until the packet reaches its destination [8]. However, the packet may get stuck in a local optimal when no neighbor exists than the current node itself.

Below is a flowchart showing all the position-based routing protocols to be discussed next (Fig. 3).

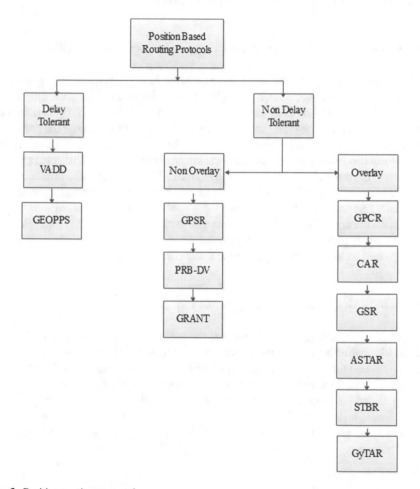

Fig. 3 Position routing protocols

2.2.1   Delay Tolerant Network Protocols

Delay Tolerant Network, DTN, is a wireless network desired to efficiently perform in frequently disconnected and large-scale networks, limited bandwidth availability, and power constraints. DTN avoids congestion and complexity with high flexibility. The nodes in this type of network may have a limited transmission range; thus, increasing delays. In DTN protocols, there is no guarantee of connected links, therefore, packets may be cached temporarily at intermediate nodes. The forward and carry method stores data packets until the network is associated [9].

Vehicle-Assisted Data Delivery in Vehicular Ad Hoc Networks (VADD)

VADD implementing the store and forward scheme handles frequently disconnected and highly mobile networks. A moving node saves the packet till forwarding it to a new node that entered its zone range. Node mobility is predictable by two factors: network traffic and route type. This protocol most probably will route the packet on the path with minimum transmission delay. Before forwarding the packet, each node adds information about its formers thus breaking the routing loop. Whenever a packet is received, the node checks the hop's information. The previous hops information is checked to avoid forwarding packets to them thus avoiding the routing loop problem. In VADD, the wireless channel is the major medium for data transmission unless when a condition requires high speed thus data packets are transmitted through wires [10]. Since this protocol is mainly applicable in highly mobile networks, changes in topology and traffic density might cause a large delay.

Geographical Opportunistic Routing (GEOPPS)

GeOpps is a forwarding protocol that uses the available navigation system. The collected information will select the closest vehicles from the destination. The protocol uses a store and forward technique and navigation system to provide efficient packet delivery. To select the next-hop every neighboring node finds the future closest node and calculates the delay for packet delivery. If the node stops the packets, then the packet should be forwarded to another neighbor. In GeOpps not all nodes are required to calculate the routes thus transmission rate depends only on the route topology and nodes' mobility.

Summary of Delay Tolerant Routing Protocols

The table below shows a summary of the above discussed delay-tolerant routing protocols showing their pros and cons (Table 3).

Table 3 Summary of delay tolerant routing protocols

Protocol	Strategy	Pros	Cons
Vehicle Assisted Data Delivery in Vehicular Ad Hoc Networks (VADD)	– Implements store and forward scheme – Route the packet on the path with minimum transmission delay	– Handles frequently disconnected and highly mobile networks	– Due to topology change and traffic density large delays might be caused
Geographical Opportunistic Routing (GEOPPS)	– Uses the available navigation system – Uses store and forward technique	– Not all nodes are required to calculate the routes	– Privacy is an issue because navigation information is disclosed to the network

2.2.2 Non-delay Tolerant Network Protocols (NON DTN)

The non-DTN protocols; also known as min-delay protocols; are suitable for high density network since they do not take into account the connectivity issue assuming there are always several nodes to achieve successful communication. In these protocols, the node should know about neighbor nodes to forward its packet to the closest neighbor. In case no closest neighbor to the destination exists other than the current node, then these protocols face problems. Non-Delay tolerant protocols are mainly used to decrease packet delivery communication time. To reduce the packet delivery ratio, the packet should pass through a minimum intermediate node and this path should be shortest and optimal. Beacon routing protocols use the "HELLO" message sent periodically to discover a neighbor node in the network and maintain the neighbor information list. Beacon routing protocols are furthermore classified into non-overlay and overlay networks.

Non-overlay

The Non-Overlay network uses greedy forwarding method to transfer data. Below are a few examples of Non-Overlay routing protocols

Greedy Perimeter Stateless Routing (GPSR)

Greedy forward is forwarding packets to neighboring nodes that are closer to the destination or uses perimeter forwarding to find the node to which the packet should be transmitted in case no closer node exists. This stateless protocol increases scalability since it saves information of its first hop neighbor's position. A link failure might occur due to high mobility and frequent topology changes. This is solved by perimeter forwarding as mentioned before. Perimeter forwarding results in increased packet loss and latency because of the numerous hops. Moreover, whenever the destination changes location, the saved location in the packet header will never be updated. Advanced Greedy Forwarding (GPSR + AGF) combines the speed, direction, and overall travel time in the packet. It also consists of packet processing time based on the current forwarding node within the data packets.

Position Based Routing with Distance Vector (PBR-DV)

The PRB-DV protocols use AODV-route strategy method if the data packets come into the local maximum range. The receiver node of any packet checks whether it is nearer to the destination or fall in local maximum, otherwise, it will store the node's details from where request packet came. The node rebroadcast the request packet or sent it back to the previous node from where received. The route request data packets consist of destination position details and node position details.

Greedy Routing with Abstract Neighbor Table (GRANT)

The extended greedy routing mechanism is used whenever every node in the network knows the information of its x hop neighbors. In this protocol, the path length is shorter as compared to the other traditional greedy routing since the next forwarding node is chosen based on the multiplication distance between the nodes.

Summary of Non-overlay Routing Protocols

The table below shows a summary of the above discussed non-overlay routing protocols showing their pros, cons, and enhancements (Table 4).

Table 4 Summary of non-delay tolerant routing protocols

Protocol	Strategy	Pros	Cons	Enhancements
Greedy Perimeter Stateless Routing (GPSR)	– Forwards packets to neighboring nodes which are closer to the destination	– Increases scalability	– Increased packet loss and latency	Advanced Greedy Forwarding (GPSR + AGF): Combines the speed, direction, and overall travel time in the packet
Position Based Routing with Distance Vector (PBR-DV)	– Use AODV-route strategy method	– Might have a better packet delivery ratio	– For non-greedy part, excessive flooding is required	
Greedy Routing with Abstract Neighbor Table (GRANT)	– Used whenever every node knows the information of its x hop neighbors	– Path length is shorter	– Possible inaccuracy in packet delivery	

Overlay Network

In an overlay network, logical links interconnect all nodes. Those links are constructed on an existing network such as city maps.

Greedy Perimeter Coordinator Routing (GPCR)

GPCR protocol is the best to be used in highly mobile environments. The greedy forwarding technique forwards packets to a neighbor node. This neighbor node must be closest to the destination. Every node must know its location, its neighbors' position, and the destination's position using location services such as GPS and periodic beaconing. Packets are forwarded until reaching the next intersection. GPCR has a maintenance process that consists of decision making. On the intersection this decision states to which intermediate node the packet will be forwarded to. On the other hand, the coordinator node states the route on which the packet will be forwarded on. In the case of coordinator node's absence, the packet will be forwarded to the furthest node in the route. This protocol does not require any global information since it is based on the destination node connectivity and the density of the road. If the node's density is low, GPCR would not connect to the destination thus increasing the transmission delay [11].

Connectivity Aware Routing (CAR)

This protocol is mainly considered for inter-vehicle communication. Mainly this protocol has four measures: route discovery, forwarding packets along the discovered path, error recovery and path maintenance. Guards are used to track the target node's current position.

Geographic Source Routing (GSR)

This protocol uses a route map to discover the shortest path. These route maps convert the junctions and routes into a graph where junction act as vertices and routes acts as edges. Packets are forwarded from junction to junction. In case there is no connectivity, recovery mode and greedy forwarding technique are used.

Anchor-Based Street and Traffic Aware Routing (ASTAR)

A-STAR protocol is mainly used in city scenarios for inter-vehicle communication. It uses a street map through which data packets are sent to its destination to calculate a series of anchors. Anchor route is calculated with traffic awareness. For route discovery A-STAR protocol uses two overlaid maps: statistically and dynamically rated map. Rated map uses a graph of the city route and makes sure that the route has stable traffic. Dynamically rated map consists of real-time traffic route information. The difference between statistical and dynamic rated maps is the traffic's condition.

Street Topology Based Routing (STBR)

Street topology-based routing protocols consist of three states: master, slave and forwarder nodes. This protocol chooses a master node in a junction and checks the links of the next junction. It navigates multiple junction for long distance routing. The master node consists of a two-level junction. The first will be via a neighbor node to its direct junction node. The second will be from a neighbor node to their junction nodes. The complexity in STBR protocols is high because if the old master node leaves the junction it must transfer the table information to the new master node in two hops.

Greedy Traffic Aware Routing Protocol (GyTAR)

GyTAR protocols follow carry and forward technique to recover from local problems. Packets are delivered between the concerned junctions by using a greedy routing strategy. GyTAR uses a digital map to identify the location of neighbor junction and gives a score to each junction neighbor based on destination distance and density of traffic. The maximum score junction can be select as the next intersection junction.

Summary of Overlay Routing Protocols

The table below shows a summary of the above discussed overlay routing protocols showing their pros and cons (Table 5).

Table 5 Summary of overlay routing protocols

Protocol	Strategy	Pros	Cons
Greedy Perimeter Coordinator Routing (GPCR)	– Forwards packets to a neighbor node closest to the destination – Maintenance process which consists of decision making	– Best to be used in highly mobile environments – Does not require any global information	– Every node must know its location, its neighbors' position, and the destination's position – Transmission delay might be increased
Connectivity Aware Routing (CAR)	– Mainly considered for inter vehicle communication	– No digital map is required	– It cannot adjust with different sub-path when traffic environment changes
Geographic Source Routing (GSR)	– Uses a route map to discover the shortest path	– In case there is no connectivity, recovery mode and greedy forwarding technique are used	– Higher routing overhead

(continued)

Table 5 (continued)

Protocol	Strategy	Pros	Cons
Anchor Based Street and Traffic Aware Routing (ASTAR)	– Mainly used in city scenario for inter vehicle communication	– Anchor route is calculated with traffic awareness – Uses two overlaid maps: statistically and dynamically rated map	– Packet delivery ratio is low
Street Topology Based Routing (STBR)	– Consist of three states: master, slave and forwarder nodes	– Traverses least spanning multiple junctions for long distance unicast communication	– High complexity
Greedy Traffic Aware Routing Protocol (GyTAR)	– Follow carry and forward technique to recover from local problems	– Throughput, delay and routing overhead is better than GSR	– Depends on roadside units

2.3   Broadcast Routing

Broadcast Routing protocols are simple and used for applications such as weather, traffic, road situation, … Broadcasting routing floods packets to all available nodes inside the broadcast domain. Packets are delivered by several nodes to achieve reliable packet transmission, but it might consume network bandwidth due to replicated packets. Each node must identify replicated packets to be discarded. Based on the road network hierarchical structure, the route is divided into virtual cells. There are two levels of hierarchy. The first level has all the nodes of the cell, and the second level is represented by reflector cells which are very close to the cell's center. Some reflectors might behave as group leaders to manage emergency messages sent from cell members or a neighbor. Broadcast routing is based on flooding where independent node re-broadcast to others with the advantage of message delivery guarantee and ease of set up. The drawback is when the number of nodes increases exponentially a decrease in network performance takes place [12].

2.3.1   Density-Aware Reliable Broadcasting Protocol (DECA)

DECA is a reliable broadcasting protocol based-on nodes' density. DECA exchange nodes' knowledge and information using beacon messages. It uses the store and forward transmission technique. The node selects the next-hop to rebroadcast the packet based on the node's information density. Thus, the next-hop node will have the highest density. Before broadcasting the packet, the sender node adds the next-hop ID. Excluding the next hop, all nodes save the packet and start a waiting timer. When the timer expires, and no rebroadcast packet is received those nodes rebroadcast the packet. Any neighboring node adds its ID to the regular beacon when receiving the broadcasted packet. Adding the ID help other nodes in determining which neighbors haven't received the broadcasted packet so that they rebroadcast the packet. This protocol is friendly to several environments since it doesn't use any global position information. An increase in network overhead and a decrease in performance transmission is due to periodic beacon transmissions and rebroadcasted packets from neighbors upon timer expiration.

2.3.2 Position-Aware Reliable Broadcasting Protocol (POCA)

POCA selects certain neighbor nodes based on their position. Those nodes will be used in packets rebroadcasting. On the other hand, the unselected nodes store the packet and start a waiting timer. In case the timer expires, and no rebroadcast is received, the unselected nodes rebroadcast the packet. Beacon messages are used to get neighbors' locations, speed, and connectivity status. POCA is highly reliable in networks with numerous nodes but flooding packets might increase congestion when waiting time is over.

2.3.3 Distributed Vehicular Broadcast Protocol (DV-CAST)

DV-CAST, a multi-hop protocol, appropriate for all traffic situations. Each node continuously checks the connectivity status of its neighbor nodes for broadcasting packets. This protocol handles many aspects of the network such as traffic and nodes' connection states. Periodic beacon messages are used to extract network topology information. The node might rebroadcast packets using nodes moving in the same direction. The source node uses the store and forward scheme in case of disconnection from the neighboring nodes. The store and forward technique states that the node stores the broadcasted packet until finding another node moving into its broadcast domain. In case there is no node, the packet will be disregarded whenever the packet live time is over. In this protocol, the nodes decide whether the packet is received before or not using a flag parameter. DV-CAST protocol minimizes the broadcasting overhead yet end to end delay in the data transmission might be increased.

2.3.4 Distribution-Adaptive Distance with Channel Quality (DADCQ)

DADCQ targets to accomplish an adaptive multi-hop broadcast protocol for large congested networks. Forwarding nodes rebroadcast packets based on their position. Upon receiving a packet, the node checks its distance from the destination. The node won't rebroadcast the packet if it was very close; since the rebroadcast will not cover a new area. On the other hand, the node rebroadcasts the packet if this distance is large. DADCQ has a minimum transmission overhead due to its dependence on the node's measurement, however large message overhead may be caused.

2.3.5 Summary of Broadcast Routing Protocols

The table below shows a summary of the above discussed broadcast routing protocols showing their pros and cons (Table 6).

Table 6 Summary of broadcast routing protocols

Protocol	Strategy	Pros	Cons
Density Aware Reliable Broadcasting Protocol DECA)	– Uses beacon messages – Uses store and forward transmission technique	– Doesn't use any global position information	– Increase in network overhead – Decrease in performance transmission
Position Aware Reliable Broadcasting Protocol (POCA)	– Selects certain neighbor nodes based on their position	– Highly reliable	– Flooding packets might increase congestion
Distributed Vehicular Broadcast Protocol (DV-CAST)	– Handles many aspects of the network such as traffic and nodes' connection state	– Appropriate for all traffic situations	– End to end delay in the data transmission might be increased
Distribution Adaptive Distance with Channel Quality (DADCQ)	– Forwarding nodes rebroadcast packets based on their position	– Minimum transmission overhead	– Large message overhead may be caused

2.4 Geocast Routing Protocols

Geocast routing protocols, location-based protocols, multicast packets from their source to their destinations. Within a specified geographic region, a directed flooding mechanism is applied to broadcast the packets [13]. Using position determining services such as GPS, every node knows its position and neighbor's geographic position. The nodes do not save any routing information. In Goecast, the network is divided into zones of relevance (ZOR). Thus, vehicles belong to the same ZOR whenever they are in the same geographical area. The vehicle does not belong to the ZOR whenever it moves out of its geographical region. In these protocols, each vehicle transmits packets to all other nodes located in the ZOR and drops the packet when it leaves the associated ZOR. Another zone is defined known as the zone of forwarding (ZOF). ZOF is the geographic region where vehicles must transmit the packets to other ZOR vehicles. Geocast decreases packet collisions avoidance and rebroadcast rate, but it is limited in waiting time while making the rebroadcasting decision. Network disconnection may cause delays in packet transmission.

2.4.1 Robust Vehicular Routing (ROVER)

ROVER, a multicast protocol, permits vehicles to deliver packets to all vehicles within a defined ZOR using on-demand routing. Control packets are flooded, and data packets are unicasted to increase consistency and efficiency. It considers vehicles to have identification numbers, digital maps, and information for global locations. ZOR is flooded using the route request packet from the source node to start a route discovery. The route request packet has source ID, location, recent ZOR, and a sequence number. The route request packet is accepted whenever received by a vehicle close to the source and located inside the ZOF and ZOR. When a route request packet is accepted a reply packet with its ID is sent to a one-hop vehicle, and the route request packet is retransmitted. If the vehicle was outside the ZOR, the route request packet is not accepted. When the vehicle sends a reply, a record containing the route request information is saved in its routing table. A tree of multicast routing can be constructed by each node. The source node will be the tree root. Due to the increased number of retransmissions, the control packet overhead and data delivery delay is increased.

2.4.2 Mobile Just in Time Multicasting Protocol (MOBICAST)

MOBICAST, a multicast geographical protocol, uses a location provider (GPS) and takes into consideration the time aspect. Packets are transmitted to vehicles inside ZOR at time t, ZORt, to provide management for spatiotemporal needs. At time t, all vehicles inside the ZOR must maintain real-time data communication. If any ZOR vehicle unexpectedly changes speed the communication of ZOR will fail. If the network is temporally fragmented, MOBICAST packets may not efficiently be received within a ZOR. This problem is solved using ZOF. ZOFt, which may not have optimal size, advertise a MOBICAST packet to all vehicles in the ZORt. Several unrelated vehicles might have to deliver the MOBICAST packets whenever the ZOFt size is larger than the optimal one. On the other hand, whenever the ZOFt size is smaller than the optimal one, a temporal network fragmentation takes place. MOBICAST estimation for ZOFt size is required to be the closest possible to the optimal size. Since MOBICAST relies on the location provider (GPS) to get global network density knowledge it may not perform well in large dynamic networks.

2.4.3   Summary of Geocast Routing Protocols

The table below shows a summary of the above discussed geoacast routing protocols showing their pros and cons (Table 7).

Table 7 Summary of geocast routing protocols

Protocol	Strategy	Pros	Cons
Robust Vehicular Routing (ROVER)	– Permits vehicles to deliver packets to all vehicles within a defined ZOR using on-demand routing	– Increase consistency and efficiency	– Control packet overhead and data delivery delay are increased
Mobile Just in Time Multicasting Protocol (MOBICAST)	– Uses a location provider (GPS) and takes into consideration the time aspect – Packets are transmitted to vehicles inside ZOR at time *t*, ZORt	– Reduced the packet overhead ratio by using the ZOF	– If any ZOR vehicle unexpectedly changes speed the communication of ZOR will fail – Packets may not efficiently be received within a ZOR – May not perform well in large dynamic networks

2.5   Cluster-Based Routing Protocols

These protocols support scalability, media access, and security infrastructure using virtual networks. These protocols have limitations on mobility and high speed which makes the implementation complex for VANET. These protocols split the network into clusters with vehicles having similar characteristics such as the same direction or velocity. Every cluster has a cluster head that manages communication inside the cluster by direct links and outside the cluster by group leaders. Group formation and cluster head selection are critical processes. Each header will be the linking node between its cluster and other clusters. The header is responsible for nodes' management inside the cluster and management between the clusters. The management of these protocols decreased the costs and delays in packet delivery. In general, these protocols may perform well in large networks but may cause more network overhead due to clusters' structure heads.

2.5.1   Clustering for Open IVC Network (COIN)

The clustering technique COIN improves network scalability by dividing the network into clusters based on the following three parameters: mobility, positions, and behavior. This protocol decreases control overhead by associating for each cluster a specified time to live to.

2.5.2   Cluster Based Routing (CBR)

In CBR each node has a preloaded digital map and GPS device. According to the geographic clusters are built, and each cluster is divided into fragments. The packets are routed along the clusters using the cluster heads (CH). The source sends the packet to its CH which chooses the next best CH based on the distance between source and destination. This process is repeated until the packet reaches the destination [14].

2.5.3   Cluster-Based Directional Routing Protocol (CBDRP)

CBDRP splits the network based on the moving direction into groups, clusters, then it selects a vehicle to be the cluster-head (CH) [15]. The selected CH has the shortest distance to the cluster's center. The cluster formation and the CH selection are the initialization phase. Following the initialization phase the data transmission phase starts which is composed of four phases: routing request, routing establishment, routing maintenance and link disconnect. A request packet is flooded whenever a source requests a route. The request only passes among cluster heads thus reduces traffic overhead. The packet is delivered by the CH to the destination whenever it arrives to CH of the cluster having the destination node. The packet size, overhead, and transmission delay will increase due to the addition of the node's information in the route in highly crowded clusters [16].

2.5.4   Summary of Cluster Routing Protocols

The table below shows a summary of the above discussed cluster routing protocols showing their pros and cons (Table 8).

Table 8 Summary of cluster routing protocols

Protocol	Strategy	Pros	Cons
Clustering for Open IVC Network (COIN)	– Dividing network into clusters	– Improves network scalability	– Decreases control overhead
Cluster Based Routing (CBR)	– Each node has a preloaded digital map and GPS device	– Routing overhead is less	– Doesn't consider the velocity and direction of a vehicle
Cluster Based Directional Routing Protocol (CBDRP)	– Splits the network based on the moving direction into groups	– Reliable and rapid data transfer	– Packet size, overhead, and transmission delay will increase due to the addition of node's information in the route

3 Internet of Vehicles

According to the European Commission Internet of Things, IoT, involves "Things having identities and virtual personalities operating in smart spaces using intelligent interfaces to connect and communicate within social, environmental, and user contexts" [17] (Fig. 4).

Fig. 4 Internet of Things (IoT) world

In transportation systems, IoT offers promising solutions in transforming the vehicles' operation through equipping them with various sensors to ensure safer travels, optimal route determination, traffic congestion avoidance, etc. [18]. By the end of 2020 decade, Digital Center expects that 250 million connected vehicles are to be on the road [19]. Vehicles are not just a moving node, but also a computer and storage center with intelligent process capability. Connecting large-scale intelligent vehicles with advanced telematics evolved traditional VANET to the Internet of Vehicles (IoV) which is expected to evolve into the Internet of Autonomous Vehicles [20].

IoV characteristics are different from those of IoT. The main difference is that in IoT most end-users have a random trajectory, whereas in IoV vehicles' trajectory is limited by the road distribution. Multi-vehicle, multi-thing and multi-network systems need multi-level collaboration in IoV [21].

According to the transmission strategy in IoV, Lin et al. classified them into three categories [22]:

- Unicast protocol
- Multicast protocol
- Broadcast protocol.

3.1 Unicast Protocol

The main goal of this protocol is to transfer packets from the source to a single destination. The transfer of packets might be one-hop or multi hops using either hop-by-hop greedy forwarding mechanism or carry-and-forward mechanism. In the greedy forwarding mechanism, intermediate vehicles must transmit the packets as soon as possible [23]. Whereas in the carry-and-forward mechanism, intermediate vehicles can carry the packets until a forward decision is made based on the routing algorithm applied.

3.2 Multicast Protocol

Multicast routing protocols save bandwidth, increase resource utilization, and reduce transmission costs compared to unicast routing. Multicast routing protocols are efficient in the dynamic environment due to its reduction in power consumption, transmission, and control overhead. This protocol simultaneously sends multiple copies of the packets to various vehicles. Therefore, packets are sent from a single sender to multiple receivers [24].

3.3 Broadcast Protocol

Broadcast routing is mostly applied in transmitting information for traffic, weather, road conditions, advertisements, and announcements. Packets are to be sent beyond the vehicle's transmission range, therefore, multi hops are used. Packets are sent to all vehicles in the network using flooding to ensure packet delivery. Due to flooding of packets bandwidth is wasted and vehicles might receive duplicate packets [25].

Based on the method for relay selection, the broadcast strategies are divided into two aspects:

- sender-based specified scheme: good at disseminating messages rapidly, rarely affected by traffic density, relies on the neighbor information
- receiver-based distributed scheme: good in sharing wireless channel features, low response, local broadcast storm [26].

4 Conclusion

This chapter discussed the VANET routing protocols and their classifications. VANET routing protocols might be classified based on several criteria. The most known classification criteria for VANET is based on the routing characteristics and techniques. This classification groups the VANET routing protocols into five groups: topology-based routing protocols, position-based routing protocols, broadcast based routing protocols, geocast based routing protocols, and cluster-based routing protocols. Several routing protocols are members of each of the above-mentioned classification. Due to VANET's high mobility, frequent topology changes, frequent disconnections, unknown vehicle trajectory, etc. routing protocols are very challenging. None of the mentioned routing protocol is perfect in all case scenarios. Every protocol has pros and cons with some protocols having enhancements protocol to improve their weakness.

In addition, this chapter pointed out the transition from VANET to IoV which is a specialized branch in IoT due to the special characteristics of vehicular communication. This chapter discussed the routing protocols applied in the IoV which are classified into three subgroups based on their transmission strategy. The three classification groups are unicast, multicast and broadcast which are discussed towards the end of this chapter.

References

1. Bourdena, A., Mavromoustakis, C.X., Kormentzas, G., Pallis, E., Mastorakis, G., Yassein, M.B.: A resource intensive traffic-aware scheme using energy-aware routing in cognitive radio networks. Future Gener. Comput. Syst. **39** (2014)
2. Mavromoustakis, C.X., Karatza, H.D.: Real-time performance evaluation of asynchronous time division traffic-aware and delay-tolerant scheme in adhoc sensor networks. Int. J. Commun. Syst. **23**(2) (2010)
3. Pandey, P.: Effect of selfish behavior on network performance in VANET. In: Fifth International Conference on Communication Systems and Network Technologies, India (2015)

4. Qureshi, K.N., Abdullah, H.: Topology based routing protocols For VANET and their comparison with MANET. J. Theor. Appl. Inf. Technol. **58**(3) (2013)
5. Mahajan, A.N., Saproo, D., Khedikar, R.: Analysis and comparison of VANET routing protocols using IEEE 802.11p and wave. Int. J. Eng. Res. Appl. **3**(4), 1556–1561 (2013)
6. Boussoufa-Lahlah, S., Semchedine, F., Louiza Bouallouche-Medjkoune, L.: A position-based routing protocol for vehicular ad hoc networks in a city environment. In: The International Conference on Advanced Wireless, Information, and Communication Technologies, Tunisia (2015)
7. Jung, S.D., Lee, S.S., Oh, H.S.: Improving the performance of AODV(-PGB) based on position-based routing repair algorithm in VANET. KSII Trans. Internet Inf. Syst. **4**(6) (2010)
8. Kaur, S., Gupta, A.K.: Position based routing in mobile Ad-Hoc networks: an overview. Int. J. Comput. Sci. Technol. **3**(4) (2014)
9. Dias, J.A., Rodrigues, J.J., Xia, F., Mavromoustakis, C.X.: A cooperative watchdog system to detect misbehavior nodes in vehicular delay-tolerant networks. IEEE Trans. Ind. Electron. **62**(12), 7929–7937 (2015)
10. Goel, N., Dhyani, I., Sharma, G.: A study of position based VANET routing protocols. In: International Conference on Computing, Communication and Automation, India (2016)
11. Anil Kumar, T., Reza, A.T., Sivakumar, T.: A survey on unicast routing protocols for VANET. Int. J. Eng. Comput. Sci. **5**(5), 16555–16565 (2016)
12. Misra, S.C., Woungang, I., Zhang, I., Misra, S.: Guide to Wireless Ad Hoc Networks. Springer, London (2009)
13. Toulni, H., Nsiri, B.: A hybrid routing protocol for VANET using ontology. Procedia Comput. Sci. **73**, 94–101 (2015)
14. Bitam, S., Mellouk, A.: Conventional routing protocols for VANETs. In: Bio-Inspired Routing Protocols for Vehicular Ad Hoc Networks (2014)
15. Senouci, O., Zibouda, A., Harous, S.: Survey routing protocols in vehicular ad hoc networks. In: Second International Conference on Advanced Wireless Information, Data, and Communication Technologies, France (2017)
16. Khayat, G., Mavromoustakis, C.X., Mastorakis, G., Maalouf, H., Batalla, J.M., Mukherjee, M.: Tuning the uplink success probability in damaged critical infrastructure for VANETs. In: IEEE International Workshop on Computer Aided Modeling and Design of Communication Links and Networks, Cyprus (2019)
17. Mavromoustakis, C.X., Mastorakis, G., Dobre, C.: Advances in Mobile Cloud Computing and Big Data in the 5G Era, vol. 22. Springer Studies in Big Data (2017)
18. Mavromoustakis, C.X., Mastorakis, G., Batalla, J.M.: Internet of Things (IoT) in 5G mobile technologies. In: Modeling and Optimization in Science and Technologies, vol. 8. Springer (2016)
19. Digital Center (2016) Connected cars: on the road to 2020. https://www.digitalcentre. technology/news/connected-cars-on-the-road-to-2020/
20. Xu, W., Zhou, H., Cheng, N., Lyu, F., Shi, W., Chen, J., Shen, X.: Internet of vehicles in big data. J. Autom. Sinica, **5**(1) (2018)
21. Yang, F., Wang, S., Li, J., Liu, Z., Sun, Q.: An overview of internet of vehicles. China Commun. **11**(10) (2014)
22. Lin, Y.-W., Chen, Y.-S., Lee, S.-L.: Routing protocols in vehicular ad hoc networks: a survey and future perspectives. J. Inf. Sci. Eng. **26**(3), 913–932 (2010)
23. Cheng, J., Cheng, J., Zhou, M., Liu, F., Gao, S., Liu, C.: Routing in internet of vehicles: a review. IEEE Trans. Intell. Transp. Syst. **16**(5), 2339–2352 (2015)
24. Farooq, W., Khan, M.A., Rehman, S., Saqib, N.A.: A survey of multicast routing protocols for vehicular ad hoc networks. Int. J. Distrib. Sens. Netw. **11**(8) (2015)
25. Nagaraj, U., Dhamal, P.: Broadcasting routing protocols in VANET. Netw. Complex Syst. **1**(2) International Institute for Science, Technology and Education (2011)
26. Wang, W., Sang, L., Luo, T., Kang, H.: A distributed broadcasting protocol based on local information sensing in IoV. In: IEEE 4th international conference on computer and communications, Chengdu, China (2018)

Towards Ubiquitous Privacy Decision Support: Machine Prediction of Privacy Decisions in IoT

Hosub Lee and Alfred Kobsa

Abstract We present a mechanism to predict privacy decisions of users in Internet of Things (IoT) environments, through data mining and machine learning techniques. To construct predictive models, we tested several different machine learning models, combinations of features, and model training strategies on human behavioral data collected from an experience-sampling study. Experimental results showed that a machine learning model called linear model and deep neural networks (LMDNN) outperforms conventional methods for predicting users' privacy decisions for various IoT services. We also found that a feature vector, composed of both contextual parameters and privacy segment information, provides LMDNN models with the best predictive performance. Lastly, we proposed a novel approach called one-size-fits-segment modeling, which provides a common predictive model to a segment of users who share a similar notion of privacy. We confirmed that one-size-fits-segment modeling outperforms previous approaches, namely individual and one-size-fits-all modeling. From a user perspective, our prediction mechanism takes contextual factors embedded in IoT services into account and only utilizes a small amount of information polled from the users. It is therefore less burdensome and privacy-invasive than the other mechanisms. We also discuss practical implications for building predictive models that make privacy decisions on behalf of users in IoT.

1 Introduction

Smartphone apps and websites increasingly ask users to make privacy decisions about personal information disclosure, e.g., to grant or deny an app permission to access their location or their phone book. However, previous research indicates that people are often unable to make these decisions due to limits in their available

H. Lee (✉) · A. Kobsa
University of California, Irvine, Irvine, CA 92697, USA
e-mail: hosubl@uci.edu

A. Kobsa
e-mail: kobsa@uci.edu

© Springer Nature Switzerland AG 2020
G. Mastorakis et al. (eds.), *Convergence of Artificial Intelligence
and the Internet of Things*, Internet of Things,
https://doi.org/10.1007/978-3-030-44907-0_5

time, motivation, and abilities to fully understand the tradeoff between utility and privacy in the disclosure of personal information. This problem will continue to grow in ubiquitous computing environments like the Internet of Things (IoT), as an array of computing devices around the user unobtrusively collects his/her personal information or infers more sensitive information from various types of sensor data collected [3, 4, 31, 33]. Clearly, rich personal information helps IoT systems better understand users, thus providing better-tailored services to them. At the same time, however, this leads to considerable increase in privacy concerns that may lead users to stop using the service [12, 37, 43]. Therefore, providing IoT services with minimized privacy risks is crucial for both protecting users' privacy and keeping IoT ecosystems sustainable.

One possible way to achieve this objective is to assist users with making better privacy decisions, by predicting decisions based on their historical decision-making behavior and recommending privacy settings accordingly (i.e., privacy decision support). Most previous research on privacy decision support has focused on personal information disclosure in online or mobile social network services. Researchers employed supervised machine learning approaches that build predictive models by utilizing users' past behavior as training data, and then utilized the trained models to recommend the most appropriate privacy decisions in the current situation [8, 14, 39–41, 44]. They also verified that privacy decision support systems alleviate users' cognitive burden, thereby allowing them to make their preferred decisions more easily. This kind of technology will become more crucial in IoT environments, not only because users will need to make privacy decisions more frequently, but also because there are no or only very limited user interfaces available for users to state their preferences or decisions to the services (e.g., there are no standard keyboards and displays).

In this chapter, we proposed a novel machine learning mechanism for predicting privacy decisions of users in IoT environments. The aim of this mechanism is to correctly predict users' decisions whether or not to allow the given personal information monitoring based on both the user's current context and personal attitudes on privacy. We tested the proposed mechanism on a privacy-related behavioral dataset collected from 172 users who were presented with descriptions of personal information tracking scenarios relating to their physical location on a university campus [26]. These scenarios are defined by five different contextual parameters such as place (*where*), type of collected information (*what*), entity (*who*), purpose (*reason*), and frequency (*persistence*) of the monitoring. The dataset also contains each user's privacy decision behavior on each scenario in terms of five reaction parameters such as willingness to be notified (*_notification*), willingness to allow (*_permission*), and subjective evaluations of comfort, risk, and appropriateness (*_comfort, _risk,* and *_appropriateness*) of the monitoring. We treated the five contextual parameters as basic features which collaboratively represent the current context in which information monitoring is performed by IoT devices. Additionally, we assigned each user to a specific privacy perception segment based on a small portion of their data (i.e., prior privacy decisions). We then used this privacy segment information as an additional feature. We used the reaction parameter *_permission* as target value (or label), since it best reflects users' substantive privacy decisions in IoT environments (namely, allow or reject the monitoring). The dataset contains 6,618 data points, and each data point (row) represents a user's privacy decisions in a specific IoT scenario.

First, we utilized a state-of-the-art machine learning model, called linear model and deep neural networks (LMDNN [10]), to make privacy decisions on behalf of users. LMDNN, which is also known as Wide and Deep Learning, jointly trains wide linear models and deep neural networks. By doing so, it can take the benefits of memorization (linear models) and generalization (neural networks) at the same time. LMDNN showed a remarkable performance on binary classification problems with sparse input features [10, 42]. Because the dataset collected in [26] is also composed of categorical data with many possible feature values (e.g., 24 values for the contextual parameter *what*), we decided to use LMDNN to build predictive models for privacy decision support in IoT. We also selected machine learning models that have been widely used in the literature (e.g., decision trees) and compared them with LMDNN in terms of predictive performance on the dataset. Experimental results indicated that LMDNN outperforms all the conventional models.

Next, we explored the most suitable combination of features for building LMDNN models with a reasonable predictive performance. We chose the five contextual parameters as basic features because these parameters are known to be related to users' privacy decisions in IoT [11, 25, 26]. In addition, we considered each user's privacy segment information as an additional feature. This is because previous research indicates that privacy segment information is helpful for machine learning models to better predict users' privacy decisions [29, 30]. We applied an unsupervised data clustering algorithm, K-modes clustering, on a subset of users' privacy decisions, in order to segment the users by their perceptions of privacy (i.e., privacy segmentation). Therefore, we tested the following feature combinations: (1) contextual parameters only (basic features), (2) contextual parameters with interactions between them, and (3) contextual parameters with interactions between them and privacy segment information. Experimental results showed that the feature combination (3) gives LMDNN models with the highest predictive performance. It also means that both interactions between contextual factors (e.g., *who* by *what*) and privacy segment information are useful for LMDNN models to predict privacy decisions of the users.

Lastly, we investigated different approaches for training machine learning models. There exist two traditional approaches in the literature: individual and one-size-fits-all modeling. Individual modeling is a process of building a user-specific predictive model based on each user's data only. This approach is known to be effective for modeling each user's unique characteristics (e.g., habits and personality). A model with a reasonable predictive performance, however, typically requires a considerable amount of training data from each individual user. In contrast, one-size-fits-all modeling utilizes multiple users' data as a single training data and constructs a universal model for all of them (including new users). This approach enables predictive models to make general predictions, therefore it can be useful for new users who did not provide data to the system yet, but want to get recommendations immediately. However, prediction results may not be personalized to each individual user. We present another approach called one-size-fits-segment modeling, a variant of one-size-fits-all, taking privacy segment information into account in building predictive models. The basic idea is to serve each user with a machine learning model trained by data collected from others who share the same notion of privacy with this user. We divided the dataset based on the results of privacy segmentation, then trained an LMDNN model for each segment of the users. By using both contextual and privacy segment information as input features, we compared the predictive performance of individual, one-size-fits-all, and one-size-fits-segment modeling. Experimental results confirmed that the proposed approach performs the best in the dataset. Final LMDNN models trained via one-size-fits-segment modeling showed an average area under curve (AUC) of 0.6782 across all users. We noticed that one-size-fits-segment modeling performs much better than individual modeling for about 80% of the users. However, it does not work well for about 20% of the users who have highly accurate individual models ($AUC > 0.7$). We elaborated more on the challenges of accurately predicting IoT users' privacy decisions in a later section.

To sum up, our proposed prediction mechanism not only showed a reasonable performance for most users, but also can cause less burden and privacy risks to users since it mainly utilizes non-personal contextual information which can often be automatically collected from the IoT environment, and only prompts each user for a small amount of privacy decisions (answers to five reaction parameters about a single scenario) to determine his/her privacy segment. We also presented some practical implications for designing and developing machine learning-based privacy decision support systems for IoT.

In summary, our work makes the following contributions to the field of privacy decision support, through the convergence of artificial intelligence and IoT:

1. We adopted a state-of-the-art machine learning model called linear model and deep neural networks (LMDNN) to make privacy decisions on behalf of IoT users, and verified that LMDNN outperforms conventional methods that have been widely used in the literature;
2. We investigated the best approach for training machine learning models in terms of input feature and training strategy, and verified that the proposed approach called one-size-fits-segment modeling outperforms preexisting methods such as individual and one-size-fits-all modeling;
3. We reported some practical implications in predicting users' privacy decisions on personal information monitoring in IoT.

2 Related Work

In this section, we first present a literature review of privacy decision prediction based on machine learning techniques. As mentioned above, a considerable research effort has been made towards developing intelligent agents which infer and recommend users' privacy decisions on diverse personal information disclosure scenarios. Researchers have also claimed that this kind of technology could help users make a correct decision, thereby minimizing potential privacy risks. In addition, we also summarized previous research about privacy segmentation. It is understood that privacy segmentation provides useful information for understanding and predicting people's privacy decision-making.

2.1 Prediction of Privacy Decision-Making

Much research on the prediction of privacy decision-making focused on online or mobile social network services (SNS). By using (semi-) supervised machine learning techniques, researchers aimed to accurately predict SNS users' sharing policies for their own contents (e.g., to allow or disallow Facebook friend John to see my photos). This is because vendor-provided privacy setting mechanisms, which ask users to manually specify their preferences, have been proven ineffective for protecting user privacy [8, 39, 44].

Fang et al. proposed a framework named Privacy Wizard that infers and recommends each user's access control policies for his/her personal information on Facebook [14]. Each policy specifies who can access specific personal information. The authors adopted a supervised machine learning approach to learn each user's policies by asking him/her a number of questions (e.g., would you like to share your birthday with Facebook friend John?). Users' answers were considered intended policies (labels) for their friends. Regarding input features for machine learning models, the authors utilized both demographic information and community membership which characterized each user. Most importantly, the framework asked users about policies that machine learning models are most uncertain about (namely, active learning with uncertainty sampling). By selectively asking users to label the most informative data first, active learning can effectively reduce manual labeling efforts. In the experiments with 45 Facebook users, Privacy Wizard showed 90% accuracy in predicting individual's privacy policies with a small amount of labeled training data (25 out of 200 friends with privileges). The authors utilized a decision tree as a machine learning model.

Shehab et al. proposed a framework called Policy Manager that predicts users' sharing policies on SNS [40]. Like [14], Policy Manager was designed to infer binary access control policies on a specific personal object posted on each user's SNS account. The authors used both user profiles (that included, e.g., gender) and social network structure (e.g., closeness among users) as input features, and asked users to manually determine policies for a subset of their friends (labeling of training data). They tried nine different machine learning algorithms for each user, and chose the best algorithm showing the highest cross-validation accuracy on his/her data. In addition, Policy Manager selected models (i.e., classifiers) trained by other users, based on their accuracy in predicting a target user's training data. It then utilized these classifiers with the target user's own classifier to produce a final classification (i.e., classifier fusion via group voting). The authors tested their framework with 200 Last.FM users, and confirmed that the proposed classifier fusion approach was effective in improving predictive performance compared to using an individual classifier only: 83 and 70% best accuracy when using an alternating decision tree (ADTree) machine learning model, respectively. The authors extended their work by applying an active learning paradigm into the semi-supervised learning framework [41]. Similar to [14], they aimed at minimizing user burden in labeling training data by selecting the most informative data points to label. Sinha et al. also developed automated tools to assist users in correctly configuring privacy policies for text-based content on Facebook [44]. The authors utilized diverse features such as the text of posts, time of creation, n-grams, the previous policy, and attachments to predict future policies (e.g., visible to only me) for Facebook posts. They also adopted a supervised learning approach based on the MaxEnt algorithm. Through an online survey study with real Facebook users, they found out that the system could predict policies with a maximum accuracy of 81%, leading to a 14% increase in accuracy compared to when users used Facebook's privacy setting mechanism for a new post.

Regarding mobile SNS, Sadeh et al. proposed a mobile social networking application, PeopleFinder, which recommends optimized location privacy policies to users [39]. To this end, the authors adopted random forests, an ensemble supervised learning method, to build a classifier predicting sharing policies for each user's current location based on his/her previous decisions. The authors proved that these machine-generated policies have better accuracy than the user-defined policies: 91 and 79% success rate in matching users' actual behavior, respectively. Recently, Bilogrevic et al. proposed a personal information sharing platform named SPISM that semi-automatically determines whether to disclose users' personal information on SNS and at what level of granularity [8]. Like other previous works, the authors used a supervised learning approach to predict SNS users' privacy decision-making. For each user, SPISM constructed a multi-level classifier based on naÃrve Bayes or support vector machine (SVM) by using his/her past behavior as training data. Training data are composed of diverse personal and contextual factors, such as the identity of requester, type of information requested, user location, co-presence of others, and time (features) and each user's past privacy decision (label). Like [14, 41], SPISM also adopted an active learning paradigm to minimize users' labeling efforts. SPISM made decisions automatically whenever the confidence (probability) in the classification result was high enough; otherwise it explicitly asked users' decisions then added them to the preexisting training data. With the updated training data, SPISM continuously learns and adapts to users' privacy behaviors. Therefore, it will require less and less user input over time. A user study with 70 participants indicated that SPISM outperforms user-defined policies; it showed a median prediction accuracy of 72% when each user provided 40 manual decisions. Even if the authors focused on building a personalized machine learning model for each user (i.e., individual modeling), they also assessed potentials of one-size-fits-all modeling. A universal model trained by all users' data showed a reasonable performance with a median accuracy of 67%. The authors claimed that one-size-fits-all modeling could be suitable for building an initial predictive model that produces default privacy settings for the new user.

All these works not only confirm the necessity of decision support systems for protecting user privacy in social network services, but also provide practical guidelines for learning people's privacy behavior which often evolves over time. Specifically, most previous works were based on supervised machine learning algorithms with an individual modeling approach. That is, researchers proposed ways to build user-specific predictive models considering each user's unique behavioral characteristics. To reduce the user burden in providing sample privacy decisions, the researchers let users respond to select privacy-invasive scenarios or situations (i.e., active learning).

2.2 Privacy Segmentation

Researchers have also investigated methodologies to segment users into several categories in terms of their privacy attitudes and behaviors. The most commonly cited methodology is Westin's privacy segmentation model [23]. Westin had conducted several surveys about privacy issues in various domains such as e-commerce, national identification systems, and e-health. To effectively summarize the survey results, Westin developed an indexing scheme that categorizes survey participants into three categories: Privacy Fundamentalists, Privacy Pragmatists, and Privacy Unconcerned. Westin treated participants' responses to several pre-defined statements (scenarios) as criteria to derive these three categories (e.g., Privacy Fundamentalists are respondents who agreed with the first statement and disagreed with the second and third statement).

Privacy segmentation has been studied in diverse contexts. Lin et al. proposed an unsupervised data clustering approach for categorizing smartphone users into distinctive groups based on their privacy preferences regarding mobile app permission (e.g., grant or deny permission to an app to access personal information) [29]. The authors utilized an agglomerative hierarchical clustering algorithm (Ward's method) on about 21,000 privacy preferences collected from 725 Android users. Each preference represents each user's willingness to grant permission to a given app for a specific purpose (i.e., app-permission-purpose triple). The authors identified four privacy profiles from this cluster analysis. They also presented default privacy settings to each user based on his/her privacy profile. This was intended to help users better control their privacy when confronted with numerous permission requests on Android platforms. In a follow-up study, the authors utilized users' privacy profile information as one of the input features for machine learning models to predict Android users' permission settings [30].

Lankton et al. clustered SNS users into four categories based on their privacy management strategies [24]. They surveyed college students' behavior on Facebook, including the use of privacy settings, degree of content disclosure, and variety and size of friend lists. The authors then conducted a two-stage cluster analysis on this dataset. Like [29], the authors first performed hierarchical clustering to determine both the correct number of clusters and initial cluster centroids. They found that a four-cluster solution is optimal. Next, they conducted non-hierarchical (K-means) cluster analysis on the dataset, using the pre-determined cluster centroids as a starting point. After statistically comparing survey responses in each cluster, the authors confirmed that the resulting clusters are distinctive enough regarding the degree of the users' privacy concerns.

Most recently, the market research firm Forrester published a report suggesting that consumers can be divided into four privacy categories: Data-Savvy Digitals, Reckless Rebels, Nervous Nellies, and Skeptical Protectionists [21]. This finding is based on large-scale online survey studies designed to capture people's behavioral reactions toward personal data collection and use by Internet companies. About 34% of the study participants were categorized as those who are not willing to share their information (Nervous Nellies and Skeptical Protectionists) since they are skeptical about corporate privacy practices.

Even though the number and type of privacy segments vary somewhat across these works, most researchers came to the same common conclusion about privacy segmentation: it is practically feasible to identify distinctive privacy segments by collecting and analyzing human behavioral data. Furthermore, privacy segment information can be used as an informative feature for understanding and predicting people's future privacy behavior because it represents users' overall perception of privacy [29, 30].

3   Dataset of Privacy Decisions

For this study we used the dataset collected in our previous work [26], which is arguably the first comprehensive collection of privacy decisions in IoT environments. We gathered people's privacy preferences (decisions) about diverse privacy-invasive scenarios in simulated IoT environments, through the experience sampling method (ESM). We used Google Glass for presenting the IoT scenarios to study participants in order to let them perceive the scenarios as realistically as possible. Specifically, we developed a Google Glass app to dynamically display scenarios based on participants' location. Participants were asked to walk around a university campus wearing Google Glass. As participants moved towards one of 130 selected locations on campus, the app presented the scenario pertaining to this location. Participants then answered several questions on their preferred privacy protection in the given scenario. This immersive spatial setup seems more suitable to gather accurate privacy behaviors from participants than a traditional online survey system, since the setup situates them in scenarios and is therefore likely to better capture the situatedness of privacy decisions [15, 35]. In addition, location has been found to be a particularly critical component in understanding people's privacy decision-making [7, 8, 39].

In order to formalize users' privacy decision-making in IoT, we defined several parameters representing both contextual characteristics of IoT scenarios (contextual parameters) and possible user reactions (reaction parameters).

In our earlier interview and online survey studies [11, 25], we had already identified the five contextual parameters that have the most influence on people's reactions toward privacy (see Table 5). These five parameters define the place where the monitoring occurs (parameter *where*), the type of information being monitored (*what*), the entity that is monitoring (*who*), the reason for monitoring (*reason*), and the frequency of the monitoring (*persistence*). Each IoT scenario can be described by an expression that includes every contextual parameter together with its respective parameter value for this scenario. We produced 130 IoT scenarios specifically related to known geographical locations on the campus.

We also identified reaction parameters that serve as proxies of people's notion of privacy, namely the desire to be notified about (parameter *_notification*) and the willingness to allow (*_permission*) the monitoring. In addition, we also found it important to measure people's opinion on each monitoring activity in terms of comfort, risk, and appropriateness (parameters *_comfort*, *_risk*, *_appropriateness*). Table 6 shows the type of reaction parameters, together with their values which are all categorical or ordinal. A single user response about a specific IoT scenario (i.e., one row in a dataset) is composed of the following attributes: participant ID, scenario ID, five contextual parameter values, and the five user-provided reaction parameter values.

We gathered 172 participants in total over a period of three months: 106 males and 65 females (one person did not disclose his/her gender), with the majority (82%) being 18–25. Because we recruited the participants on campus, most of them have some university affiliation (about 2/3 undergraduates and 1/3 graduates). Participants answered 39 scenario descriptions on average. The collected dataset contains a total of 33,090 privacy decisions for 6,618 IoT scenarios. The dataset will be referred to as IoTP in this chapter.

4 Privacy Decision Prediction

By using the IoTP dataset, we investigated mechanisms to learn and predict users' privacy decision-making in IoT environments. Specifically, we aimed to predict the value of the reaction parameter *_permission* (R_2) based on the current context and privacy segment of the user. Even though this parameter denotes four possible privacy decisions (see Table 6), we focused on binary decisions by converting $R_2 = 1, 2$ into $R_2 = 1$ (*allow*) and $R_2 = 3, 4$ into $R_2 = 0$ (*reject*). Therefore, prediction for this parameter can be formalized into binary classification problems (*allow* or *reject* the monitoring).

In this vein, we conducted a series of machine learning experiments, varying the models (algorithms), features, and training strategies. First, we tested multiple machine learning models to find the most suitable for making predictions about privacy decisions in IoT. We first tried LMDNN since it is known to be very effective for processing categorical data with high sparsity. We then compared the performance of LMDNN with several machine learning models that have been extensively used in earlier research. Thereafter we also assessed the impact of input features, consisting of contextual and privacy segment information, on the predictive power of the machine learning model. This assessment allowed us to determine which features should be used for building classifiers. Last, and most importantly, we conducted a comparative evaluation of the predictive performance of well-known model training strategies, such as individual and one-size-fits-all modeling, and our proposed approach called one-size-fits-segment modeling. The results informed us of practical implications for developing a privacy decision support system for IoT environments.

4.1 Machine Learning Model

In this sub-section, we explained in detail why we chose the LMDNN machine learning model for realizing privacy decision support in IoT. We also presented experimental results showing that LMDNN can provide the most reasonable predictive performance compared to conventional machine learning models used in the literature.

4.1.1 LMDNN

Linear model and deep neural networks (LMDNN), also known as Wide and Deep Learning, has been proposed by [10] to solve the problem of recommending apps on Google Play. Generalized linear models like logistic regression are widely used for large-scale regression and classification problems as they are simple, scalable, and interpretable. The models are often trained on binarized sparse input features with one-hot encoding. Memorization of diverse feature interactions can be efficiently achieved by feeding a wide set of cross-product feature transformations with a target value (or label) into the model. While linear models are effective for learning relationships between categorical features and a target value, they cannot generalize the relationships to identify feature-target patterns that do not exist in training data. In contrast, deep neural networks (DNN) can better generalize to the previously unseen patterns by using low-dimensional dense embedding vectors learned from the sparse input features (i.e., transforming a categorical feature value into a vector of continuous values). This means that DNN can make a reasonable prediction for new observations based on preexisting training data. At the same time, DNN also can over-generalize when the underlying feature-target matrix is too sparse (e.g., rare interactions between features and target value). LMDNN is a mixture of logistic regression (wide learning) and DNN (deep learning) to achieve both memorization and generalization in a single model. By jointly performing wide and deep learning, LMDNN complements the weakness of deep learning (i.e., over-generalization) by letting the wide learning take some cross-product feature transformations into account in producing final classifications. This Wide and Deep Learning paradigm shows a remarkable predictive performance on diverse classification problems [9, 10, 42].

Figure 1 shows the LMDNN model structure we used in this study. The bottom (input) layer receives training data composed of categorical features (contextual

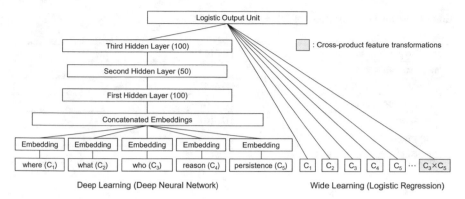

Fig. 1 Architecture of the LMDNN model

parameters) together with a target value (binarized reaction parameter _permission_). Next, the model can generate cross-product feature transformations and/or dense embedding vectors from the inputted data. As explained above, feature transformations and embedding vectors are used by wide and deep learning, respectively. For the deep learning, an 8-dimensional embedding vector is learned from each categorical feature. The model also combines all embedding vectors into a single dense embedding vector. The resulting concatenated embedding vector is then fed into three hidden layers with the ReLU activation function, and finally the logistic output unit (here, the sigmoid function). We configured 100, 50, and 100 units in consecutive hidden layers, respectively. We determined this network structure based on internal performance benchmarking on the IoTP dataset. In training LMDNN models, we followed the default mechanism (e.g., backpropagation, mini-batch stochastic optimization, AdaGrad regularization) described in [10]. We utilized a TensorFlow [1] implementation of LMDNN for all our experiments.

4.1.2 Predictive Performance Evaluation

To verify whether the LMDNN model is suitable for making predictions based on many categorical input features, we compared its predictive performance on the IoTP dataset against other conventional machine learning models: recursive partitioning tree [46], conditional inference tree (CTree) [16], random forests and bagging ensemble based on CTree (conditional random forests), SVM, naÃrve Bayes, and logistic regression. We chose these models because they are widely used in privacy decision support systems based on machine learning (e.g., decision trees [14, 40], random forests [39], and SVM [8]). For each machine learning model, we measured its predictive performance on the dataset, through 10-fold cross validation (CV). We only utilized the five contextual parameters as input features in these experiments, without cross-product feature transformations for memorization. This is because we first needed to assess the generalization capability of these models since the dataset

Table 1 Predictive performance of ML models (10-fold CV)

Machine learning model	AUC
Recursive partitioning tree	0.6142
Conditional inference tree	0.6293
Conditional random forests	0.6353
Support vector machine	0.6107
Naïve bayes	0.6161
Logistic regression	0.6208
Deep neural networks (of LMDNN)	0.6421

is small (6,618 rows), thereby potentially leading to overfitting. Therefore, we evaluated a deep model of LMDNN (see the left side of Fig. 1) in this experiment.[1] Regarding a performance metric, we primarily used the area under the ROC curve (AUC) because it is unaffected by the class imbalance problem (there were 64% allow and 36% reject decisions in the dataset) and is independent of the threshold applied to compute the probability of the binary classification results. Additionally, AUC itself is comprehensible; a random classifier has an AUC score of 0.5 while a perfect classifier has an AUC score of 1.0.

Experimental results indicated that LMDNN outperforms all other models (see Table 1). However, the performance difference is not large (up to 3%). Conditional random forests yield competitive performance because it has proven effective at generalizing to variants not in the training set [6], just as deep neural networks do. As mentioned before, however, LMDNN is known to further enhance the deep model by efficiently memorizing feature-target patterns that are rarely observed in a sparse dataset (wide learning). For this distinctive feature, we decided to utilize LMDNN as the machine learning model for this study.

4.2 Features

Here we explained how we identified the most useful features for training LMDNN models. Specifically, we presented the reasons why we chose the five contextual parameters with interactions between them as underlying input features. In addition, we described how we conducted privacy segmentation on the study participants in [26] and why we utilized the resulting privacy segment information as an additional feature.

---

[1] The TensorFlow implementation of LMDNN provides an API that enables programmers to selectively configure a wide, deep, or wide and deep model.

4.2.1 Contextual Information

We decided to use the five contextual parameters as basic features for the following reasons: (1) our previous research [26] indicated that all these contextual parameters impact people's privacy decision-making, and (2) we aimed to make predictions based on contextual information which can be automatically collected by IoT environments, thereby avoiding as much as possible asking users to manually enter additional information. We also considered interactions between the contextual parameters because people's privacy decisions about specific contextual information could be influenced by other factors. For instance, the monitoring of personal information (e.g., face photos: $C_2 = 11$) can be perceived differently depending on who is performing this monitoring (e.g., unknown vs. employer/school: $C_3 = 1$ vs. 6). Furthermore, LMDNN models can make predictions for unusual feature-target patterns by considering cross-product feature transformations (i.e., memorization). Because the contextual parameters *what* (C_2) and *who* (C_3) have the most significant influence on people's privacy decisions [25, 26], we determined that we should feed feature transformations based on *what* and *who* parameters (e.g., $C_2 \times C_1, C_2 \times C_3$, $C_2 \times C_4, C_2 \times C_5$) into the LMDNN models.

4.2.2 Privacy Segment

As discussed before, privacy segmentation is known to provide useful information for understanding and predicting people's privacy behavior. By applying the K-modes clustering algorithm on the IoTP dataset, we identified distinctive privacy segments and then assigned each user into one of the segments. Even though K-means clustering is the most famous data mining technique, we could not directly apply K-means to the categorical IoTP dataset because it can only process continuous numerical values as its input. As a variant of K-means, the K-modes clustering algorithm was designed to cluster categorical (or ordinal) values without data conversions. It modifies the original K-means by (1) replacing cluster means with cluster modes, (2) using the simple matching dissimilarity function instead of Euclidean distance to calculate the distance between categorical objects, and (3) updating modes with the most frequent categorical values in each iteration of the clustering [17, 18]. We used klaR [38], an R implementation of K-modes, for the task of privacy segmentation. In [26], we had already performed cluster analysis on the same dataset. However, in that study, we clustered 130 IoT *scenarios* in terms of users' privacy concerns about contextual factors (e.g., *what* and *who*). In this study, we aimed to cluster *users* into privacy segments based on each user's expectation of privacy in a single IoT scenario.

To that end, we selected scenario #60[2] as a base scenario that all participants had responded to, and filtered the dataset by considering the base scenario only. All

---

[2]A device of ICS ($C_3 = 6$) takes a photo of you ($C_2 = 11$). This happens once ($C_5 = 0$), while you are in DBH ($C_1 = 3$), for safety purposes ($C_4 = 1$), namely to determine if you are a wanted criminal.

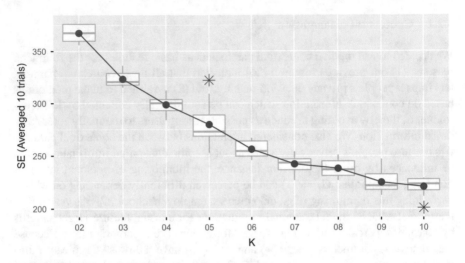

Fig. 2 Clustering errors (*SE*) and the size of clusters (*K*)

participants responded to scenario #60 because it is related to the location where each experiment started [26]. By analyzing this partial dataset, we expected to understand how individual users perceive and react differently in the same scenario.

Determining the number of clusters (K) is the first step in the data clustering process. We need to find a balance between maximum data compression by assigning all data points into a single cluster ($K = 1$) and maximum accuracy by assigning each data point into an individual cluster ($K = n$). Thus, we heuristically searched for the optimal K by utilizing the well-known Elbow method [32]. First, we computed the sum of errors (*SE*) of the K-modes clustering with a maximum of 50 iterations, while increasing K from 2 to 10. The *SE* is defined as the sum of the distance between each instance of the cluster and the cluster's centroid (mode):

$$SE_K = \sum_{i=1}^{K} \sum_{x \in c_i} dist(x, c_i)$$

where x is a data point belonging to the ith cluster and c_i is the mode of the ith cluster. We repeated this procedure 10 times, and took average values of *SE* for each value of K. Next, we calculated the values for the mean difference between SE_K and SE_{K-1}, and found that the largest decrease in errors occurred when we increased K from 2 to 3 (see Fig. 2). Therefore, we chose 3 as a suitable number of clusters ($K = 3$), and used it as a parameter (modes) for running the K-modes clustering algorithm on the dataset.

Table 2 summarizes the resulting cluster modes, which are composed of both contextual and reaction parameter values. Note that all contextual parameter values are identical across the clusters because we fixed them to describe the base scenario. The clusters are quite distinct from each other, primarily in the reaction parameters

Table 2 Modes of the resulting privacy segments (clusters)

Contextual param $\{C_1, C_2, C_3, C_4, C_5\}$	Reaction param $\{R_1, R_2, R_3, R_4, R_5\}$	Privacy segment (cluster mode)
3, 11, 6, 1, 0	1, **1**, 4, 6, 6	Indifferent (M_1)
3, 11, 6, 1, 0	1, **0**, 3, 3, 3	Some sensitive (M_2)
3, 11, 6, 1, 0	1, **0**, 1, 1, 1	Sensitive (M_3)

_comfort (R_3), _risk (R_4, reverse-coded), and _appropriateness (R_5): each mode has a unique combination of values for R_3, R_4, and R_5. As shown in Table 6, these parameters represent people's privacy attitudes about IoT scenarios on a scale of 1 to 7. For example, cluster mode 3 (M_3) contains $R_3 = 1$, $R_4 = 1$, and $R_5 = 1$, indicating that the given scenario is perceived by participants as *very uncomfortable*, *very risky*, and *very inappropriate*, respectively. For M_3, the value of the reaction parameter _permission (R_2) is zero. This means that users belonging to cluster 3 are likely to reject the base scenario because they have negative views on privacy in this scenario. Therefore, we marked the clusters (privacy segments) using these three reaction parameters. We labeled privacy segment 1 (PS_1) as *Indifferent to Privacy* since its mode contains the second highest value for R_4 and R_5 (namely, 6 on a 7-item scale). Likewise, we labeled privacy segment 2 (PS_2) as *Somewhat Sensitive to Privacy* (the values of R_3, R_4, and R_5 fall slightly below the scale average), and privacy segment 3 (PS_3) as *Sensitive to Privacy*. As a result, 74%, 15%, and 11% of the participants were assigned into *Indifferent to Privacy*, *Somewhat Sensitive to Privacy*, and *Sensitive to Privacy* segments, respectively. Finally, we repeated this clustering on additional scenarios (#20, #73, #93, #111)[3] which were the next most frequently visited scenarios after #60. We arrived at the same conclusions regarding the number ($K = 3$) and labels of the resulting privacy segments.

To validate the distinctiveness of the resulting privacy segments, we performed two Welch's t-tests on the R_3 parameter between the following pairs of privacy segments: (PS_1, PS_2) and (PS_2, PS_3). The reason for using Welch's t-test is that all privacy segments have different variances in the R_3 parameter. The tests confirm that the difference in the means of the R_3 parameter between each pair of the segments is statistically significant ($p < 0.025$, Bonferroni-corrected for two comparisons). Next, we also conducted Welch's t-tests on the R_4 and R_5 parameters and drew the same conclusion. Thereby, we verified that the privacy segments are sufficiently distinct from each other in terms of participants' reactions to the given scenario.

Because privacy segment information for all users was available from such a cluster analysis, we then utilized it as an additional feature for building predictive models. This is because privacy segment information is known to be useful for quantifying an individual's judgement about privacy and utility in various circumstances [22, 24].

---

[3]Number of respondents (scenario ID): 140 (#20), 138 (#73), 136 (#93), 162 (#111).

Table 3 Predictive performance of LMDNN models (10-fold CV)

Feature combination	AUC
(1) C_1, C_2, C_3, C_4, C_5	0.6421
(2) $C_1, C_2, C_3, C_4, C_5, C_2 \times C_{\{1,3,4,5\}},$ $C_3 \times C_{\{1,4,5\}}$	0.6528
(3) $C_1, C_2, C_3, C_4, C_5, C_2 \times C_{\{1,3,4,5\}},$ $C_3 \times C_{\{1,4,5\}}, PS$	0.6725

4.2.3 Predictive Performance Evaluation

As described in Sect. 4.1.2, we treated the five contextual parameters as basic features for training the deep learning part of LMDNN. We then considered interrelated parameters, especially based on the *what* (C_2) and *who* (C_3) parameters, in conducting both wide and deep learning via LMDNN. This is not only because both contextual parameters have a significant impact on people's privacy decisions, but because the memorization of some interactions between parameters (i.e., cross-product feature transformations) can improve the performance of the deep model of LMDNN. As an additional feature, we adopted privacy segment information because it differentiates users according to their perceptions of privacy. We tested the following combinations of features to assess their influence on the predictive performance of LMDNN models.

1. Contextual parameters (*deep* learning)
2. Contextual parameters with interactions (*wide* and *deep* learning)
3. Contextual parameters with interactions and privacy segment information (*wide* and *deep* learning).

Using the whole IoTP dataset, we performed 10-fold CV on the LMDNN model trained with each of these feature combinations. As expected, the AUC score gradually improves as we added cross-product feature transformations and privacy segment information to the five basic contextual features (see Table 3). We then concluded that both the contextual parameters (including interactions) and privacy segment information can act as informative features for predicting the binary value of the reaction parameter _*permission* (R_2) through LMDNN.

4.3 Training Strategy

Based on the selected machine learning model (LMDNN) and features (contextual and privacy segment information), we investigated the best strategy to build a predictive model (i.e., classifier) for each individual user. First, we reviewed two commonly used strategies, individual and one-size-fits-all modeling. Individual modeling typically utilizes a single user's instances as training data, and will therefore result in a highly personalized user-specific model if a sufficient amount of training data is available. One-size-fits-all modeling, in contrast, trains a single model based on all users' data, so that reasonable predictions can be made about new users for whom insufficient data is available. Next, we proposed our strategy that we dub one-size-fits-segment modeling. It was designed to utilize the one-size-fits-all paradigm for making predictions for users grouped by privacy segmentation. We analyzed the overall and per-user performance of the predictive models trained through these three different strategies, and then draw some practical implications.

4.3.1 Individual Modeling

This is the most popular and straightforward approach for building user-specific machine learning models for privacy decision support systems, especially targeted at predicting users' decisions about the disclosure of personal information on social network services [8, 14, 39–41, 44]. It constructs a distinctive predictive model per each user by using his/her data only. This assumes that each user has a very different point of view regarding privacy, therefore the others' data are not useful for modeling and predicting his/her own privacy decision-making. As most previous works have adopted supervised machine learning approaches, each user will need to provide a certain amount of labeled training data (e.g., historical decisions for privacy-invasive scenarios). For instance, Bilogrevic et al. [8] stated that they needed 40 manual decisions from each user to build personalized models with a reasonable predictive performance, which is a quite burdensome amount. The more training data a user provides, the more the performance of the individual predictive model tends to improve. For these reasons, an individual modeling strategy is suitable for situations in which service providers could acquire enough training data from each individual user. Like [8], the study participants in [26] made about 40 privacy decisions (reaction parameter _permission) on average. Therefore, we first applied individual modeling for constructing personalized LMDNN models to check whether this strategy is adequate for solving our own problem. Specifically, we trained and evaluated a single LMDNN model for a specific user based on his/her data. We repeated it for all users in the dataset. As input features, we utilized solely the five contextual parameters with their interactions. Since each individual LMDNN model is exclusively trained using each user's data, we do not use privacy segment information in this experimental setup.

4.3.2 One-Size-Fits-All Modeling

One-size-fits-all contrasts with an individual modeling strategy. Instead of building multiple user-specific predictive models, it generates a single universal model based on data collected from a crowd of users. Because a one-size-fits-all predictive model is trained using a larger dataset (multiple users' data), it typically represents a wider range of common feature-target patterns than an individual model. Therefore, researchers have often utilized a one-size-fits-all modeling strategy for building a predictive model that makes initial predictions (e.g., default privacy settings) for new users who did not provide training data [8]. Recently published works [34, 45] also adopted one-size-fits-all modeling to make their privacy decision support systems generalizable to a wide range of users. The authors in [34] collected IoT-related privacy decisions from 1,007 Amazon Turkers through an online survey, and built one-size-fits-all predictive models for randomly chosen 50 participants. They reported that the overall accuracy of these trained models ranges from 76 to 80%, depending on whether most (75%) or all (100%) of the other participants' responses are used as training data. To verify the applicability of the one-size-fits-all modeling strategy to privacy decision support for real-world IoT environments (using IoTP dataset), we built a one-size-fits-all LMDNN model for each user based on data collected from all other users, utilizing both contextual and privacy segment information as input features. Each user's data was used thereafter to assess the predictive power of his/her LMDNN model.

4.3.3   One-Size-Fits-Segment Modeling

We proposed a novel approach called one-size-fits-segment modeling. This is a modified version of one-size-fits-all modeling. It builds multiple universal models for several groups of users rather than for the entire user population. Here, we utilized the result of privacy segmentation (see Sect. 4.2.2) as a criterion to cluster users. We intended to improve the performance of one-size-fits-all modeling by constructing per-user predictive models based on data collected from same-minded users in terms of privacy. We believe that each individual user will be benefitted if a predictive model is trained on large volumes of data provided by others similar to him/her. To verify the proposed approach, we conducted experiments as follows. First, we divided users in the IoTP dataset into three groups according to our privacy segmentation. For each single user, we determined the privacy segment to which he/she belongs, and built a one-size-fits-segment LMDNN model by utilizing data collected from other users in the corresponding privacy segment. Each user's data was used as test data for measuring the predictive performance of his/her LMDNN model. Unlike one-size-fits-all modeling, we only utilized contextual parameters (including interactions between them) as input features. As explained, privacy segment information was utilized to split users with regard to the subset of their stated privacy decisions, thereby constructing one-size-fits-segment LMDNN models for every single user. In comparison with individual modeling and one-size-fits-all modeling (described in [26]), the proposed mechanism does not burden new users because it just asks five questions (reaction parameters) about a single base scenario to perform privacy segmentation.

Table 4 Predictive performance of LMDNN models

Training strategy	Mean AUC	Std. dev.
Individual	0.4806	0.2179
One-size-fits-all	0.6699	0.1721
One-size-fits-segment	0.6782	0.1668

4.3.4 Predictive Performance Evaluation

We compared the predictive power of these three different training strategies as follows. Regarding individual modeling, we trained and assessed the predictive performance of user-specific LMDNN models via 10-fold CV. Each of these models was trained on all data instances from a different user in the IoTP dataset. We repeated this procedure for all users and took the average of their AUC scores as the overall predictive performance of individual modeling. For one-size-fits-all or one-size-fits-segment modeling, we constructed a target user's LMDNN model based on all other users' data and tested the trained model using the target user's data. Unlike [34], we repeated this for all users and calculated the average AUC score for each strategy. Finally, we compared these mean AUC scores for three training strategies to choose the best.

Table 4 summarizes the experimental results showing that both the one-size-fits-all and the one-size-fits-segment modeling strategy significantly outperform the individual modeling strategy. Furthermore, one-size-fits-segment shows a slightly better performance than one-size-fits-all. This is because one-size-fits-segment LMDNN models are trained by the data of same-minded users, thereby predicting each user's privacy decisions more accurately. This finding was unexpected since most previous research about privacy decision support has reported that individual modeling is better than one-size-fits-all modeling for inferring users' privacy decision-making. One possible explanation is that the size of per-user training data was not large enough to learn sparse high-dimensional feature spaces. Participants in the dataset responded to an average of 39.58 scenarios (std. dev: 14.65), or about 33% of all available scenarios. Each scenario received 50.9 responses on average (std. dev: 40.16). When we transformed the dataset into a user-scenario matrix (i.e., each cell represents an observed feature-target pattern), the sparsity[4] of this matrix was 0.6566. This low response rate may also be found in future real-world situations since the number of applications and services in IoT environments is likely to increase over time. As a result, it could be difficult to collect enough training data from each IoT user to build individual predictive models from scratch.

Figure 3 presents the per-user predictive performance of two model training strategies: individual and one-size-fits-segment modeling. Results have been filtered out by the threshold of $AUC > 0.5$ and sorted by the AUC score of individual LMDNN models in ascending order. As can be seen, one-size-fits-segment LMDNN models

---

[4]1—(nonzero entries/total entries in a user-scenario matrix).

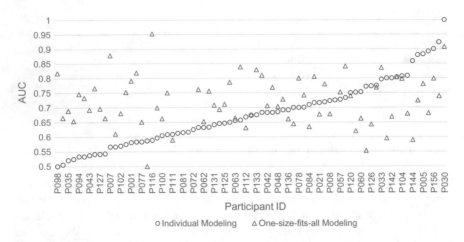

Fig. 3 Per-user predictive performance

show a better predictive performance than individual LMDNN models for about 80% of the users (left side of the chart). However, there is also the opposite effect for about 20% of users (far right side of the chart) who have highly accurate individual LMDNN models ($AUC > 0.7$) trained solely on their data. To check the causes of this per-user difference, we performed Pearson's chi-square tests for independence between personal attributes (gender, age, type of university affiliation)[5] and the fitted training strategy. For instance, we constructed a 2×2 contingency table using two binary variables, age (18–25 or older than 25) and training strategy (one-size-fits-segment or individual modeling), and assessed the significance of the difference between the two proportions (i.e., users aged 18–25 and older users, both have superior one-size-fits-segment LMDNN models). The test confirms that there is no statistical evidence of an association between these two variables at the .05 significance level. We iterated this test for other personal attributes such as gender (male or others) and the type of university affiliation (undergraduate students or others), and reached the same conclusion. Identifying latent factors (e.g., cultural background in privacy decision-making [28]) that cause this difference is one possible future research direction. If we were to find these factors, we could determine the most appropriate modeling strategy for each individual user in advance, and then accordingly utilize it for better predictive performance.

4.4 Implications

We confirmed that privacy segment information is helpful for understanding and predicting people's privacy decision-making about various IoT services. Specifically,

---

[5]Mode values of these attributes are male, 18–25, and undergraduate students, respectively.

one-size-fits-segment modeling shows the best predictive performance (mean AUC of 0.6782; N = 172) compared with the previously proposed model training strategies. However, even a single *false positive* may cause undesired information monitoring in IoT, making users reluctant to use privacy decision support systems. At its current accuracy, we should therefore use the prediction results only to give recommendations to users that they can still inspect and override; we should not use the prediction results for automated disclosure decisions. Moreover, we should strive for further improvement of the predictive performance. In the following, we explained how the current predictive performance could be improved.

4.4.1 Role of Privacy Segment Information

To better interpret the performance of one-size-fits-segment modeling, we measured the predictive performance of the trained LMDNN models for each privacy segment. Figure 4 indicates that users who are assigned to the *Sensitive to Privacy* segment have the most accurate LMDNN models (mean AUC of 0.729, std. dev: 0.074). Likewise, users belonging to the *Somewhat Sensitive to Privacy* segment also tend to have LMDNN models with a reasonable performance (mean AUC of 0.708, std. dev: 0.101). Users in the *Indifferent to Privacy* segment however have relatively inaccurate LMDNN models compared to those in other privacy segments (mean AUC of 0.665, std. dev: 0.184). Here, we also could not identify any inter-segment differences in terms of personal characteristics such as gender, age, and the type of university affiliation. One possible explanation is that participants in both the *Sensitive to Privacy* and *Somewhat Sensitive to Privacy* segments presumably responded to IoT monitoring scenarios more carefully than participants in the *Indifferent to Privacy* segment, thereby providing more consistent privacy decisions (i.e., high-quality training data). We might need to consider strategies for improving the data collection process for the *Indifferent to Privacy* segment. For instance, users who are indifferent about privacy

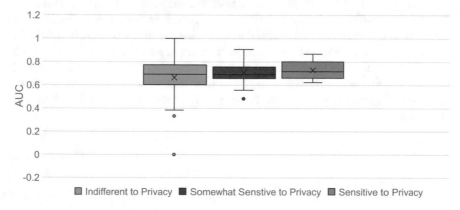

Fig. 4 Per-segment predictive performance

could be asked for more training data than the other users. This is because a higher amount of training data would reduce uncertainty in predictive models, and therefore likely improve performance. It might also be possible to utilize privacy nudges [30], which can make privacy-insensitive users aware of unexpected outcomes of personal information monitoring, for steering users towards more consistent privacy behavior.

4.4.2 Size of Individual Training Data

Regarding individual modeling, 28 users had highly accurate ($AUC > 0.7$) individual LMDNN models trained on their own data. Furthermore, individual models can be continuously improved as each user provides additional training data. To verify this claim, we built a linear regression model with the number of user responses (i.e., size of training data) as the independent variable and AUC scores (i.e., predictive performance) as the dependent variable. According to the fitted regression model, a statistically significant positive relationship exists between these two variables ($p < 0.005$). Therefore, the collection of extra training data might further improve the performance of individual machine learning models. Yet, asking for additional data can also be burdensome and privacy-invasive.

4.4.3 Hybrid Modeling

For these reasons, it might be desirable to gradually morph one-size-fits-segment models into individual models. Like [8, 34, 45], we verified that our one-size-fits-segment LMDNN models can make reasonably accurate predictions for two segments, even for new users. Therefore, we can utilize them to produce *default* privacy settings that are applicable to the general population (grouped by privacy segmentation). As mentioned before, however, these general predictive models would need to be tailored to each user to make more accurate predictions. One possible avenue would be to gradually transform the pre-trained one-size-fits-segment LMDNN models into *personalized* individual LMDNN models for each user. Specifically, it is necessary to continuously re-train (update) the base one-size-fits-segment LMDNN model with privacy decision samples collected from a specific user, while increasing the weights of this user-specific data (i.e., transfer learning [20, 27, 36]). This should then better fit the updated LMDNN model with the user's unique behavioral patterns. Active learning paradigms can also be applied to reduce users' labeling efforts as much as possible.

5 Discussion and Future Work

We studied machine learning mechanisms for modeling and predicting users' privacy decisions regarding personal information tracking in IoT environments. After investigating various machine learning algorithms, input features, and model training strategies, we proposed a novel mechanism called one-size-fits-segment modeling for privacy decision support systems in IoT. We showed that the proposed mechanism exhibits a reasonable predictive performance on privacy behavioral data captured in the field. Our work goes far beyond the limited machine learning approach in [26], where contextual parameters (without consideration for interaction effects) and cluster membership information of IoT service scenarios were used as features for building a decision tree-based predictive model. We believe this approach is not very practical because it requires users to specify reaction parameter values for dozens of scenarios in advance. Another difference between [26] and this work lies in the examination of the model training strategies; we only tested one-size-fits-all modeling (i.e., single classifier) in [26], and hence the results may provide limited implications (e.g., lack of consideration for per-user predictive performance). Yet, our work still has some shortcomings that will be addressed in the next few sub-sections.

5.1 Representability of Data

First, we need to consider the representativeness of the dataset we used. The study participants were predominantly university students aged 18–25 (82%), since we recruited them on campus. This may induce a sampling bias that makes our results less generalizable. In other words, we do not know how the proposed prediction mechanism will work on datasets collected from other populations (e.g., older users who are not familiar with IoT). Previous researchers repeatedly stated that the machine learning model trained on a specific user group's data does not perform well for other user groups due to the discrepancies between different user populations [5]. In this regard, we plan to validate our mechanism with more diverse and representative samples, thereby confirming a future direction of this research (e.g., continuous updates of one-size-fits-segment LMDNN models with user-provided data). It will also be necessary to apply the proposed mechanism to different domains (e.g., privacy decisions in healthcare settings), and verify its expandability. To that end, we need to collect or get access to users' privacy behavioral data regarding such a domain.

5.2 Reliability of Privacy Segmentation

We utilized the results of privacy segmentation to improve the performance of our predictive models. As discussed, we performed privacy segmentation by clustering users into several groups based on their responses toward a single scenario. Here, we utilized scenario #60 as a base scenario since all users responded to this scenario; this enabled us to assign privacy segment information to each of the users. We also tried four different base scenarios to validate whether our clustering methodology (i.e., determining the correct number and labels of the clusters) is sound, and the result was positive. However, it is also important to check the invariance of the clustering results. As future work, we plan to collect all users' privacy decisions regarding more than one base scenario and conduct pair-wise comparisons on the clustering results so as to confirm the resulting privacy segments are stable no matter which base scenario is used for privacy segmentation. If a user was assigned to the same privacy segment regardless of the base scenario, we can consider the results of privacy segmentation as ground truth. Otherwise, we need to devise a way to decide on the most accurate privacy segment for this user. One possible approach is a majority vote among the segmentation results from all base scenarios.

5.3 Privacy Paradox

The privacy paradox is the phenomenon that people's stated privacy preferences or decisions often seem inconsistent with their actual behaviors [2, 13, 19]. As explained before, we utilized stated privacy decisions collected through ESM in a simulated IoT environment [26] as training data, and not actual behavior observed in a working IoT environment. Although we had tried to make participants perceive they were in a real situation, we do not know how they would actually behave in real-world situations. Therefore, we need to construct a working IoT environment, let participants freely interact with the environment, and then collect the corresponding privacy decisions, possibly with sensor data. To that end, we plan to conduct an experimental study for collecting (large-scale) privacy behavioral data from real users in an operational IoT system based on open protocols for IoT discovery and interaction, such as the Open Connectivity Foundation (OCF), Web of Things, or Physical Web. By using this dataset, we need to confirm the effectiveness of the proposed mechanism.

6 Conclusion

In this chapter, we proposed a novel machine learning mechanism for predicting people's privacy decisions in IoT environments. We aimed to predict binary privacy decisions for each user, namely whether to allow or reject a given personal information monitoring scenario in IoT. To begin with, we adopted linear model and deep neural networks (LMDNN) as the machine learning model for our study. Using a privacy behavioral dataset ($N = 172$) collected from our earlier study, we confirmed that LMDNN provides better predictive performance than conventional machine learning models that have been widely used in the literature. Next, we utilized both contextual and privacy segment information as input features for training LMDNN models. We adopted a wide range of contextual factors comprising diverse IoT scenarios. We then generated users' privacy segment information by clustering their privacy decisions about a single selected IoT scenario. Lastly, we proposed a new model training strategy called one-size-fits-segment modeling, and compared its performance with two commonly used strategies: individual and one-size-fits-all modeling. Experimental results indicated that one-size-fits-segment outperforms other modeling strategies. We also presented some practical implications regarding the design and development of privacy decision support systems for IoT environments. Future work will focus on collecting privacy-related human behavioral data from more representative samples of users interacting with working IoT systems, and validating the proposed mechanism on this new dataset.

Acknowledgements This research was funded by NSF Grant SES-1423629. The human subjects research described herein is covered under IRB protocol #2014-1600 at the University of California, Irvine.

Appendix

See Appendix Tables 5 and 6.

Table 5 Contextual parameters

Parameter (id)	Values
where (C_1)	(0) your place
	(1) someone else's place
	(2) semi-public space (e.g., restaurant)
	(3) public space (e.g., street)
what (C_2)	(1) phoneID
	(2) phoneID $\Rightarrow$ identity
	(3) location
	(4) location $\Rightarrow$ presence
	(5) voice
	(6) voice $\Rightarrow$ gender
	(7) voice $\Rightarrow$ age
	(8) voice $\Rightarrow$ identity
	(9) voice $\Rightarrow$ presence
	(10) voice $\Rightarrow$ mood

(continued)

Table 5 (continued)

Parameter (id)	Values
	(11) photo
	(12) photo $\Rightarrow$ gender
	(13) photo $\Rightarrow$ age
	(14) photo $\Rightarrow$ identity
	(15) photo $\Rightarrow$ presence
	(16) photo $\Rightarrow$ mood
	(17) video
	(18) video $\Rightarrow$ gender
	(19) video $\Rightarrow$ age
	(20) video $\Rightarrow$ presence
	(21) video $\Rightarrow$ mood
	(22) video $\Rightarrow$ lookingAt
	(23) gaze
	(24) gaze $\Rightarrow$ lookingAt
who (C_3)	(1) unknown
	(2) colleague/fellow
	(3) friend
	(4) own device
	(5) business
	(6) employer/school
	(7) government
reason (C_4)	(1) safety; (2) commercial; (3) social; (4) convenience; (5) health; (6) none
persistence (C_5)	(0) once
	(1) continuously

Table 6 Reaction parameters

Parameter (id)	Values
_notification (R_1)	(1) notify me, always
	(2) notify me, just this time
	(3) don't notify me, just this time
	(4) don't notify me, always
_permission (R_2)	(1) allow, always
	(2) allow, just this time
	(3) reject, just this time
	(4) reject, always

(continued)

Table 6 (continued)

Parameter (id)	Values
_comfort (R_3)	(1) very uncomfortable
	(2) uncomfortable
	(3) somewhat uncomfortable
	(4) neutral
	(5) somewhat comfortable
	(6) comfortable
	(7) very comfortable
_risk (R_4)	(1) very risky
	(2) risky
	(3) somewhat risky
	(4) neutral
	(5) somewhat safe
	(6) safe
	(7) very safe
_appropriateness (R_5)	(1) very inappropriate
	(2) inappropriate
	(3) somewhat inappropriate
	(4) neutral
	(5) somewhat appropriate
	(6) appropriate
	(7) very appropriate

References

1. Abadi, M., Agarwal, A., Barham, P., Brevdo, E., Chen, Z., Citro, C., Corrado, G.S., Davis, A., Dean, J., Devin, M., et al.: Tensorflow: Large-scale machine learning on heterogeneous distributed systems. arXiv preprint arXiv:1603.04467 (2016)
2. Acquisti, A., Grossklags, J.: Privacy attitudes and privacy behavior. In: Economics of Information Security, pp. 165–178. Springer (2004)
3. Atzori, L., Iera, A., Morabito, G.: The internet of things: a survey. Comput. Netw. **54**(15), 2787–2805 (2010)
4. Batalla, J.M., Gajewski, M., Latoszek, W., Krawiec, P., Mavromoustakis, C.X., Mastorakis, G.: ID-based service-oriented communications for unified access to IoT. Comput. Electr. Eng. **52**, 98–113 (2016)
5. Beel, J., Breitinger, C., Langer, S., Lommatzsch, A., Gipp, B.: Towards reproducibility in recommender-systems research. User Model. User-Adap. Inter. **26**(1), 69–101 (2016)
6. Bengio, Y., Delalleau, O., Simard, C.: Decision trees do not generalize to new variations. Comput. Intell. **26**(4), 449–467 (2010)
7. Benisch, M., Kelley, P.G., Sadeh, N., Cranor, L.F.: Capturing location-privacy preferences: quantifying accuracy and user-burden tradeoffs. Pers. Ubiquit. Comput. **15**(7), 679–694 (2011)
8. Bilogrevic, I., Huguenin, K., Agir, B., Jadliwala, M., Gazaki, M., Hubaux, J.P.: A machine-learning based approach to privacy-aware information-sharing in mobile social networks. Pervasive Mob. Comput. **25**, 125–142 (2016)

9. Burel, G., Saif, H., Alani, H.: Semantic wide and deep learning for detecting crisis-information categories on social media. In: International Semantic Web Conference, pp. 138–155. Springer (2017)

10. Cheng, H.T., Koc, L., Harmsen, J., Shaked, T., Chandra, T., Aradhye, H., Anderson, G., Corrado, G., Chai, W., Ispir, M., et al.: Wide and deep learning for recommender systems. In: Proceedings of the 1st Workshop on Deep Learning for Recommender Systems, pp. 7–10. ACM (2016)

11. Chow, R., Egelman, S., Kannavara, R., Lee, H., Misra, S., Wang, E.: HCI in business: a collaboration with academia in IoT privacy. In: International Conference on HCI in Business, pp. 679–687. Springer (2015)

12. Christin, D., Reinhardt, A., Kanhere, S.S., Hollick, M.: A survey on privacy in mobile partici-patory sensing applications. J. Syst. Softw. **84**(11), 1928–1946 (2011)

13. Connelly, K., Khalil, A., Liu, Y.: Do I do what I say?: Observed versus stated privacy prefer-ences. Hum. Comput. Interact. **2007**, 620–623 (2007)

14. Fang, L., LeFevre, K.: Privacy wizards for social networking sites. In: Proceedings of the 19th international conference on World Wide Web, pp. 351–360. ACM (2010)

15. Hine, C.: Privacy in the marketplace. Inform. Soc. **14**(4), 253–262 (1998)

16. Hothorn, T., Hornik, K., Zeileis, A.: Unbiased recursive partitioning: a conditional inference framework. J. Comput. Graph. Stat. **15**(3), 651–674 (2006)

17. Huang, Z.: A fast clustering algorithm to cluster very large categorical data sets in data mining. DMKD **3**(8), 34–39 (1997)

18. Huang, Z.: Extensions to the k-means algorithm for clustering large data sets with categorical values. Data Min. Knowl. Disc. **2**(3), 283–304 (1998)

19. Jensen, C., Potts, C., Jensen, C.: Privacy practices of Internet users: self-reports versus observed behavior. Int. J. Hum. Comput. Stud. **63**(1), 203–227 (2005)

20. Karayev, S., Trentacoste, M., Han, H., Agarwala, A., Darrell, T., Hertzmann, A., Winnemoeller, H.: Recognizing image style. arXiv preprint arXiv:1311.3715 (2013)

21. Khatibloo, F.: It's Here! Forrester's Consumer Privacy Segmentation. Available at https://go. forrester.com/blogs/its-here-forresters-consumer-privacy-segmentation/. 15 Dec 2016

22. Knijnenburg, B.P., Kobsa, A., Jin, H.: Dimensionality of information disclosure behavior. Int. J. Hum. Comput. Stud. **71**(12), 1144–1162 (2013)

23. Kumaraguru, P., Cranor, L.F.: Privacy indexes: a survey of Westin's studies. Carnegie Mellon University, Pittsburgh, PA (2005)

24. Lankton, N., McKnight, D., Tripp, J.: Privacy management strategies: an exploratory cluster analysis. In: Proceedings of the 22nd Americas Conference on Information Systems (AMCIS 2016), pp. 1–10 (2016)

25. Lee, H., Kobsa, A.: Understanding user privacy in Internet of Things environments. In: Internet of Things (WF-IoT), 2016 IEEE 3rd World Forum on, pp. 407–412. IEEE (2016)

26. Lee, H., Kobsa, A.: Privacy preference modeling and prediction in a simulated campuswide IoT environment. In: Pervasive Computing and Communications (PerCom), 2017 IEEE International Conference on, pp. 276–285. IEEE (2017)

27. Lee, H., Upright, C., Eliuk, S., Kobsa, A.: Personalized object recognition for augmenting human memory. In: Proceedings of the 2016 ACM International Joint Conference on Pervasive and Ubiquitous Computing: Adjunct, pp. 1054–1061. ACM (2016)

28. Li, Y., Kobsa, A., Knijnenburg, B.P., Nguyen, C., et al.: Cross-cultural privacy prediction. Proc. Priv. Enhancing Technol. **2017**(2), 113–132 (2017)

29. Lin, J., Liu, B., Sadeh, N., Hong, J.I.: Modeling users' mobile app privacy preferences: restoring usability in a sea of permission settings. In: Proceedings of the 10th Symposium on Usable Privacy and Security (SOUPS 2014), pp. 199–212 (2014)

30. Liu, B., Andersen, M.S., Schaub, F., Almuhimedi, H., Zhang, S., Sadeh, N., Acquisti, A., Agarwal, Y.: Follow my recommendations: a personalized privacy assistant for mobile app permissions. In: Proceedings of the 12th Symposium on Usable Privacy and Security (SOUPS 2016), pp. 27–41 (2016)

31. Liu, Y., Yang, C., Jiang, L., Xie, S., Zhang, Y.: Intelligent edge computing for IoT-based energy management in smart cities. IEEE Netw. **33**(2), 111–117 (2019)

32. Madhulatha, T.S.: An overview on clustering methods. arXiv preprint arXiv:1205.1117 (2012)
33. Mavromoustakis, C.X., Batalla, J.M., Mastorakis, G., Markakis, E., Pallis, E.: Socially oriented edge computing for energy awareness in IoT architectures. IEEE Commun. Mag. **56**(7), 139–145 (2018)
34. Naeini, P.E., Bhagavatula, S., Habib, H., Degeling, M., Bauer, L., Cranor, L., Sadeh, N.: Privacy Expectations and preferences in an IoT world. In: Proceedings of the 13th Symposium on Usable Privacy and Security (SOUPS 2017), pp. 399–412 (2017)
35. Norberg, P.A., Horne, D.R., Horne, D.A.: The privacy paradox: personal information disclosure intentions versus behaviors. J. Consum. Aff. **41**(1), 100–126 (2007)
36. Pan, S.J., Yang, Q.: A survey on transfer learning. IEEE Trans. Knowl. Data Eng. **22**(10), 1345–1359 (2010)
37. Perera, C., Ranjan, R., Wang, L., Khan, S.U., Zomaya, A.Y.: Big data privacy in the Internet of Things era. IT Prof. **17**(3), 32–39 (2015)
38. Roever, C., Raabe, N., Luebke, K., Ligges, U., Szepannek, G., Zentgraf, M.: Package klaR. Available at https://cran.r-project.org/web/packages/klaR/klaR.pdf/. 20 Feb 2015
39. Sadeh, N., Hong, J., Cranor, L., Fette, I., Kelley, P., Prabaker, M., Rao, J.: Understanding and capturing people's privacy policies in a mobile social networking application. Pers. Ubiquit. Comput. **13**(6), 401–412 (2009)
40. Shehab, M., Cheek, G., Touati, H., Squicciarini, A.C., Cheng, P.C.: User centric policy management in online social networks. In: Policies for Distributed Systems and Networks (POLICY), 2010 IEEE International Symposium on, pp. 9–13. IEEE (2010)
41. Shehab, M., Touati, H.: Semi-supervised policy recommendation for online social networks. In: Advances in Social Networks Analysis and Mining (ASONAM), 2012 IEEE/ACM International Conference on, pp. 360–367. IEEE (2012)
42. Shi, S., Zhang, M., Lu, H., Liu, Y., Ma, S.: Wide and deep learning in job recommendation: an empirical study. In: Asia Information Retrieval Symposium, pp. 112–124. Springer (2017)
43. Sicari, S., Rizzardi, A., Grieco, L.A., Coen-Porisini, A.: Security, privacy and trust in Internet of Things: the road ahead. Comput. Netw. **76**, 146–164 (2015)
44. Sinha, A., Li, Y., Bauer, L.: What you want is not what you get: predicting sharing policies for text-based content on facebook. In: Proceedings of the 2013 ACM Workshop on Artificial Intelligence and Security, pp. 13–24. ACM (2013)
45. Spyromitros-Xioufis, E., Petkos, G., Papadopoulos, S., Heyman, R., Kompatsiaris, Y.: Perceived versus actual predictability of personal information in social networks. In: International Conference on Internet Science, pp. 133–147. Springer (2016)
46. Therneau, T.M., Atkinson, E.J., et al.: An Introduction to Recursive Partitioning Using the RPART Routines. Tech. rep, Mayo Foundation (1997)

Energy-Efficient Design of Data Center Spaces in the Era of IoT Exploiting the Concept of Digital Twins

Spyros Panagiotakis, Yannis Fandaoutsakis, Michael Vourkas, Kostas Vassilakis, Athanasios Malamos, Constandinos X. Mavromoustakis, and George Mastorakis

Abstract Research on the power management of server units and server room spaces can ease the installation of a Data Center, inferring cost reduction, and environmental protection for its operation. This type of research focuses either on the restriction of power consumption of the devices of a data center, or on the minimization of energy exchange between the data center space as a whole and its environment. The work presented here focuses mainly on the latter. In particular, we attempt to estimate the effect that can infer on the energy performance of a data room, various interventions on the structural envelop of the building that hosts the data center. The performance of the data center is evaluated in this work by measuring its PUE index. Our modeling methodology includes the selection of a proper simulation tool for the creation of the digital twin, that is of a digital clone, of our real data center, and then building its thermal model via energy measurements. In this work, the researchers exploit

S. Panagiotakis (✉) · Y. Fandaoutsakis · M. Vourkas · K. Vassilakis · A. Malamos
Department of Electrical and Computer Engineering, Hellenic Mediterranean University,
Heraklion 71004, Crete, Greece
e-mail: spanag@hmu.gr

Y. Fandaoutsakis
e-mail: fandaty@gmail.com

M. Vourkas
e-mail: vourkas@teicrete.gr

K. Vassilakis
e-mail: kostas@hmu.gr

A. Malamos
e-mail: amalamos@hmu.gr

C. X. Mavromoustakis
Department of Computer Science, University of Nicosia and University of Nicosia Research
Foundation (UNRF), 46 Makedonitissas Avenue, 24005, 1700 Nicosia, Cyprus
e-mail: mavromoustakis.c@unic.ac.cy

G. Mastorakis
Department of Management Science and Technology, Hellenic Mediterranean University, 72100
Agios Nikolaos, Crete, Greece
e-mail: gmastorakis@hmu.gr

© Springer Nature Switzerland AG 2020
G. Mastorakis et al. (eds.), *Convergence of Artificial Intelligence and the Internet of Things*, Internet of Things,
https://doi.org/10.1007/978-3-030-44907-0_6

the potential of EnergyPlus software for modelling and simulating the behavior of a
data center space. The energy measurements were conducted for a long period, using
low cost IoT infrastructure, in the data center space of an Educational building in
Heraklion Crete. This paper describes the procedures and the results of this work.

1 Introduction

The term Green Information and Communication Technologies (Green ICT)
describes the study and practice of environmentally sustainable technologies in the
operation of Information Technoloy (IT) systems. Modern IT systems rely on a complex mix of people, software, networks and hardware. Therefore, a green initiative
should cover all these areas. One solution may also need to address end-user satisfaction, management restructuring, regulatory compliance, and Return on Investment
(ROI). There are also considerable fiscal incentives for companies to take control
of their own energy consumption. Literature review reveals that 44% of all energy
consumed in the EU is used in buildings and in the U.S., 40 billion kilowatt-hours
(kWh) of electricity are used annually for cooling. In parallel, power consumption
in Data Centers, nowadays, is constantly increasing. Data Centers provide the world
community with an invaluable service: almost unlimited access to all kind of information imaginable with the support of most Internet services, such as Web Hosting
and e-commerce services. Data Centers include, sometimes thousands of Servers and
cooling infrastructure support. Energy efficient Data Center design should contribute
to a better use of the hosting space increasing power performance and efficiency. The
huge energy consumptions of data centers demonstrate their significant potential
for energy saving making them the desired target of energy conservation measures
[1, 2]. An energy efficient Data Center design should address all aspects of energy use
in a Data Center: from IT and HVAC equipment to the actual location, configuration
and construction of the building. In that context, hosting a data center in a building
that assures minimal energy requirements, can save huge amounts of energy (and
money) providing green-friendly administration.

The work presented here focuses mainly on the minimization of energy exchange
between the data center space as a whole and its environment. In particular, we
attempt to estimate the effect that can infer on the energy performance of a data
room various interventions on the structural envelop of the building that hosts the data
center. The knowledge of the energy behavior of the building's envelop can assure
better performance and power management over the operation of the data center. The
performance of the data center is evaluated in this work by measuring its Power Usage
Efficiency index (PUE) and its variations under various interventions at the building
envelop. Scope of this work is to propose interventions that can reduce PUE. In order
to test the thermal behavior of a data center under different structural envelops, we
need at first a simulation environment to create the digital twin, that is the digital clone,
of our real data center, and then a proper methodology for creating its thermal model.
For the former, several simulation tools have been proposed in bibliography able to

evaluate the thermal performance and energy efficiency of buildings or building sub-sectors. In this work, the researchers exploit the potential of EnergyPlus software for modeling and simulating the behavior of a data center space. EnergyPlus is a software platform that can be used in relation to other tools for the creation of a digital twin, offering a complete and user-friendly simulation environment for the buildings' thermal modeling. In order to develop a modeling procedure especially for data center rooms, the researchers studied these tools and used them to create the digital twin of the Educational building in Heraklion Crete that hosts the data center of Hellenic Mediterranean University. In this context, we describe the development of the 3-D geometry of the case study building in Sketchup, and the selection of input parameters and generic modeling procedure using openstudio. In order to evaluate the developed model, a set of measurements was conducted in situ for a long period using low cost IoT infrastructure, so the indoor air temperature of the Educational building's spaces is logged. Main objective of this work was to assure that the model built for the Data Center space confirms the measurements obtained about temperature variation. When this was achieved, the device loads and the required temperature as measured in the space of the Data Center was set in our model, and through the EnergyPlus the required cooling energy and therefore the associated PUE index were calculated. Finally, various interventions were evaluated over the building envelope in terms of PUE reduction for upgrading the energy efficiency of the Data Center space. The following sections elaborate further on the procedures and the results of this work. In particular, Sect. 2 presents the current recommendations for data center spaces and discusses some similar works. Section 3 illustrates our methodology and introduces to the software and hardware tools we used for thermal modeling. Section 4 evaluates the produced thermal model and calculate the PUE index of data center for various structural interventions. Finally, Sect. 5 concludes the paper.

2 Related Work

2.1 Energy Consumption in Data Centers

Nowadays Data Centers provide the world community with an invaluable service: almost unlimited access to all kind of information imaginable with the support of Internet services, such as Web hosting and e-commerce services. The power and energy consumption are key concerns for Internet Data Centers. A benchmark of 22 Data Center buildings, determined that Data Center can be over 40 times energy inten-sive as conventional office buildings. The huge energy consumptions of data centers demonstrate the significant potential of energy saving making them the desired target for energy conservation measures [1]. Data Centers include, sometimes thousands of Servers and cooling infrastructure support. Research on power management in

Servers can ease the installation of a Data Center, inferring cost reduction, and environmental protection. Given these benefits, researchers have made great strides in energy savings in Server rooms.

The electricity that has been used in Data Centers across the World doubled from 2000 to 2005, but that rate of growth slowed significantly from 2005 to 2010. This slowing was the result of the 2008–2009 economic crisis, the increased prevalence of virtualization in Data Centers, and the industry's efforts to improve efficiency of these facilities since 2005. If we take the midpoint between the Low and High cases in this analysis, US and world electricity use in Data Centers grew by about 36% and 56%, respectively, from 2005 to 2010, totaling about 1.3% of World electricity use and 2% of US electricity use in 2010. The US Data Center market appears to have been hit harder than the world market by the economic crisis, with growth slowing more noticeably in that market than in the world as a whole. Finally, while Google is a large user of Data Centers, that company's facilities represent less than 1% of all Data Center electricity use worldwide [3].

The first step in prioritizing opportunities for energy saving is to get a solid understanding of Data Center energy consumption and energy indicators in Data Centers. The indicators can be used to shape strategies for saving energy, and to determine the effectiveness of the measures taken to reduce energy consumption. In a Data Center, there are many elements that consume power [4]. The visible elements that consume power include Servers, network equipment and UPS, but what is not visible is the function of CPUs that run programs and the continuous flow of information in and out of the system [5].

Figure 1 illustrates the distribution of power consumption in a typical Data Center. This is a high availability Data Center with N +1 air conditioners (CRAC), operating at a typical load of 30% of design capacity [4].

Determining the requirements for developing an IT architecture in terms of hardware and software, the thermal model of a Data Center can be obtained. Operations in a typical Data Center, as presented in Fig. 2, comprise those of IT equipment, such as Servers, routers, etc., and non-IT equipment, like air conditioning, UPS, etc.

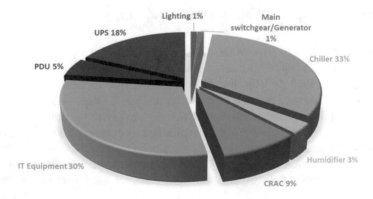

Fig. 1 Energy consumption in a typical Data Center

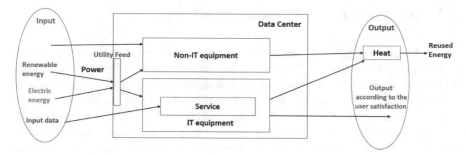

Fig. 2 Operations of a typical Data Center [6]

As input to the Data Center are considered both data and energy, while as output is obtained data that bring satisfaction to the customer. The input energy includes energy from both grid and renewable energy sources. The output is also accompanied by heat that can be reused. In the output the energy losses are also included. Since the natural environment (for example, the position in a cold area) is a large factor of the input power to the system it is also incorporated in the model [4].

We can say that the two parties involved in the application development are the users and the Data Center. Between the two is approved a Service Level Agreement (SLA). An SLA defines the expectations between the consumer and the provider. It helps determining the relationship between the two parties. It is the cornerstone of how the service provider sets and retains an obligation for the consumer of the service. A good SLA sets five key aspects according to [7]:

- The obligations of the provider.
- The way the provider offers its services.
- Who measures the quality of provision, and which way?
- What happens if the provider fails to deliver as promised?
- How the SLA changes over the time.

The SLA encapsulates many secondary factors that contribute to energy consumption. Also, the SLA represents user satisfaction. The power consumption is not constant over time but varies depending on various parameters. The main factor affecting the consumption is a Data Center workload and the external environment. The modelling of energy efficiency and the losses of Data Center equipment is a complex task and important cases generate large errors [8]. First of all, the assumption that the losses associated with power and cooling equipment is stable over time is wrong. It has been observed that energy consumption is a function of IT load and non-IT equipment (devices responsible for cooling and power supply) [6].

2.2 Degrees of Freedom in Energy Efficiency

We call degrees of freedom, the number of independent pieces of information that can be found in an equation. The plant benefits from degrees of freedom from: energy transmission and transformation, cooling and draining heat from the construction of the Data Center. On the other hand, many degrees of freedom can be secured from its use. Then, we will analyze each degree of freedom [8]. When searching for the highest yield in the cooling system, it is imperative for the Data Center to operate at the correct temperature. The pursuit of this objective should include assessments of the energy spent in the cooling process and the power supply fan and should also be considered the best cooling solution flows and management techniques. One of the degrees of freedom that had been proposed in the literature is the increasing temperature of the Data Center [8].

Some of the major manufacturers of Servers and performance experts for Data Centers share the view that Data Centers can operate at much higher temperatures than those operating today without sacrificing the uptime but enabling huge savings in both cooling costs and CO_2 emissions. According to [8] Data Centers can save 4–5% in energy costs for every degree increase in single inlet temperature in Server. The air temperature must be raised but only after consideration of the effects on each piece of IT equipment. If a Data Center is successful in raising the temperature, there is a big saving in energy with the chiller, potential savings in humidity control, with the possibility of increasing the number of hours that economizers can be used.

There are several different methods for free cooling in Data Centers, which can be used instead of/in conjunction with traditional refrigeration for full or partial free cooling [9]:

1. Direct air free cooling: Ambient air is treated (filtered, humidified and refrigerated where necessary) and brought into the data hall. The cooling system (and therefore electrical plant) needs to be sized for the worst-case maximum refrigeration load.
2. Indirect air free cooling: Ambient air is passed through a heat exchanger and hot/cold air transferred with the data hall internal air stream. Adiabatic humidification on the external air stream allows evaporative cooling at hot dry ambient conditions. Any refrigeration capacity required is minimal (supplementary only), allowing reduced sizing of mechanical and electrical plant and therefore capital cost saving.
3. Quasi-indirect air free cooling (thermal wheel): Similar to indirect air free cooling except that a constant volume of fresh air is brought into data hall which requires treatment.
4. Water side free cooling: Heat is rejected to ambient without the use of a refrigeration cycle, for example using cooling towers or dry coolers. This is often an easier retrofit option compared with air side free cooling designs as this method usually requires modification to external plant only and is typically less demanding in terms of plant footprint.

According to [2] the following best practices have emerged through the study of 22 data centers:

- Improved air management, emphasizing control and isolation of hot and cold air streams.
- Right-sized central plants and ventilation systems to operate efficiently both at inception and as the data center load increases over time.
- Optimized central chiller plants, designed and controlled to maximize overall cooling plant efficiency, including the chillers, pumps, and towers.
- Central air-handling units with high fan efficiency, in lieu of distributed units.
- Airside or water-side economizers, operating in series with, or in lieu of, compressor-based cooling, to provide "free cooling" when ambient conditions allow.
- Alternative humidity control, including elimination of simultaneous humidification and dehumidification, and the use of direct evaporative cooling.
- Improved configuration and operation of uninterruptible power supplies.
- High-efficiency computer power supplies to reduce load at the racks.
- On-site generation combined with adsorption or absorption chillers for cooling using the waste heat, ideally with grid interconnection to allow power sales to the utility.
- Direct liquid cooling of racks or computers, for energy and space savings.
- Reduced standby losses of standby generation systems.
- Processes for designing, operating, and maintaining data centers that result in more functional, reliable, and energy-efficient data centers throughout their life cycle.

Other authors [10], show how the equipment is placed in the Data Center in order the release of heat to be as effective as possible. The way in which the Data Center was constructed affects the energy efficiency of the Data Center. The goal of all researchers in this area is to find the best location and construction of the Data Center space to ensure efficient release of heat that produced by the electrical equipment installed and operated. In conclusion, the development of a smoother cooling controller and additionally with intelligent placement of equipment can minimize cooling consumption and can support energy optimization and management process in a Data Center.

If we look at performance, we can state that a promising approach to increasing energy efficiency could be virtualization and Server Consolidation [11]. The latter may introduce additional degrees of freedom that are not available in traditional operating models [12]. Furthermore, another degree of freedom can be achieved if the integration of Servers is associated with scalable voltage and frequency provided by conventional processors, for even better results. Remarkable work has also been done in terms of data storage including the increase in the speed of accessing discs. This allows saving energy in fatigue and also rapid recovery in terms of performance [12].

2.3 Power Usage Effectiveness

The Power Usage Effectiveness index (PUE) is a Unit of measurement that was established by the Green Grid organization, and expresses the energy performance of a Data Center, namely a central computing facility, in terms of basic infrastructure conditions. PUE expresses the ratio of the total energy required to the energy actually consumed by the computing resources (Fig. 3). Total energy is calculated from all the resources needed to support the load under normal conditions, such as UPS, air conditioners, etc. For example, in a Data Center where 1000 W of total power are consumed and 500 W on IT equipment, PUE obtained has a value of 2 (1000/500 = 2). The closer the value is to 1, ideally 100% of the energy supplied is directed to the computing infrastructure, the more energy efficient is the Data Center, which in turn means lower operating costs. The precise measurement of PUE allows for the evaluation, control and better compliance to even better ("greener") energy technologies.

PUE is calculated by taking energy measurements close to the utility meter. If the Data Center is a mixed-use Data Center or office, the measurement should be taken only to the extent that supplies the Data Center. If the Data Center is in a separate utility meter, the amount of energy consumed by the non-given center portion is estimated and removed from the equation. The IT equipment load should be measured after the power conversion, switching, and the air conditioning. According to The Green Grid, the more likely measurement point would be the output of computer room Power Distribution Units (PDUs). This measurement should represent the total power delivered to the Server racks in the Data Center.

The Uptime Institute has an average PUE of 2.5. This means that for every 2.5 W measured at the meter, only one watt is delivered to the IT load. Uptime estimates that more facilities could achieve 1.6 PUE using the most efficient equipment and best technologies. These include measures to reduce the IT load, such as decommission or repurposing of Servers, powering down Servers when they are not in use, enabling of power management, replacement of inefficient Servers, virtualization

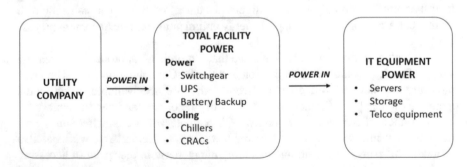

PUE = Total Facility Power / IT Equipment Power

Fig. 3 PUE parameter flow

and consolidation. According to [13] PUE is a great tool for the facilities side of the Data Center. It allows facility engineers to measure the impact of changes they make to the infrastructure, things like raising the Data Center temperature, upgrading to a higher efficiency UPS, increasing voltage to the rack and so on. In [13] it is stated that Virtualization and consolidation, while reducing the overall energy usage, will actually increase the PUE unless the power and cooling infrastructure are downsized to align with the IT load. Raising the Server inlet temperature may reduce the PUE, but the overall energy usage may actually increase if the increased power required for Server fans is greater than the cooling savings. Hence, we conclude that PUE should not be the only metric for energy efficiency in a Data Center. Making the IT equipment operation more efficient, would not decrease PUE, if the power and cooling equipment are not optimized to adapt to the new IT load. Improving Power Usage Effectiveness (PUE)—Tripp Lite [14] also argues that to decrease PUE one should concentrate first on cooling, which is often the biggest consumer. Cooling load can be reduced by organizing cables, installing banking panels, isolating and removing hot air.

Worldwide according to the Uptime Institute's Data Center Survey [15] the average PUE of the largest Data Center ranges between 1.8 and 1.89. Best PUE is achieved by companies such as Yahoo, Google, Facebook, and Microsoft that have values close to unity and less than 1.2. The reason why they have achieved such values is that their specially designed facilities are often placed in very favorable climates for operation of an efficient Data Center. In many cases, the external cold air is used to entirely cool the Data Center avoiding one of the most intensive parts of the traditional Data Center, air conditioning, while in specially designed Data Centers water cooling uses the cold ocean water with positive effects also on the value of PUE. Cold Isle Containment systems, i.e. passive separation of air flow between the cold and hot isle, are used for PUE enhancement. The hot and cold isle ensures both the introduction of cold conditioned air from the front of the system and the extraction of hot air from the back of the Rack and other more efficient system operation. Simultaneously, air conditioners positioning can be set for more efficient channeling of cold air in the cold area at the lowest possible cost. Also, climatic sensors can be installed and configured that are responsible for the immediate notification and the calculation of useful metrics to forecast future improvements in the infrastructure. Virtualization/consolidation reduces costs, since it reduces capacity requirements and the number of physical machines required. Typical Server utilization ranges from 5 to 30%, which allows accommodating more than one Servers on one physical machine. Another popular metric, namely performance per watt attempts to capture the performance-power tradeoff. However, a related problem is that characterizing the impact of power on performance is often very challenging and there is little progress in the way formal models can address this gap in a virtualized environment. It should be possible to estimate power and thermal effects of individual VM's, but in a virtualized environment this can be very challenging since the VMs can interact in complex ways, for example, a poorly behaved VM can increase the power consumption of other VMs [16].

2.4 Data Centers and Building Envelopes

Green technologies of information—American Society of Heating, Refrigerating and Air-Conditioning Engineers (ASHRAE) published in 2011 the whitepaper "Thermal Guidelines for Data Processing Environments". The purpose of thermal guidelines is to develop recommended envelopes as to give guidance to data center operators on maintaining high reliability and also operating their data centers in the most energy efficient manner [17]. Any choice outside of recommended envelops should be a balance between the additional energy savings of the cooling system considering the operating temperature, versus the deleterious effects in performance, reliability, or acoustics (see Fig. 4).

In the same context, [18] categorizes green computing performance metrics in Data Centers as it is illustrated in Fig. 5 and also presents the formal process for benchmarking a Data Center using developed software tools, including EnergyPlus, for designing energy efficient buildings and Data Centers.

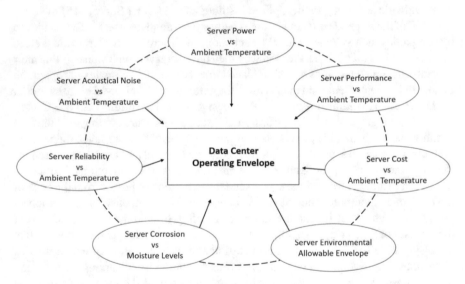

Fig. 4 Server Metrics for Determining Data Center Operating Environmental Envelope

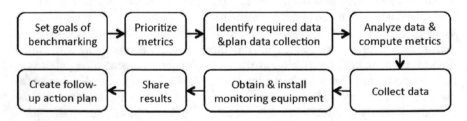

Fig. 5 Formal process for designing an energy efficient Data Center [18]

Similarly, Pan et al. in [1] have developed an energy model with EnergyPlus for two office buildings in a R&D center in Shanghai, China to evaluate the energy cost saving of green building design options compared to the baseline building.

Recently, a number of efforts have devised models and simulation tools for capturing thermal effects using temperature measurement facilities at various levels (starting from chip level), which can be useful for control purposes [16]. According to [9] the refrigeration periods can be modelled, based on the weather data for different locations, and hence, the total energy consumption of the Data Center equipment for a given IT load is estimated.

3 Proposed Methodology

3.1 Modeling Procedure

The building of interest in this study (see Fig. 6) is a building of Hellenic Mediterranean University (HMU) in Heraklion, Crete. It is a 3-story building that has study rooms, classrooms, offices and IT rooms. On the 2nd floor there are two structurally identical rooms, one is used as Data Center and the other as Hardware Laboratory (see Fig. 7). Since the data center space is permanently in use, we chose to start our methodology for building the thermal model of the data center space using the opposite Hardware Lab, which is totally empty from people and operating computer

Fig. 6 The case study building

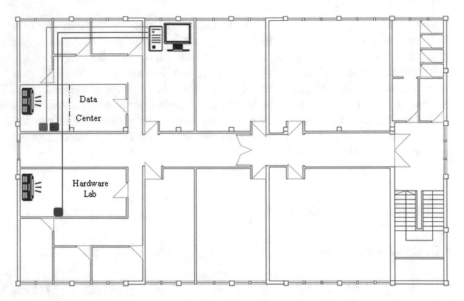

Fig. 7 Third floor plan with air conditioner, temperature sensor (red), electricity consumption sensors (blue) and computer with Emoncms software

and cooling devices during the Christmas holiday period. The idea is that the thermal model of the data center room, when it is empty from people and operating devices, will follow the same temperature variation with Hardware Lab, considering that both spaces share the same structural envelop and have identical dimensions and configuration. Hence, we measured along this period the variation of the indoor air temperature at the Hardware Laboratory. In parallel, we measured the variation of the indoor air temperature at the data center and also the electric power consumption of the non-IT equipment (energy consumed for cooling the data center), along with the total energy consumption from the computing and cooling devices. Following this measuring step, we had to develop at Energy Plus the digital twins of our rooms, so we can simulate their thermal behavior. The next step was to compare simulation results from Energy Plus with measurements for indoor air temperature variation for the Hardware Laboratory in order to see if our simulation model is validated. If it is true, we can assume that we have successfully simulated the thermal behavior of our structural envelopes for both spaces. As the Data Center and Hardware Laboratory are both spatially and structurally the same spaces, we can then start adding passive energy loads (computing and cooling devices) in our model similar to the real loads in the data center. The aim is to define in our model the characteristics of the real Data Center space in order to have a report about the electric energy and cooling capacity that is required to cool the space regarding the temperature set point at the Data Center. This new simulated thermal model is again validated towards the measurements we have collected from the operating data center during the winter holidays. Then, having measured the electric energy needed for cooling the data center and the total energy consumed in the data center, the PUE index of the data center can be estimated. Considering that our thermal model for the data center is successfully validated, we will be able to propose ways for the reduction of the PUE index of Data Center trying various structural changes on the envelope of our digital twin. We note that the HVAC system of the data center comprises two inverting wall-mounted 24.000 Btu minisplit air conditioners.

3.2 Software Simulation Tools

EnergyPlus [19] is a thermal analysis software that allows engineers to simulate the thermal behavior of buildings. EnergyPlus is a Dynamic Simulation Modelling tool (DSM) and is used to simulate the thermal performance of new and existing buildings. Dynamic simulation traces the building's thermal state through a series of hourly snapshots, providing a detailed picture of the way the building performs under extreme design conditions and throughout a typical year. Results consist of several data points for each variable that is output. EnergyPlus allows for preparation of a model in 3-D geometry in a third party software (Sketchup) based on already existing building plans and detailed photos and then guides the user to set up the internal condition zones and structural element characteristics that will describe the building model (by relative plugins). The simulation process calculates annual hourly data that can be examined in detailed and filtered for each zone and variable.

Sketchup [20] is a 3D modeling software for a wide range of drawing applications. There is an online open source library of free model assemblies (e.g., windows, doors, automobiles, etc.), 3D Warehouse, to which users may contribute models. The program includes drawing layout functionality, allows surface rendering in variable "styles", supports third-party "plug-in" programs (e.g., openstudio), and enables placement of its models within Google Earth. SketchUp 4 and later support software extensions written in the Ruby programming language, which add specialized functionality. SketchUp has a Ruby console, an environment which allows experimentation with Ruby. The free version of SketchUp also supports Ruby scripts, with workarounds for its import and export limitations.

The Legacy OpenStudio Plugin-in for SketchUp [21] facilitates the creation and editing of building geometries in EnergyPlus input files. This free plug-in also allows the launching of EnergyPlus simulations and viewing of the results which leaving the SketchUp 3D drawing program. The Legacy OpenStudio Plug-in was created by the National Renewable Energy Laboratory of the U.S. Department of Energy (www.nrel.gov) as an interface to the EnergyPlus simulation engine. Active development has moved from the Legacy OpenStudio Plug-into the more comprehensive OpenStudio suite of tools and libraries.

The OpenStudio Application Suite [22] combines all the plug-ins for modeling, simulation and results viewing with EnergyPlus and SketchUp. It is a cross-platform (Windows, Mac, and Linux) collection of software tools to support whole building

energy modelling using EnergyPlus. OpenStudio is an open source project to facilitate community development, extension, and private sector adoption. OpenStudio
includes graphical interfaces along with a Software Development Kit (SDK). It is the
next version of Legacy OpenStudio plug-in that incorporates all the necessary tools
for EnergyPlus simulations. OpenStudio enable potential developers to write their
OpenStudio measure (henceforth referred to as a "measure"). Measure is a program
(or 'script', or 'macro') that can access the leverage the OpenStudio model and API
to create or make changes to a building energy model, as defined by an OpenStudio
model (.osm). Measures are written in Ruby, which allows the measure author to
access OpenStudio directly as well as through the SketchUp plugin. Measures can
be created from scratch, but existing measures may also be used as a starting basis.

xEsoView [23] is a file viewer for EnergyPlus.eso output files that gives to the
user a very fast overview of the simulation results. xEsoView use the Qwt extension
of the Qt toolkit from Trolltech Inc. xEsoView presents all reported variable names
in a list, which can be sorted and filtered, on the left-hand side. On the right side
it presents the graphical representation of the selected variable. The time axis can
be changed using predefined ranges (hour, day, week, month, total) with zooming
support. With a selection box you can switch between the available environments,
e.g. summer design day and run-period. A simple CTRL + C copies the variable
data to the clipboard. So you can easily paste the data into a spreadsheet for further
analysis.

Elements software [24], Version 1.0.4, is a free, open-source, cross-platform software tool for creating and editing custom weather files for building energy modeling.
Elements software tool was developed by Big Ladder Software with the generous
funding and collaboration of Rocky Mountain Institute.

3.3   Creating the 3D Geometry in SketchUp

In this work, we used the SketchUp 15 trial version for 3D modelling. We created
the thermal 3D model from scratch, and all surfaces was "matched" and linked to
each other according to the boundary conditions (Fig. 8).

Before creating the geometry, a basic template of a Large office building was
imported in the design in order to allow for the basic construction and thermal
characteristics to be determined automatically. Changes and adaptions were made to
account for the specificities of our application later in the modeling process using
OpenStudio Application Suite. The basic design of the floor was expanded using the
"Create Spaces from Diagram" command. Three floors were created, 1st and 2nd
floor. In the 2nd floor, one space was created, which account to the Data Center.

The basic space types and thermal zones can be set in the 3D geometry, while
further details are imported later in the .osm model. First the space types are set as
large open office areas for all the areas except the corridors, the stairs and the spaces
which of interest for this study. All office areas are imported in the same thermal
zone. The stairs area of the ground floor is a separate thermal zone. The areas of the

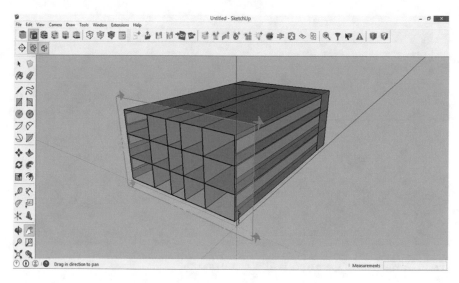

Fig. 8 Rendering by boundary conditions view of SketchUp 3D model

Data Center and the Hardware Laboratory are in different thermal zones. The Data Center and was set as IT space in the .osm model. Hence five different thermal zones are set in the .osm file that is created (Figs. 9 and 10). The thermal simulation and all further data are imported in the OpenStudio suite environment.

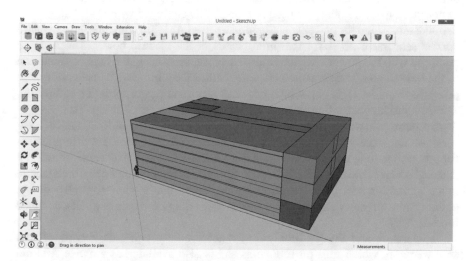

Fig. 9 Assignment of thermal zones in Sketchup 3D model

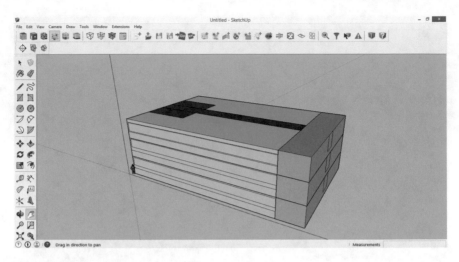

Fig. 10 Assignment of space types in Sketchup 3D model

3.4 Measurements

For temperature variation and power consumption measurements we used the low-cost Arduino technology. The open-source web application from Emoncms [25] was used for storage, real time representation and treatment of the archived measurement data, which was collected from the ethernet-connected Arduino. In particular, for temperature measurements we used Arduino UNO equipped with DS18b20 + temperature sensor and for power consumption we used Arduino with three non-invasive current sensors of 30 A for the three phases of the electric power. The Arduino equipped with the power consumption sensors was measuring the total consumption of the Data Center. The measurement data for the IT equipment's consumption were provided from the UPS unit which provided electrical power to the IT equipment. The Arduino layout for temperature measurement is presented in Fig. 11.

Emoncms [25] was used for storage, real time representation and treatment of measurement data received from arduinos. Emoncms is an open-source web application for processing, logging and visualizing energy, temperature and other environmental data and is part of the OpenEnergyMonitor project. In brief the emoncms architecture is a combination of a front controller on the Server, model-view-controller design pattern and a directory structure that makes easy adding features in a self-contained modular way.

Temperature and Energy Monitoring is achieved through a dashboard, offered as a web app, which runs on Emoncms and is also available as a native Android app. Through this dashboard we can Explore—visualize Site temperature in Celsius Degrees, power consumption in Watts, daily energy consumption in KWh and PUE calculations (see Figs. 12 and 13).

Fig. 11 Arduino layout for temperature measurement

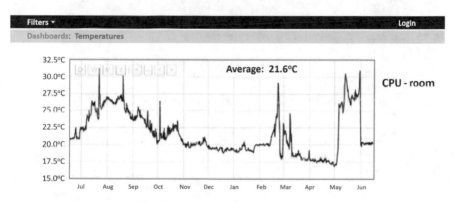

Fig. 12 Graphical representation of ambient Temperature

Figure 14 presents measurements obtained for the Data Center's (a) and the hardware lab's (b) air temperature respectively, from December 25th to December 30th. Measurements show that Data Center's temperature deviates in the range of 21.4–22.7 °C and at the same time, hardware lab's air temperature deviates in the range of 19.4–20.8 °C.

3.5 Setting Model Parameters in OpenStudio .OSM File

In the Site tab (see Fig. 15) we set the path of the .epw weather file for simulation. The weather data has been provided by a Weather Station located in HMU, about

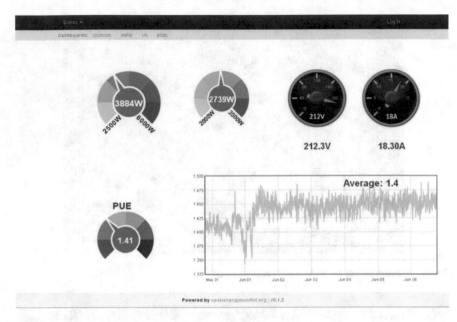

Fig. 13 Real time visualization of the power consumption

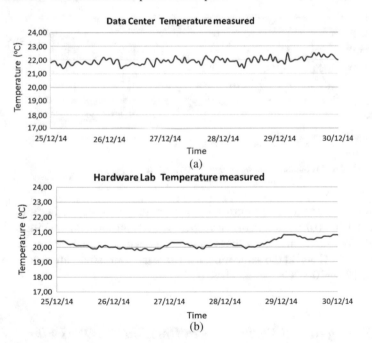

Fig. 14 Data Center's (**a**) and hardware lab's (**b**) temperatures measured

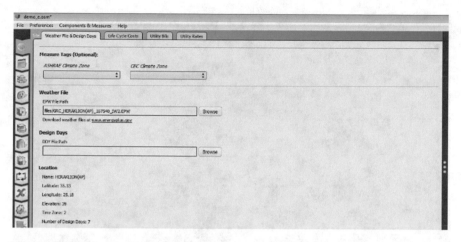

Fig. 15 OpenStudio Site tab

700 M away from the case study building. Weather file has to be in .epw format for simulations in EnergyPlus. Also, a weather file in .epw format was purchased from White Box Technologies Company [26]. The data of the weather file of White Box Technologies Company was edited and updated by the Elements software with the weather data from the Weather Station of HMU.

In the .osm file we have to make some changes in order to adapt to the case study of the building that we are referring to. We select the appropriate materials from the Library tab and drag and drop them to the relative position (Fig. 16). The materials and their attributes were set according to Technical Guideline 20701–2/2010 of Technical Chamber of Greece [27].

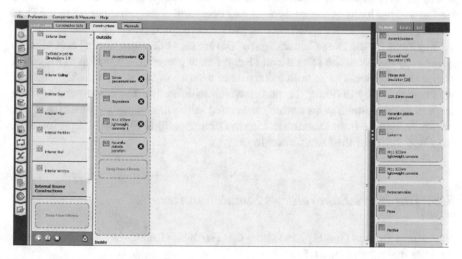

Fig. 16 Constructions and their materials

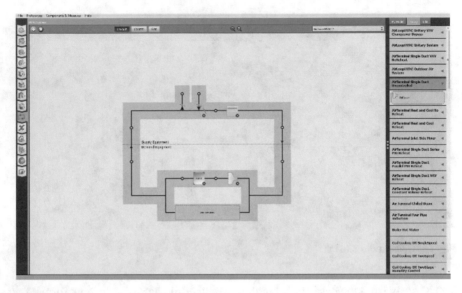

Fig. 17 Building up a simple split AC system from standard components

Apart from the materials, we set the HVAC system for heating and cooling (see Fig. 17). As the openstudio does not provide a template for a standard split air condition, standard components from the library were used to build up a simple split AC system. As the thermal zone of Data Center was set to conditioned thermal zone the other thermal zones were set to not conditioned and left drifting during simulation.

The loads for the Data Center (Space 306) are set according to measurements about the IT and non IT equipment power consumption in the closet of the Data Center and the loads of the equipment was operating at the same period in the Data Center. The Data Center space was set as IT Room space type.

The loads for the Data Center (Space 306) are set as IT room Equip" according to measurements about the IT and non IT equipment power consumption in the closet of the Data Center. The schedule consists of a series of fraction rules. The fraction was calculated by dividing the power which measured in the corresponding time step, to the maximum power which measured in the total simulate time. Each rule or profile applied from December 25th to December 30th. Figure 18 represents the fractional rules for the Data Center loads.

3.6 Hardware Laboratory's Simulation Results

Simulation Results (see Fig. 19) show that Hardware Laboratory's air temperature deviates in the range of 21.3–22.7 °C.

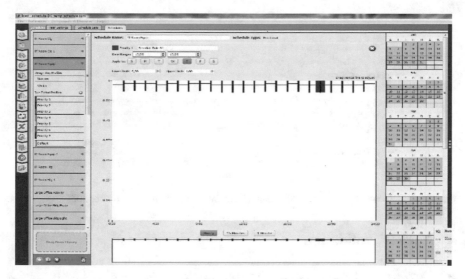

Fig. 18 Fractional rules for the Data Center loads

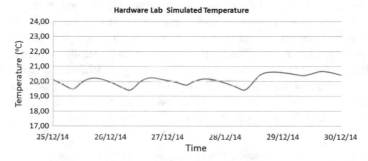

Fig. 19 Hardware Laboratory hourly simulated results from December 25th to December 30th

3.7 Data Center Simulation Results

Simulation Results (see Fig. 20) show that Data Center's air temperature deviates in the range of 21,3–22,7° C.

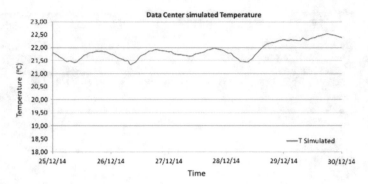

Fig. 20 Data Center hourly simulated results from December 25th to December 30th

4 Models' Validation

4.1 Hardware Laboratory's Validation Results

The validation of the model for Hardware Laboratory is realized during Christmas holidays, so that no activity or internal gains are perturbing the results. Figure 21 presents the measurements obtained for the hardware lab air temperature, the temperature from simulation and the outdoor temperature from December 25th to December 30th, 2014. Measured temperatures deviate in the range of 19.8–20.8 °C and the simulated temperatures deviate in the range of 19.31–20.62 °C. The standard deviation of the difference between simulated and measured temperatures is roughly 0.2 °C. So, we can say that the EnergyPlus model gives satisfying results.

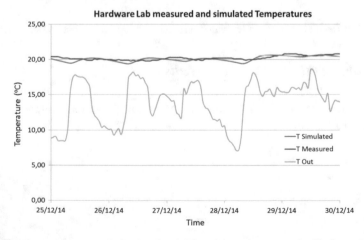

Fig. 21 Temperatures measured, simulated and the outdoor temperatures for Hardware laboratory

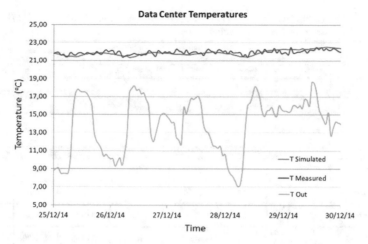

Fig. 22 Temperatures measured, simulated and the outdoor temperatures for Data Center

4.2 Data Center's Validation Results

The validation of the model for Data Center is realized during Christmas holidays, so that no activity or internal gains are perturbing the results. Figure 22 presents measurements obtained for the Data Center air temperature, the temperature from simulation and the outdoor temperature from December 25th to December 30th, 2014. Measured temperatures deviate in the range of 21.3–22.7 °C and the simulated temperatures deviate in the range of 21.4–22.5 °C. The standard deviation of the difference between simulated and measured temperatures is roughly 0.2 °C. So, we can say that our model in EnergyPlus gives satisfying results.

4.3 PUE Calculations for Various Structural Interventions

In this section we consider the possible interventions that can be done in our building in order to improve energy performance of the Data Center space. A distinct simulation was made for each proposed intervention of the building model according to measured temperatures in comparison with the recommended envelope proposed by the ASHRAE for Data Processing Environments (Table 1). Power consumption, and therefore PUE, comparison between the previous and new situation was made.

According to ASHRAE TC 9.9—Thermal Guidelines for Data Processing Environments, typical Data Centers are operated in a temperature range of 20 to 21 °C with a common notion of "cold is better". The cooling setpoint of the Data Center we are considering is set from the administrator at 19 °C and this is an indication that this applies in respect to the cooling setpoint. The upper limit temperature

Table 1 2011 and 2008 Thermal Guideline Comparisons Source: ASHRAE TC 9.9 [17], Thermal Guidelines for Data Processing Environments

Classes (a)	Equipment Environmental Specifications								
	Product Operations (b)(c)					Product Power Off (c) (d)			
	Dry-Bulb Temperature (°C) (e)(g)	Humidity Range, non-Condensing (h) (i)	Maximum Dew Point (°C)	Maximum Elevation (m)	Maximum Rate of Change(°C/hr) (f)	Dry-Bulb Temperature (°C)	Relative Humidity (%)	Maximum Dew Point (°C)	
	Recommended (Applies to all A classes; individual data centers can choose to expand this range based upon the analysis described in this document)								
A1 to A4	18 to 27	5.5ºC DP to 60% RH and 15ºC DP							
	Allowable								
A1	15 to 32	20% to 80% RH	17	3050	5/20	5 to 45	8 to 80	27	
A2	10 to 35	20% to 80% RH	21	3050	5/20	5 to 45	8 to 80	27	
A3	5 to 40	-12°C DP & 8% RH to 85% RH	24	3050	5/20	5 to 45	8 to 85	27	
A4	5 to 45	-12°C DP & 8% RH to 90% RH	24	3050	5/20	5 to 45	8 to 90	27	
B	5 to 35	8% RH to 80% RH	28	3050	NA	5 to 45	8 to 80	29	
C	5 to 40	8% RH to 80% RH	28	3050	NA	5 to 45	8 to 80	29	

value for the recommended envelope proposed by the ASHRAE for Data Processing Environments is at 27 °C.

Simulating the proposed innervations, we exported the results illustrated in Fig. 23 (a) and (b) for the PUE of the December and August period.

Simulating the proposed innervations, we exported the results in Table 2 for the average reduction of cooling consumption for the December and August period. The results are sorted in increasing order.

As one can see the most beneficial option is to increase the cooling temperature limit according to the envelope proposed by ASHRAE, without incurring the life of the equipment. Definitely, the applicable and most beneficial option should be considered among the others based on the prices of the local electric power market.

5   Conclusions

This work investigated thermal performance of buildings with emphasis on data center spaces. In order to evaluate the thermal performance and energy efficiency of buildings and building sub-sectors, several simulation tools have been proposed. In this work, the researcher exploited the potential of EnergyPlus software. EnergyPlus is a software platform that is used in relation to other tools in order to create a user-friendly simulation environment. In order to develop a modeling procedure, the researcher studied the simulation tools and used them in order to simulate the Educational building in Heraklion, Crete that hosts the data center of HMU. In order to evaluate the developed model, measurements were performed at specific periods, and

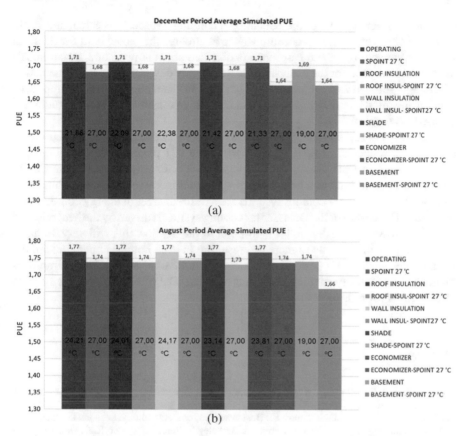

Fig. 23 Average Simulated PUE for December (**a**) and August (**b**) periods for all interventions and relevant average ambient temperatures

Table 2 Exported results for average cooling consumption reduction for the December and August period, sorted in increasing order

Roof Insulation—Setpoint at 19 °C	0,00%
Wall Insulation—Setpoint at 19 °C	0,00%
Shade—Setpoint at 19 °C	0,00%
Economizer—Setpoint at 19 °C	0,09%
Install Equipment in a Basement—Setpoint at 19 °C	6,99%
Wall Insulation and ASHRAE recommended envelope—Setpoint at 27 °C	7,47%
Roof Insulation and ASHRAE recommended envelope—Setpoint at 27 °C	9,01%
ASHRAE recommended envelope—Setpoint at 27 °C	9,32%
Shade and ASHRAE recommended envelope—Setpoint at 27 °C	10,66%
Economizer and ASHRAE recommended envelope—Setpoint at 27 °C	17,97%
Install Equipment in a Basement and ASHRAE recommended envelope—Setpoint at 27 °C	33,44%

the indoor air temperature of the Educational building's spaces was calculated. Main objective of the evaluation procedure was to assure that the model built for the Data Center space confirms the measurements obtained about temperature variation. All the software utilized for the analysis were thoroughly described: the implementation of the 3-D geometry of the case study building in Sketchup using Sketchup 2015 version, the input parameter selection and modeling procedure using openstudio and EnergyPlus respectively.

The huge energy consumptions of the data centers demonstrate the significant potential for reduction of their energy cost. Green ICT policies and regulations put more pressure on Data Center operators to consider Energy-efficient Data Center design, so to contribute to a better use of space and increase performance and efficiency. The Data Center cooling systems are a desired target of energy conservation measures. The design of the Data Center cooling system is driven by the heat released from IT hardware it serves, and it is difficult to predict how this will vary. Exhaust temperatures will continue to increase, resulting in even hotter temperatures in the Data Center spaces. Air performance metrics can be used to benchmark improvements and to quantify inefficiencies and calculate Data Center's cooling systems effectiveness towards optimization.

References

1. Pan, Y., Yin, R., Huang, Z.: Energy modeling of two office buildings with data center for green building design. Elsevier (2007)
2. Greenberg, S., Mills, E., Tschudi, B.: Best Practices for Data Centers: Lessons Learned From Benchmarking 22 Data Centers. Lawrence Berkeley National Laboratory|U.S. Department of Energy (2006)
3. Koomey, J.G.: Growth in Data Center Electricity Use 2005–2010. Analytics Press, NY Times (2011)
4. Rasmussen, N.: Calculating Total Cooling Requirements for Data Centers. American Power Conversion, West Kingston (2003)
5. Fichera, R.: Power and Cooling Heat up the Data Center. Forrester Research, Cambridge (2006)
6. Green IT Promotion Council. https://home.jeita.or.jp/greenit-pc/e/index.html. Accessed 20 Sep 2019
7. McFarlane, R.: Let's Add an Air Conditioner, SearchDataCenter news article (2005). http://searchDataCenter.techtarget.com/columnItem/0,294698,sid80_gci1148906,00.html. Accessed 20 Sep 2019
8. Dunlap, K., Rasmussen, N.: The Advantages of Row and Rack-Oriented Cooling Architectures for Data Centers. American Power Conversion, TCO, West Kingston (2005)
9. Flucker, S., Tozer, R.: Data Centre Cooling Air Performance Metrics, Technical Symposium. DeMontfort University, Leicester UK (2011)
10. Patel, D., Shah, A.: Cost Model for Planning, Development, and Operation of a Data Center. Internet Systems and Storage Laboratory. HP Laboratories, Palo Alto (2005)
11. Hughes, R.: The data center of the future—part 1—current trends. Data Center J. News Article (2005). http://www.DataCenterjournal.com/News/Article.asp?article_id=315. Accessed 20 Sep 2019
12. Anthes, G.: Data Centers Get a Makeover. Computerworld News Article (2005). http://www.computerworld.com/databasetopics/data/DataCenter/story/0,10801,97021,00.html?SKC=home97021. Accessed 20 Sep 2019

13. Geng, H.: Data Center Handbook. Wiley, Hoboken (2015)
14. Improving Power Usage Effectiveness (PUE)—Tripp Lite (2014). https://blog.tripplite.com/improving-power-usage-effectiveness-pue/. Accessed 20 Sep. 2019
15. 2019 Data Center Industry Survey Results, Uptime Institute (2019). https://uptimeinstitute.com/2019-data-center-industry-survey-results. Accessed 20 Sep 2019
16. Kant, K.: Data center evolution: a tutorial on state of the art, issues, and challenges, Intel Corporation—Elsevier (2009)
17. ASHRAE TC 9.9, Thermal Guidelines for Data Processing Environments, American Society of Heating, Refrigerating and Air-Conditioning Engineers (ASHRAE) (2011)
18. Wang, L., Khan, S.U.: Review of performance metrics for green data centers: a taxonomy study. Springer (2011)
19. U.S. Department of Energy's (DOE) Building Technologies Office (BTO), National Renewable Energy Laboratory (NREL), EnergyPlus. https://energyplus.net. Accessed 20 Sep 2019
20. Trimble Navigation Limited, Sketchup, http://www.sketchup.com. Accessed 20 Sep 2019
21. National Laboratory of the U.S. Department of Energy, Office of Energy Efficiency and Renewable Energy, Legacy OpenStudio Plug-in for SketchUp. https://github.com/NREL/legacy-openstudio. Accessed 20 Sep 2019
22. National Laboratory of the U.S. Department of Energy, Office of Energy Efficiency and Renewable Energy, OpenStudio Application Suite, https://www.openstudio.net. Accessed 20 Sep 2019
23. Christian Schiefer, xEsoView. http://xesoview.sourceforge.net. Accessed 20 Sep 2019
24. Big Ladder Software, Rocky Mountain Institute, Elements software. http://bigladdersoftware.com/projects/elements
25. OpenEnergyMonitor, Emoncms. https://emoncms.org
26. White Box Technologies. http://weather.whiteboxtechnologies.com
27. Technical Chamber of Greece: Technical Guideline 20701–2/2010 (2010)

In-Network Machine Learning Predictive Analytics: A Swarm Intelligence Approach

Hristo Ivanov, Christos Anagnostopoulos, and Kostas Kolomvatsos

Abstract This chapter addresses the problem of collaborative Predictive Modelling via in-network processing of contextual information captured in Internet of Things (IoT) environments. In-network predictive modelling allows the computing and sensing devices to disseminate only their local predictive Machine Learning (ML) models instead of their local contextual data. The data center, which can be an Edge Gate- way or the Cloud, aggregates these local ML predictive models to predict future outcomes. Given that communication between devices in IoT environments and a centralised data center is energy consuming and communication bandwidth demanding, the local ML predictive models in our proposed in-network processing are trained using Swarm Intelligence for disseminating only their parameters within the network. We further investigate whether dissemination overhead of local ML predictive models can be reduced by sending only relevant ML models to the data center. This is achieved since each IoT node adopts the Particle Swarm Optimisation algorithm to locally train ML models and then collaboratively with their network neighbours one representative IoT node fuses the local ML models. We provide comprehensive experiments over Random and Small World network models using linear and non-linear regression ML models to demonstrate the impact on the predictive accuracy and the benefit of communication-aware in-network predictive modelling in IoT environments.

H. Ivanov (✉) · C. Anagnostopoulos · K. Kolomvatsos
Essence: Pervasive & Distributed Intelligence Research Group, School of Computing Science, University of Glasgow, G12 8QQ Glasgow, UK
e-mail: 2205747i@student.gla.ac.uk

C. Anagnostopoulos
e-mail: christos.anagnostopoulos@glasgow.ac.uk

K. Kolomvatsos
e-mail: kostas.kolomvatsos@glasgow.ac.uk

© Springer Nature Switzerland AG 2020   145
G. Mastorakis et al. (eds.), *Convergence of Artificial Intelligence and the Internet of Things*, Internet of Things,
https://doi.org/10.1007/978-3-030-44907-0_7

1 Introduction

This section introduces the aims of the chapter, the motivations behind it and the structure of this chapter.

1.1 Motivation

Internet of Things is defined as a "proposed development of the Internet in which everyday objects have network connectivity, allowing them to send and receive data [21]." Every IoT device has an IP address and is able to transfer data over a network. IoT devices become more popular with each passing year. In 2018 their number hit 7 billion and there are no signs of them slowing down [18].

Nowadays, IoT devices have numerous applications and almost every industry uses some sort of smart devices to automate processes, gather data or communicate with other devices. On the other hand, there are many challenges that they face. Some of them are security, connectivity, compatibility and privacy, but the main talking point of this chapter would be energy consumption and how we can reduce it (Fig. 1).

This chapter focuses on one use case of Internet of Things devices **Wireless Sensor Networks** (WSN). In this case, energy conservation is everything. The longer a sensor can be kept alive, the more data you can gather and less often you will have to recharge the device (which may be difficult, depending on the conditions sensors are located at). One way to address that problem is **Predictive Modelling**. Predictive Modelling is a process that uses statistics and machine learning models to predict outcomes (closely related to Predictive Analytics). It can never predict the future, but it can look at existing data and determine a likely outcome [11]. The benefits of predictive

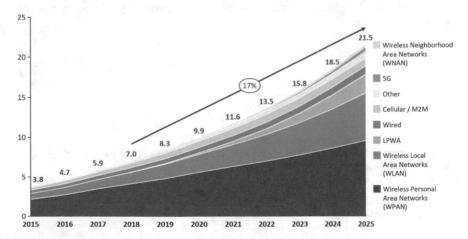

Fig. 1 Number of global IoT Connections in Billions [18]

modelling are that it brings the computation to the data and not vice versa. Sensing and computing devices do not have to send their entire local data over the network to the **Data Center** (or **sink node**), but just have to generate a local machine learning model trained on that data and send that instead. In the long term, this approach should greatly benefit the battery life of the sensing and computing devices. Moreover, it would allow the data center (DC) to study the received machine learning models, predict possible outcomes of that data and recreate part of the datasets. Of course, there are some drawbacks to this approach—adaptability of the ML models to new data (the models will not always be up to date) and models cannot be 100% accurate (if accuracy is critical this might not be a suitable approach).

1.2 Aim

The objective of this chapter is to investigate whether we could reduce the communication overhead of the sensing and computing devices (nodes) by implementing different network models and reducing the number of nodes communicating with the back-end data center system, e.g., sink node. Each device has a unique dataset and will use the **Particle Swarm Optimisation** (Swarm Intelligence) algorithm to calculate the regression coefficients (linear or non-linear) that best fit its data.

This chapter will implement and analyse two network types—Random network and Small World network. These network models will be used to generate node neighbourhoods. For each neighbourhood, we will generate a fused regression model which will 'generalise' all the local regression models in that group. One representative of each group will be responsible for collecting the local ML models of its neighbours and using the proposed methodology of this chapter to generate that fused model. After that, each of those models will be sent to the sink node, to evaluate their accuracy using a test dataset. Furthermore, the proposed methodology's performance will be evaluated against a baseline solution, in terms of energy consumption and prediction accuracy of each machine learning model. This communication overhead—prediction accuracy tradeoff will be an important talking point in the chapter.

1.3 Outline

This section introduced the motivations and aims behind the chapter. The remainder of this chapter is structured as follows:

- **Section** 2—**Background** introduces the reader to topics that are relevant to this chapter and reviews relevant literature.
- **Section** 3—**Analysis** defines the problem this chapter is trying to solve, outlines the baseline solution, proposed methodology and network models in greater detail.

- **Section 4—Implementation** describes the approach taken to solve the challenges
  mentioned in the Analysis section and outlines the implementation details of this
  chapter's baseline solution, proposed methodology and network models.
- **Section 5—Performance and Comparative Assessment** evaluates the performance of the proposed methodology against the baseline solution and discusses
  the tradeoff between communication overhead and prediction accuracy of the
  regression models.
- **Section 6—Conclusion** provides reflections on the proposed methodology, draws
  conclusions on its effectiveness and performance, identifies opportunities for
  future work and summarises the whole chapter.

2 Background

This section summarises several books, research papers and topics that I had to get
familiar with and algorithms that had to be implemented as part of the chapter.

2.1 Particle Swarm Optimisation Algorithm

The Particle Swarm Optimisation (PSO) algorithm is a population-based search and
optimisation algorithm based on the simulation of the social behaviour of birds within
a flock [9]. It was developed by Dr. Eberhart and Dr. Kennedy in 1995. The PSO
algorithm has many advantages including simplicity of implementation, reliability,
robustness, and in general, is considered an effective meta-heuristic optimisation
algorithm [23]. Kennedy and Eberhart [17] describe the algorithm as a very simple
concept, for which paradigms can be implemented in a few lines of code. It requires
only primitive mathematical operators and is computationally inexpensive in terms
of both memory requirements and speed.

The algorithm works by having a population of candidate solutions (particles) and
moving them around in the search-space according to simple mathematical formulae
over their position and velocity. Each particle's movement is influenced by its local
best known position but is also guided toward the best known position in the search
space, which is updated as better positions are found by other particles. This is
expected to move the swarm toward the best solutions [33] (see plot 2).

The PSO algorithm is used in many fields because optimisation is found every-
where. In the production of a new device, in a new artificial intelligence technique,
in a big data application or in a deep learning network, optimisation is the most
important part of an application. The PSO algorithm is used in the chapter to find
the optimal coefficients in the regression model equations (see 4 or 2.5).

Özsoy and Örkcü [23] used the PSO algorithm in their paper to estimate nonlinear regression model parameters. They note that the algorithm exhibits a rapid

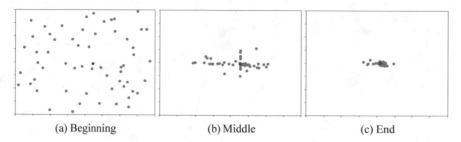

| (a) Beginning | (b) Middle | (c) End |

Fig. 2 Particle positions at different stages of the PSO algorithm

convergence tendency and it specifically converged after at most 20–25 iterations for all the models with the number of parameters ranging from 2 to 9. They conclude that the algorithm is a very efficient method for handling the problems of parameter estimation of nonlinear regression models (Fig. 2).

2.2 Particle Representation

Kennedy and Eberhart converted the social simulation algorithm into an optimisation paradigm. The idea was to let the agents (birds) find an unknown favourable place in the search space (food source) through capitalising on one another's knowledge. Each agent is able to remember its best position and knows the best position of the whole swarm. The extremum of the mathematical function to be optimised can be thought of as the food source [15].

In the algorithm each bird is represented by a particle. In this chapter each particle represents a possible estimation of coefficients (e.g. $p = [w0, w1]$). Particles move around the search coefficient space (by adding a calculated velocity to their position at every iteration) to find the optimal coefficients $[w0*, w1*]$. The following two equations calculate the velocity and the position of a particle, respectively:

$$v_{i,j}^{k+1} = wv_{i,j}^k + c_1r_1\left(xbest_{i,j}^k - x_{i,j}^k\right) + c_2r_2\left(gbest_j^k - x_{i,j}^k\right) \tag{1}$$

$$x_{i,j}^{k+1} = x_{i,j}^k + v_{i,j}^{k+1}, \tag{2}$$

where $x_{i,j}$ and $v_{i,j}$ are the position and velocity of the j component of the x particle at iteration i; $gbest_j$ and $xbest_{i,j}$ show the best position achieved so far by the whole swarm in dimension j and the personal best for particle i, respectively; r_1 and r_2 are random numbers in the interval $(0,1)$ sampled from a uniform distribution; c_1, c_2 are the cognition factor (particle's confidence in itself) and the social factor (particle's confidence in the swarm) [15]; w is the inertia weight (weighs the contribution of the velocity from the previous iteration).

Fig. 3 The movement of a
single particle based on
Eqs. 1, 2

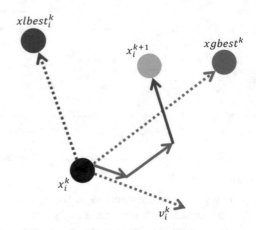

Particle positions need to be randomly initialised in order to cover as much search space as possible and to distribute them so that there is a greater chance the optimum is found. The initialisation of particles plays an important role in the performance of the algorithm. If it is not appropriate then the particles may end up searching in the wrong area, get stuck in a local minimum or find a suboptimal solution. A good technique that is used in this chapter is the following: if each dimension j is defined by the two vectors, $x_{min, j}$ and $x_{max, j}$, which respectively represent the minimum and maximum positions, then the particle's position is initialised to: (Fig. 3)

$$x = x_{min,j} + r_j(x_{max,j} - x_{min,j}), \forall j = 1, \ldots, n, \tag{3}$$

where $r_j \sim U(0, 1)$ and n is the number of dimensions [9].

2.3 Parameters

The performance of PSO depends greatly on its parameters. The optimal values for these parameters are dependent on the problem itself, the search space and the convergence time required. The control parameters of the algorithm are namely the dimension of the problem, number of particles, inertia weight, number of iterations and the cognitive and social components. Additionally, if velocity clamping (setting a limit on particles' velocity) or constriction (an approach similar to adding inertia weight) is used, the maximum velocity and constriction coefficient also influence the performance of the PSO [9]. This chapter has experimented with different values for these parameters which will be discussed in detail in Sect. 4—Implementation.

2.4   The PSO Algorithm

Each position in the n-dimensional search space corresponds to the values of n regression coefficients. During each iteration of the algorithm, each solution is evaluated by an objective function to determine its fitness. All particles will be moving through the search space to find the position for which the fitness value is best (minimum or maximum, depending on the problem). Algorithm 1 shows the pseudo code for PSO. The Particle Swarm Optimisation algorithm will be discussed in greater detail later in this work—Sect. 4—Implementation where it will be used to generate regression coefficients for the local machine learning models.

Algorithm 1: Pseudo code for the PSO algorithm.

Data: *P*, an array of particles
N the number of particles
D the number of dimensions
F the fitness function
gbest the global best
begin

   $gbest \longleftarrow MAXINT$
   for $i < N$ **do**

      for $j < D$ **do**

         Initialise $P_{i,j}$ .position
         Initialise $P_{i,j}$ .velocity

      end

   end

   for $i < N$ **do**

      $P[i].best = F(P[i].position)$
      if $P[i].best < gbest$ **then**
      $gbest = P[i].best$
      end

   end
   while *max iteration not reached or termination condition not met* **do**

      for $i < N$ **do**

         if $F(P[i].position < P[i].best$ **then**

            $P[i].best = F(P[i].position)$
            if $P[i].best < gbest$ **then**

               $gbest = P[i].best$

```
end
end
end
for i < N do
    for j < D do
        Update Pᵢ,ⱼ .velocity
        Pᵢ,ⱼ .position = Pᵢ,ⱼ .position + Pᵢ,ⱼ .velocity
    end
end
end
end
```

2.5 Regression

Regression is a statistical measurement that attempts to determine the strength of the relationship between one dependent variable (usually denoted by **y**) and a series of other changing variables (known as independent variables, denoted by **x**) [5]. Regression plays an important part in this chapter, as each device would have to compute its local regression coefficients.

Linear Regression Model

There are two types of linear regression: simple and multiple. Simple linear regression is useful for finding a relationship between two continuous variables. One is a predictor (independent variable) and the other is a response (dependent variable). It looks for a statistical relationship but not deterministic relationship [28]. In the simple linear regression model, there are two parameters—slope and intercept. The slope is the ratio of the change in the y-value over the change in the x-value and the intercept is the y-value of the point where the line intersects the y-axis [20]. The general form of a simple linear regression model looks like this:

$$y = w0 * x + w1,\tag{4}$$

where $w0$ is the slope, $w1$—the intercept, x is the independent variable and y— dependent variable. Linear regression is a linear approach to model the relationship between a scalar response and one or more explanatory variables. For more than one explanatory variable, the process is called multiple linear regression [34] (Fig. 4).

The core idea of linear regression is to obtain a line that best fits the data. The best fit line is the one for which total prediction error (of all data points) is as small as possible. The error is the distance between the point and the regression line [28]. Sometimes the relationship between variables is not linear and needs a

Fig. 4 Linear regression line

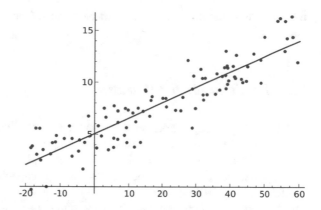

more complicated nonlinear equation, that is why this chapter also experiments with nonlinear regression models.

Nonlinear Regression Model

While a linear regression has one form, nonlinear regression can take many forms. If the equation does not satisfy the requirements for linearity, then it is not linear. Nonlinear models are used to model complex interrelationships among variables and play an important role in various scientific disciplines and engineering [23]. A nonlinear regression equation looks like this:

$$y = f(\beta, x) + \varepsilon, \tag{5}$$

where x is a vector of p predictors, β is a vector of k parameters, f is some known regression function, and s is an error term whose distribution may or may not be normal [31]. Nonlinear regression is characterised by the fact that the prediction equation depends nonlinearly on one or more unknown parameters [23]. Here are just a few nonlinear regression models that have been used in the implementation of this chapter—Logistic, Jennrich and Meyer 1 models respectively:

$$\frac{\beta_1}{1 + exp(\beta_2 - \beta_3 x)} \tag{6}$$

$$exp(\beta_1 x) + exp(\beta_2 x) \tag{7}$$

$$\frac{\beta_1 \beta_3 x_1}{1 + \beta_1 x_1 + \beta_2 x_2} \tag{8}$$

Nonlinear models make the estimation of parameters and the statistical analysis of parameter estimates more difficult and challenging. Difficulties arise due to the large number of parameters and the multimodal nature of the objective function. It is very difficult to minimise a sum of squared errors function using ordinary optimisation techniques. In order to overcome those difficulties, the use of a powerful meta-heuristic method such as the Particle Swarm Optimisation (PSO) algorithm is needed [23].

2.6 Prediction Error Metrics

Predictive modelling works on the constructive feedback principle. You build a model, get feedback from metrics, make improvements and continue until you achieve a desirable accuracy. Evaluation metrics explain the performance of a model. An important aspect of evaluation metrics is their capability to discriminate among model results [27]. Since our models will produce an output given any input or set of inputs, we can then check these estimated outputs against the actual values that we tried to predict. The difference between the actual value and the model's estimate is called a residual. The residual can be calculated for every point in the data set, and each of these residuals will be of use in the assessment. These residuals will play a significant role in judging the usefulness of a model. If the residuals are small, it implies that the model that produced them does a good job of predicting the output of interest. Conversely, if these residuals are generally large, it implies that the model is a poor estimator [24].

This chapter will use the **Root Mean Square Error** (RMSE) as an error metric to evaluate all regression models. The RMSE is the distance, on average, of a data point from the fitted line, measured along a vertical line (see Eq. 9). RMSE is directly interpretable in terms of measurement units, and so is a better measure of goodness of fit than a correlation coefficient [29]. The RMSE is calculated as follows:

$$F = \sqrt{\sum_{i=1}^{n} (y_i - \hat{y}_i)}, \tag{9}$$

where y_i is actual y in the dataset and $\hat{y}_i$ is the predicted y (Fig. 5).

Moreover, RMSE has the benefit of penalising large errors and avoids the use of taking the absolute value, because of those reasons it is more appropriate than other error metrics, e.g. MAE (Mean Absolute Error) [13].

2.7 Network Modelling

Random Network
A random network (or random graph) is a graph where edges are created by a random process. It is obtained by starting with a set of n isolated vertices and adding successive edges between them at random. Different random graph models produce

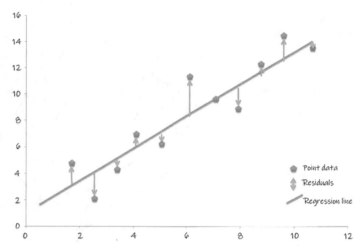

Fig. 5 Visualisation of the Root Mean Square Error [30]

different probability distributions on graphs. Most commonly studied is the one proposed by Edgar Gilbert, which is used in this chapter, denoted $G\,n, p$, in which every possible edge occurs independently with probability $0 < p < 1$. The probability of obtaining any one particular random graph with m edges is:

$$p^m(1-p)^{N-m}, N = \binom{n}{2}[35]. \qquad (10)$$

Therefore, by increasing the probability p the network becomes denser and the size of the node neighbourhoods increases. By having large neighbourhoods, we expect to have increased communication costs (node is communicating with more neighbours) and increased accuracy (the model of the neighbourhood is trained on more data). In this type of networks the difference between node degrees is not big and therefore there are rarely outliers, which makes it suitable for this chapter [4] (Fig. 6).

The connectivity links in the random network implementation used in this chapter are unidirectional. This is done because the goal is to have different neighbourhoods per node, in which neighbourhood that node receives the models from its neighbours and generates the fused model of the group.

Small World Network

The small world phenomenon, also known as six degrees of separation states that if you choose any two individuals anywhere on Earth, you will find a path of at most six acquaintances between them [4]. The Watts and Strogatz model small world networks are used in this chapter. This model is a random graph generation model that produces graphs with small-world properties, including short average path lengths and high

Fig. 6 An example of a
standard random network [8]

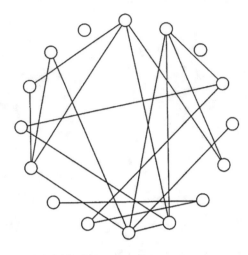

clustering. It was proposed by Duncan J. Watts and Steven Strogatz in their joint
1998 Nature paper [32].

The algorithm for creating a small world network is the following:

- Create a ring structure from the n nodes.
- Connect every node to its k closest neighbours (k 1 when k is odd).
- For every node n_i take every edge n_i, n_j and rewire it with probability p.
- Rewiring is done by removing the edge n_i, n_j and replacing it with n_i, n_k, where
  k is chosen at random.

p is the rewiring probability of the network. As we can see from the plot above
(Fig. 7) when p is equal to 0, there are no rewirings and each node is connected to
its k closest neighbours. By increasing the probability, we increase the chance that a
certain node will be disconnected with one of its neighbours and become connected
to another random node. The mean number of neighbours per node—k stays the
same, regardless of the value of p. As the value of p increases, the network starts
behaving like a random network with a constant mean number of neighbours per
node.

2.8 Mica2 Wireless Sensor Platform

The energy consumption model from the Mica2 sensor board was adopted in this
chapter. Mica2 is a wireless platform and serves as a foundation for various wireless
server network applications. We will take into account the energy costs for instruction
execution, transmitting and receiving bits when we evaluate the energy consumption
of regression models in Sect. 5—Performance and Comparative Assessment.

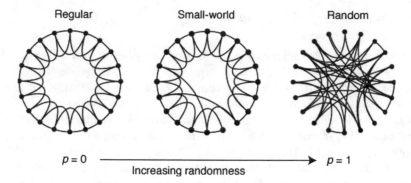

Fig. 7 Watts Strogatz small world networks with different rewiring probabilities p

Table 1 Energy costs of some Mica2 operations

Node operation mode	Energy cost
Instruction execution	4 nJ/instruction
Transmitting	720 nJ/bit
Receiving	110 nJ/bit

This energy model assumes the energy is coming out of two AA batteries that approximately supply 2,200 mAh with an effective average voltage of 3 V. It consumes 20 mA if running a sensing application continuously, which leads to a lifetime of 100 h [1]. The energy costs of the Mica2 operations are summarised in Table 1.

The packet header of the communication protocol adopted by Mica2 is 9 bytes (MAC header and CRC[1]) and the maximum payload is 29 bytes. Therefore, the perpacket overhead equals 23.7% (lowest value). This is quite high, thus necessitating a multifield transmission solution (i.e., packing multiple contextual values in the same WSN[2] message) [1]. We assume the sensors will be sensing with a lot of precision. Consequently, for each contextual value, the assumed payload is 8 bytes (high precision floating point number). Therefore, we would be able to fit two float values in each packet. This will be taken into account when calculating the energy costs of each machine learning model.

3 Analysis

This section talks about the problem this chapter is trying to solve, the baseline solution and its limitations and the proposed solution.

---

[1] Cyclic Redundancy Check.
[2] Wireless Sensor Network.

3.1 Problem Definition and Analysis

Advances in microelectronics technology have made it possible to build inexpensive, low-power, miniature sensing devices. Equipped with a microprocessor, memory, radio, and battery, such devices can now combine the functions of sensing, computing, and wireless communication into miniature smart sensor nodes [19].

Challenges in sensor networks include high bandwidth demand, high energy consumption, quality of service (QoS) provisioning and data processing. However, the energy constraint is unlikely to be solved soon due to slow progress in developing battery capacity. Moreover, the untended nature of sensor nodes and hazardous sensing environments preclude battery replacement as a feasible solution [12]. Predictive Modelling is an efficient way of reducing those communication costs, by generating a local machine learning model of the data within the sensing nodes and sending that model to the data center, which will be able to predict possible future outcomes of that data.

At its core, this chapter tries to solve the contradiction between service quality and survival time of IoT sensor networks. We are aiming to increase the survival time of the sensing and computing devices by significantly reducing the number of nodes communicating with the sink node (very costly) by exploiting node neighbourhoods. Only a representative of each neighbourhood will send the regression model, that is trained on all the data in the group, to the data center. This will lead to trading off communication overhead and accuracy which we will talk about in more detail in Sect. 5—Performance and Comparative Assessment. The base case network will not be generating neighbourhoods and every sensing device will communicate with the data center directly. This baseline solution will be useful for comparing the performance of the proposed methodology against it (Fig. 8).

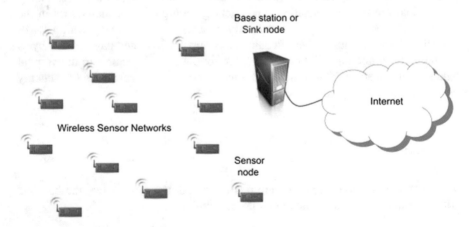

Fig. 8 A simple wireless sensor network

3.2 Baseline Methodology

In our base case solution, every node talks to the sink node. We do not have any neighbourhoods of nodes and we assume no collaboration between them. There are two ways of predictive modelling in this case: **Global Model** and **Average Model**.

Global Model
Every node n_i sends its dataset d_i (via wireless links) to the sink node. In this ideal case (very energy expensive) the sink node collects all the data and trains a machine learning model on that data and the result is the global model. This model is expected to be the most accurate because it is trained on the entire dataset (aggregation of the local datasets of all sensing devices) (Fig. 9).

Average Model
Another reference model is the Average Model. In this case, we assume that the sink node is not collecting all the data. Instead, devices are running the PSO algorithm on their own datasets in order to generate local regression coefficients—f_i. After that, each device sends its local model to the sink node. Then, the sink node is averaging all the local models to generate an aggregate model which is equal to:

$$\frac{1}{n} \sum_{1}^{n} f_i, \tag{11}$$

where n is the number of nodes in the network. This model takes the naïve approach of averaging all the parameters (in the linear regression case—local slopes and local offsets) and generates the new aggregate model. E.g., if we have two local linear models: $y_1 = 3x + 4$ and $y_2 = 4x - 3$ then the average model is equal to: $y = (3x + 4x) * 0.5 + (4 - 3) * 0.5 = 3.5x + 0.5$. The average model is going to be our point of

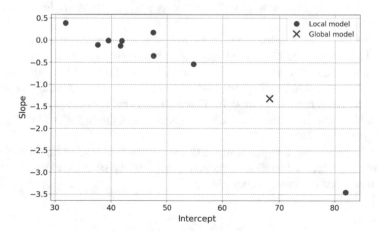

Fig. 9 Simple Linear Regression models of each node, including the Global model for comparison

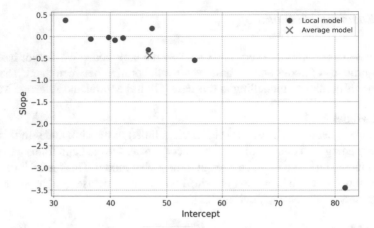

Fig. 10 Simple Linear Regression models of each node, including the Average model for comparison

reference for energy efficiency, because very low energy consumption is required to send only local predictive models per device, instead of contextual data (Fig. 10).

3.3 Limitations

There are certain limitations associated with the two reference models mentioned above. The global model could prove very costly for each sensing device because it means they have to send their entire datasets over the network, which is too expensive and very much hurts scalability. This approach will require the devices to read data from disk, load it into memory and send it over the network to the data center which overall will be very expensive in terms of energy cost (see transmission costs per bit from Table 1).

The average model introduces less communication overhead than the global model. This is because the model generation phase is performed locally at each node and only those local models are sent to the receiver. On the other hand, its accuracy is expected to be worse because of the naive approach of taking the average of all the received regression models. This method is naive because it does not take into consideration the size of the different datasets and the 'mass' of (x, y) pairs in the 2D plane (Fig. 10).

Overall, generating the global and average models will lead to high energy consumption and decreased prediction accuracy, respectfully. Both have their limitations and neither is the obviously better choice.

3.4 Proposed Methodology

This section lists the required steps of the proposed methodology.

Generate Neighbourhoods

The proposed methodology will exploit node networking neighbourhoods. Therefore, the first step will be to generate those neighbourhoods. Depending on the network type, initialise an adjacency matrix $adj[][]$ where $adj[i][j]== 1$ if and only if node j is part of the neighbourhood of node i. In that way, each node creates a neighbourhood of its own, which will be exploited to generate a fused regression model that fits the aggregated dataset from the nodes in the group.

The Local PSO Model

After that, every sensing device needs to individually analyse its contextual data—d_i and generate its local regression coefficients. Each node evidently derives a different regression model and in order to generate that model, it runs the Particle Swarm Optimisation algorithm and gets the optimal local regression coefficients of the model that best fits its local data—$PSO(d_i) \Rightarrow (w_0, w_1,.., w_n)$.

Regression Parameters Dissemination to Neighbours

Each sensing device sends information to the representative node in the neighbourhood. For example, if the device has learnt the linear model—$y = w_0 * x + w_1$, it will send the set of parameters—w_0, w_1, the x_{min}, x_{max} ranges for the input data x and the number of rows of data—N it contains to the representative node (Fig. 11).

Generate Sampling Datasets

After the representative receives the required information from all neighbours n_i, the next step is to generate 'sampling' datasets—d_i for each set of local ML models received. The goal of this process is to generate input and output data pairs that can

Fig. 11 Neighbours sending their regression models

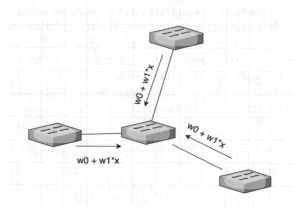

reproduce the datasets within the neighbouring nodes. Therefore avoiding transmission of the full datasets within the neighbouring nodes, which would be highly energy consuming. For each neighbouring sensing device n_i the process is the following:

- Take a random value x from the interval x_{min}, x_{max}. To do that, we need a uniform generator which gives us a uniformly distributed value p in the range [0, 1]. Then the random input x within the interval is:

$$x_{rand} = x_{min} + p * (x_{max} - x_{min})$$ (12)

  This is a uniformly generated input x from this interval.
- Before we calculate the output y we need to add Gaussian noise, in order to reduce overfitting and improve model generalisation. Gaussian noise is a value with zero mean and specific variance. We assume the variance, is relatively small, e.g., 0.1. Hence, we generate a Gaussian value q with $\mu = 0$ and $\sigma = 0.1$. Therefore, the final output looks like this: $y = w_0 * x + w_1 + q$.
- Calculate the output y.
- Generate N of these x, y pairs, run the PSO algorithm again and we will get the same model we received from node n_i. This proves that we have successfully, recreated the dataset in n_i, without needing to send its entire dataset.

The Fused PSO Model

In every node n_i, we combine the local data d_i with the aggregated generated 'sampling' datasets. The result is the fused dataset D which represents all the data of devices in the neighbourhood. We run the PSO algorithm again, this time on the fused dataset D. The final result is the fused regression model of that neighbourhood, which represents all the devices in that group $PSO\ D = f_i$.

After the generation of each fused model, they all need to be sent to the data center, which in turn has to pick the top-k performing models (in terms of energy consumption and prediction accuracy), for Predictive Modelling purposes. The end goal is to have at least k fused models outperforming the average and global models from the baseline solution so that they will be selected by the sink node.

In this chapter, we will assume that at least 70% of the nodes need to be represented by the top-k fused models so that they can represent the entire network. Therefore, a sensible value needs to be assigned to k, depending on the network type and the neighbourhood sizes. For a dense network $k = 1$ might be enough, however for a sparser network $k > 1$. The k values per network model will be set in the Performance and Comparative Assessment.

3.5   Network Model Impact

In the **Random Network** model, increasing the connectivity probability p increases the sizes of the neighbourhoods. Therefore, we expect larger neighbourhoods to have more accurate fused models, but have increased communication costs.

The **Small World** network is dependant on the order of nodes in the ring, the mean number of neighbours for each node and the rewiring probability β. Increasing the number of neighbours per node should guarantee more accuracy but at the cost of increased energy consumption. The rewiring probability and the order of the nodes, might add some unpredictability to this network model.

4   Implementation

This section discusses the software development aspects of this chapter in detail—the used datasets, the implementations of the Particle Swarm Optimisation algorithm and each of the network models.

4.1   Programming Language

The first task of the chapter was to choose a suitable programming language for the implementation. After some thought the circle was narrowed down to two programming languages—Java and Python. Some of the reasons being their ease of use, a large number of libraries available, object-oriented nature and the familiarisation at the time with those languages. Learning a new programming language would have introduced a lot of complexity and would have been time-consuming, therefore endangering the successful completion of the chapter. In the end, the chosen language was Python because of its suitability for quick prototyping, its versatility and convenience for plotting data, which is really important for analysing and evaluating results in this chapter.

Because of its versatility Python is dubbed as the Swiss Army Knife in the data science community [6]. However, the Standard Python Library alone was not sufficient for the implementation of this chapter. Additional third-party libraries were used for that additional functionality, that was missing in the standard module. SciPy and NumPy are one of the most used libraries in the Python data science community. They were useful, because of their large collection of high-level mathematical functions, ability to manipulate data and efficient operations on arrays [22]. The Matplotlib plotting library has played an important part in the chapter's performance assessment phase. This library made it straightforward to create quality graphics, which helped properly assess the performance of the machine learning models [25].

4.2   Data

Datasets Used
This chapter has used two datasets to distribute across the simulated sensing nodes

- Appliances Energy Prediction Dataset—this dataset was originally used to present and discuss data-driven predictive models for the energy use of appliances [2]. It includes measurements of temperature and humidity sensors from a wireless network. The readings are from 9 rooms in a house and the data has been collected at every 10 min for about 4.5 months. The house temperature and humidity conditions were monitored with a ZigBee wireless sensor network. Each wireless node transmitted the temperature and humidity conditions at every 3.3 min. Then, the wireless data was averaged for 10 min periods. Two random variables have been included in the dataset for testing the regression models and to filter out non-predictive attributes (parameters) [7].
  For the purposes of the experiments in this chapter, each of the 9 rooms from the dataset is considered as a separate node with different temperature and humidity readings.
- GNFUV Unmanned Surface Vehicles Sensor Dataset - this dataset comprises 4 sets of mobile sensor readings data (humidity, temperature) corresponding to a swarm of four Unmanned Surface Vehicles (USVs) floating over the sea surface in a coastal area of Athens, Greece [10]. It contains readings about the humidity and temperature of a sea surface during two months—March and July. Each USV set contains records in the following format: { 'USV-ID'; 'humidity-value'; 'temperature-value'; 'experiment-id'; 'sensing-time'}. The swarm of the USVs is moving according to a GPS predefined trajectory.
  This dataset has been collected by 4 sensing devices for 2 different months. Therefore, using this dataset we can simulate 4 * 2 = 8 nodes (Fig. 12).

Fig. 12 A picture of the USVs used in the experiments [10]

Data Diversity

After the datasets were downloaded, the next step was to analyse them and see if they are diverse enough for the purposes of the chapter. After all, the data had to be diverse, so that each node computed a unique set of regression coefficients, which is expected in a real wireless sensor network. The regression lines of the relationships between temperature and humidity (where the temperature is the independent variable and humidity the dependent one) in both datasets are seen in Fig. 13. From that, we can conclude that both datasets contain diverse linear regression models and each node would compute unique regression coefficients. As we can see from the above plots, in the Appliances dataset almost all sensing nodes have approximately the same median values of their temperature and humidity readings. This is not the case with the USV dataset, where there are more distinguishable differences between the medians. Overall, both datasets will be useful in the experimental phase of the chapter (Figs. 14, 15).

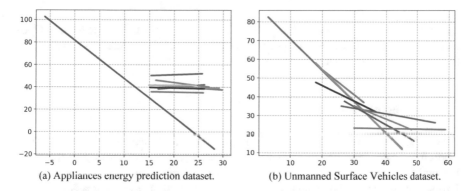

(a) Appliances energy prediction dataset.　　(b) Unmanned Surface Vehicles dataset.

Fig. 13 Regression lines of the data within each sensing device

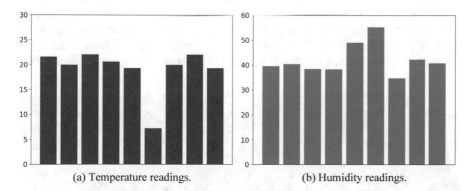

(a) Temperature readings.　　(b) Humidity readings.

Fig. 14 Median values per node in the Appliances dataset

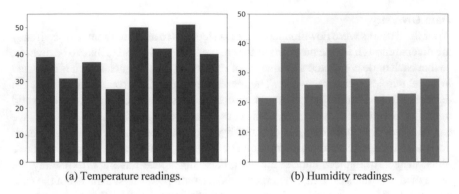

(a) Temperature readings.   (b) Humidity readings.

Fig. 15 Median values per node in the USV dataset

4.3   Particle Swarm Optimisation

In order to generate the local ML models, this chapter uses the Particle Swarm Optimisation (Swarm Intelligence) algorithm to calculate the local regression coefficients that best fit the data in each node.

4.3.1   Implementation

One of the numerous advantages of the algorithm is its ease of implementation. Therefore, implementing PSO was straightforward (see Algorithm 1). The algorithm maintains a swarm of particles, where each particle represents a potential solution. I took advantage of the object-oriented nature of Python to create a Swarm and Particle classes. The swarm class contains information that concerns all particles, such as global best fitness value and positions, the number of maximum iterations and particles (Fig. 16). It is responsible for updating the global best value and stopping the algorithm when it has converged. The particle class, as the name suggests, represents a single particle (see Fig. 17). It contains the boundaries of the search space, the position, current velocity, maximum velocity and the personal best position as fields.

```python
class Swarm:
    def __init__(self, iters, stopping_condition, number_of_parameters):
        self.particles = []
        self.gbest = sys.maxsize
        self.gbest_pos = [0.0] * number_of_parameters
        self.iters = iters
        self.stopping_condition = stopping_condition
```

Fig. 16 Swarm class implementation

```
class Particle:
    def __init__(self, lower_boundary, upper_boundary, number_of_parameters):
        self.number_of_parameters = number_of_parameters
        self.pos = [0.0] * number_of_parameters
        self.lower_boundary = -60.0
        self.upper_boundary = 60.0
        self.pbest_pos = [0.0] * number_of_parameters
        self.pbest = -1
        self.v = [0.0] * number_of_parameters
        self.v_max = (self.upper_boundary - self.lower_boundary) * 0.05
        self.initialize_position()
```

Fig. 17 Particle class implementation

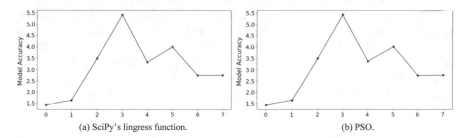

(a) SciPy's lingress function.   (b) PSO.

Fig. 18 Accuracy of generated regression coefficients for each sensing device's data

Its position is initialised as a uniform random value in the search space interval. The particle class looks like the following:

The objective function used in this chapter is the **Root Mean Square Error** (see 2.6) of the current position of the particle in the n-dimensional space (the parameters this particle represents) and the node's local dataset. It is evaluated for every particle at every iteration. All particles are aiming for the lowest value of the objective function (lower error means higher accuracy of the regression coefficients). Therefore the lowest RMSE value wins and the particle's parameters become the global best.

This implementation of the base PSO algorithm (without any optimisations or parameter tweaking) was tested on generating linear regression coefficients. It was tested against SciPy's lingress function, which also computes linear regression parameters. Figure 18 shows that their performances are exactly identical. This proves that while the algorithm may be very simple to implement, it produces fantastic results.

4.3.2 Algorithm Optimisations

Inertia Weight
Inertia weight adds a fraction of the last iteration's particle velocity to the current one. It was introduced by Shi and Eberhart as a mechanism to control the exploration

and exploitation abilities of the swarm, and as a mechanism to limit the maximum velocity a particle can reach and prevent all particles from leaving the search space [9]. Eberhart and Shi indicated that the use of the inertia weight w, which decreases linearly from about 0.9 to 0.4 during a run, provides improved performance in a number of applications [15]. This technique is used in this chapter and it enables the particles to explore a larger area at the start and as the algorithm progresses, slow down and converge to an optimum.

Velocity Clamping
Sometimes particles may reach high velocity and risk skipping appropriate search areas. They may jump over good solutions, and continue to search in fruitless regions of the search space. That is why particle velocity needs to be limited in order to find the balance between global exploration (explore different regions) and local exploitation (concentrate the search). If a particle's velocity exceeds a specified maximum velocity, the particle's velocity is set to that maximum velocity. Let $V_{max, j}$ denote the maximum allowed velocity in dimension j. If $V_{max, j}$ is too small, the swarm may not explore sufficiently beyond locally good regions [9]. On the other hand, too large values of $V_{max, j}$ risk the possibility of missing a good region. Particle velocity is adjusted before the position update using the following equation:

$$v_{i,j}(t+1) = \begin{cases} v_{i,j}(t+1) & \text{if } v_{i,j}(t+1) < V_{max,j} \\ V_{max,j} & if v_{i,j}(t+1) \geq V_{max,j} \end{cases} \quad (13)$$

4.3.3 Parameter Tuning

Swarm Size
The more particles there are in the swarm, the larger the search area that can be covered per iteration. This means that the search for optimal parameters may take fewer iterations. On the other hand, increasing particles increases greatly the computational complexity. The suitable swarm size mostly depends on the type of problem being solved. At the start of the chapter, a different number of particles has been experimented with, as can be seen from Fig. 19. Depending on the complexity of the regression model the number of particles needed to provide optimal accuracy may vary. The PSO implementation in this chapter uses between 50 and 100 particles depending on the regression models tested with (some models require particles to cover more searching space in order to find the optimal coefficients).

Number of Iterations
The number of iterations to reach a good solution is also problem-dependent. Too few iterations may terminate the search prematurely. A too large number of iterations has the consequence of unnecessary added computational complexity [9]. The maximum number of iterations needs to be set appropriately, in case another stopping condition

Fig. 19 PSO run with 30 iterations and different swarm sizes plotted against the resulting accuracy of the found parameters (linear regression)

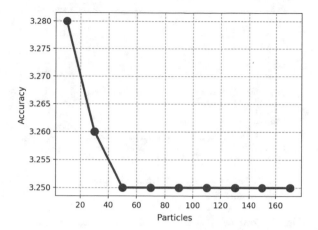

Fig. 20 PSO run with 50 particles and different maximum iterations (linear regression)

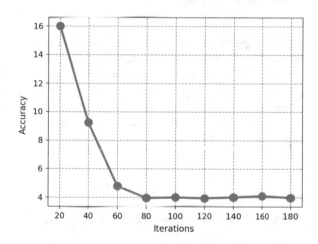

is not met. The maximum number of iterations experimented with varies from 30 to 70. As we can see from Fig. 20 the increasing number of iterations does not provide an increase in accuracy. Therefore, to avoid the additional computational overhead, for the task of finding the optimal regression coefficients, we just need to find the minimum number of iterations that provide the best accuracy.

4.4 Convergence

At the start, the PSO implementation had a fixed maximum number of iterations as the only terminating condition. The issue with that was after the particles had converged to a global minimum, the algorithm was still running and wasting iterations. An additional stopping condition was introduced to remove unnecessary iterations.

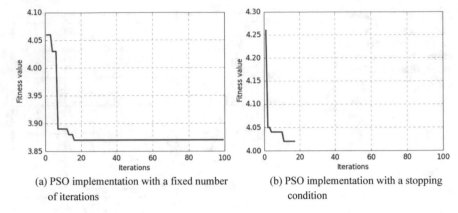

(a) PSO implementation with a fixed number of iterations

(b) PSO implementation with a stopping condition

Fig. 21 **a** Shows that the algorithm has converged but is still iterating, **b** shows a smarter implementation with a stopping condition

The stopping condition used in this implementation is to terminate the algorithm when little improvement is observed in the particles' positions over a number of iterations. If the change in some particle positions is relatively small (less than a predefined s), this would mean that the swarm has converged to a solution. This solution introduces some complexity because sensible values need to be found for those two parameters: the window of iterations w and the difference in particle positions s . Therefore, if for every dimension in the n-dimension space the following statement is true for a particle p:

$$\| p.oldpos - p.newpos \| < \varepsilon \tag{14}$$

in consecutive w iterations, then we know the algorithm has converged (Fig. 21).

4.5 Network Models

Random Network
In the random network, every node has a probability p to be connected to every other node. The implementation of the random network requires an adjacency matrix, which generates the node neighbourhoods.

Algorithm 2: Pseudo code to generate a random network

Data: N an array of nodes p connectivity probability A adjacency matrix
begin

    for $i < size\ (N)$ **do**

       for $j < size\ (N)$ **do**

          if $i == j$ **then**

             continue

          end

          if $r\ and() < p$ **then**

             $A[i][j] = 1$

          Else

             $A[i][j] = 0$

          end

       end

    end

end

Figure 22 shows the implemented adjacency matrix. The random network takes in as parameters the dataset and connectivity probability. After that, it initialises the adjacency matrix and looks it up for every node to assign its neighbours.

Small World Network

The NetworkX python library has been used to obtain an implementation of the Watts Strogatz network. NetworkX is a Python package for the creation, manipulation, and study of the structure, dynamics, and functions of complex networks [3]. The parameters that the watts_strogatz_graph() method takes in are listed in Table 2.

Fig. 22 Adjacency matrices for random networks with different connectivity probabilities

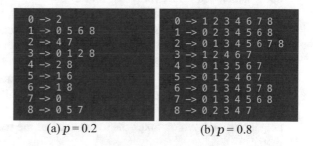

(a) $p = 0.2$ (b) $p = 0.8$

Table 2 Parameters needed to obtain a small world network from the NetworkX python library

Parameter	Description
n	Number of nodes
k	Each node is joined with k nearest neighbours in the ring topology network
p	Probability of rewiring each edge

The method first creates a ring over n nodes. Then each node in the ring is joined with its k nearest neighbours (or k 1 neighbours if k is odd). Then new edges are created by replacing some already existing edges as follows: for each edge (u, v) in the underlying "n ring with k nearest neighbours" with probability p replace it with a new edge (u, v) with uniformly random choice of existing node w [3].

5   Performance and Comparative Assessment

This section goes into detail how well the proposed methodology performs against the baseline solution and discusses the pros and cons of each network model. Each network type is evaluated in two cases: when nodes generate linear and nonlinear regression coefficients.

5.1   Prerequisites

Types of Models
This section will compare the performance of the proposed fused model (discussed in detail in Sect. 3.4) against 4 other types of models:

- **Global Model**—all sensing devices send their entire datasets to the remote sink, which in turn generates a regression model based on all the data.
- **Average Model**—all sensing devices run the PSO algorithm to generate their local regression models. After that, every device sends their model to the sink node, which takes the average of them.
- **Local-Global Model**—similar to the Global model, but just for a specific neighbourhood—all the nodes from the neighbourhood send their entire datasets to the representative node of the neighbourhood, which generates a regression model based on the received data plus its local data.
- **Local-Average Model**—similar to the Average model, but just for a specific neighbourhood—all nodes from the neighbourhood generate their local regression models. Each device sends its local model to the representative of the neighbourhood, which takes the average of all the received models plus its local model.

Linear and Non-linear Regression Models
The proposed methodology in this chapter has experimented with linear and nonlinear regression models. In the linear case, both networks were tested with simple linear regression models of the form: $y = w_0 x + w_1$, where x is the temperature reading and y is the humidity. In the other case, several nonlinear regression models have been experimented with but for the purposes of the performance assessment, we will

use the one that was most accurate for the two datasets used in this chapter—the **Misrald** model. It takes the following form:

$$y = \frac{\beta_1 \beta_2 x}{1 + \beta_2 x},$$ (15)

where again x is the temperature and y is the humidity data.

Evaluation

In order to evaluate the proposed fused models, we need to calculate their accuracy and energy cost and compare their performance against the other model types. The energy consumption is calculated using the energy cost of the Mica2 instructions mentioned in Table 1. Moreover, the energy consumption of the global and average models is calculated with respect to the average diameter of the network (the number of hops the packets need to go through to reach the sink node). Moreover, we assume that the communication between nodes of the same neighbourhood is direct (0 hops). Meanwhile, the accuracy of each model is evaluated using a test dataset that consists of a 5% random sample of the full dataset (4.2). Depending on the network type, the experiments will show if the top-k performing models are the fused models (k depends on the connectivity of the network and the network model itself).

5.2 Random Network Assessment

This section will discuss the results of the proposed methodology when run in a random network. The connectivity probabilities experimented with range from 0.1 to 0.9. The diameter of the random network is $d = \frac{n*p}{\log n}$, where n is the number of nodes and p is the connectivity probability [26]. Therefore, when a device wants to send packets to the sink node, they would have to go through d other nodes, to reach their destination. This means that these d nodes, as well as the source node will pay those communication costs. When the connectivity probability increases, so do the neighbourhoods and sensing devices need to send their regression coefficients to more neighbours. This is illustrated in Fig. 23. The energy cost of the fused models is lowest at $p = 0.1$ and highest at the opposite end—$p = 0.9$.

Depending on the connectivity of the network, the data center needs to pick a different number of fused models k. The increase in p reduces the top-k models that need to be selected, in order to represent at least 70% of the devices in the network. Figure 24 shows that increasing the density of the network, decreases the top-k models, the sink node needs.

Linear Regression Models

In this case, the PSO algorithm will be used to generate linear regression coefficients. As we can see from Fig. 25 as the connectivity in the random network increases, the fused linear models become more accurate, which is to be expected, since adding

Fig. 23 Connectivity probability and energy cost of the fused models

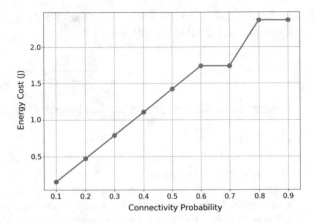

Fig. 24 Number of top-k models per connectivity probability

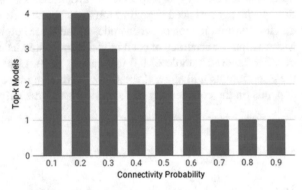

Fig. 25 Connectivity probability and median accuracy of the linear fused models

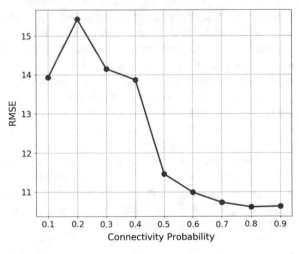

more nodes to the neighbourhoods means adding more local data to train the models. After $p = 0.7$, there is no significant improvement in the prediction accuracy, therefore we can assume that that is our optimal p value. Therefore, from Figs. 23 and 25 we can conclude that we can sacrifice that little extra energy consumption to achieve relatively greater accuracy. If we assume that $p = 0.7$ is the optimal value the following plot shows the result of the experiment for each model type.

Firstly, we can see that each model type is clustered into a group and there are almost no outliers. The Global model is the most energy consuming and the local average models are least accurate. In terms of prediction accuracy, the proposed fused models are on par with the local-global and Global models, which are our point of reference for accuracy, since they are trained on the entire dataset in the neighbourhood/network. At the same time, the fused models are two orders of magnitude more energy efficient. Some local-average models consume even less energy but have, on average, 24.4% higher error estimate. From Fig. 24 it follows that for $p = 0.7$ the sink node would select only the best performing model. Therefore, if the sink is looking for the best accuracy-energy cost tradeoff, it would pick one of the fused models (Fig. 26).

Nonlinear Regression Models

This time, each device in the Random network will compute its local nonlinear regression model. As we can see from the plot above, the results are much different from the ones in the linear case. We cannot make out any trends in the accuracy of the nonlinear fused models and there is no evident relationship between connectivity probability and prediction accuracy in this case. Therefore, we can use the p value where the fused accuracy is best -0.6 and use that to test how the proposed methodology performs in this case (Fig. 27).

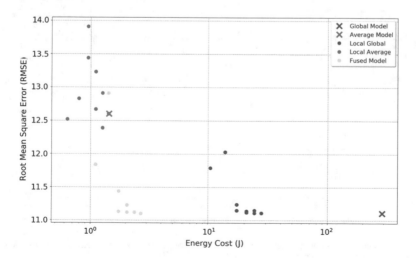

Fig. 26 Accuracy and energy consumption of each linear model in a Random network with connectivity probability $= 0.7$

Fig. 27 Connectivity
probability and median
accuracy of the nonlinear
fused models

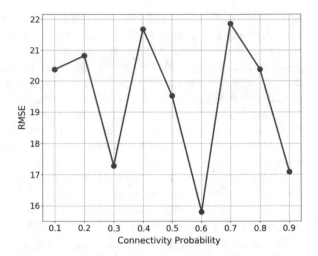

From Fig. 28 it follows that when each node is calculating nonlinear regression coefficients, there are more outliers. Half of the fused models have the same prediction accuracy as the 'naive' average models. On the contrary, the other half have on average 30% higher prediction accuracy. In that regard, they are on par with the global models, but at the same time, they consume between 10 and 100 times less energy. Some local-average models consume a little less energy but are much more inaccurate. In this case, the connectivity probability is equal to 0.6. Consequently, the sink node has to select *top k* models where $k = 2$. Overall, there are more than 2

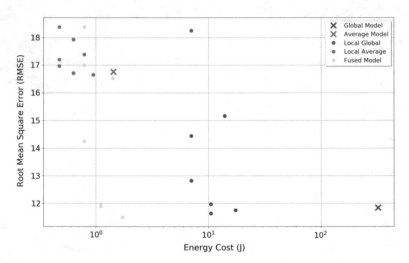

Fig. 28 Accuracy and energy consumption of each nonlinear model in a Random network with connectivity probability $= 0.6$

fused models that by far outperform the baseline solution. Therefore, this experiment is successful.

5.3 Small World Network Assessment

The datasets used in this chapter contain data for 8 (USV dataset) and 9 sensing devices (Appliances dataset). A suitable number of neighbours had to be chosen in order to appropriately assess the small world network. At the start, every device is connected to its k closest neighbours ($k - 1$ if k is odd). Therefore, every node needs to have an even number of neighbours and for the given datasets, the options are 2, 4 and 6. Increasing the neighbours per node from 2 to 4 lowers the fused model error value by 30%. Having 6 neighbours does not provide any increase in the fused model accuracy and it further increases the communication costs. Therefore, 4 neighbours will lead to a great increase in the models' accuracy, without being energy expensive. Consequently, for the purposes of this assessment, we will choose the neighbours per node to be 4 (Fig. 29).

The number of top performing models—k needs to be established for this network model. The number of neighbours—N, is the parameter that controls the size of the neighbourhoods. If we set $N = 4$, for the purposes of the experiments, it follows that the mean size of the neighbourhoods will be 5. If the datasets that are used in this chapter have been collected from 8 and 9 sensing devices, then one neighbourhood will consist of 62.5 and 55.6% of all devices respectively. Therefore, the data center will need $k = 2$ models to involve at least 70% of all the nodes' data in the generation of the models. Consequently, for the proposed methodology to be successful in this network model, the top 2 best performing regression models have to be fused models.

Fig. 29 The fused model accuracy for a specific number of neighbours

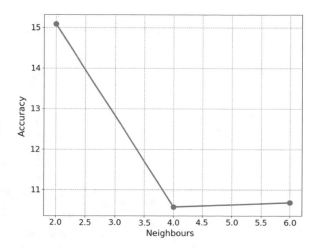

The small world network offers "six degrees of separation", that means the maximum distance between any 2 nodes is 6 hops. This is true for social media as well-everyone in the world is connected to everyone else by no more than six degrees of separation [16]. This property will be useful when calculating the energy consumption of the Global and Average models because the packets from the source devices would have to go through 6 other devices (in the worst case) before reaching their destination - the sink node.

Linear Regression Models

In this case, the sensing and computing devices in the small world network will have to compute their local linear regression models. As we can see from the above plot we cannot make a conclusion about the relationship between the fused model prediction accuracy and the rewiring probability of the network. For that reason, for the purposes of the experiment, we will use the rewiring probability for which the fused error is minimum—0.6. The models are mostly grouped together in clusters (Figs. 30 and 31).

In terms of accuracy, the fused models perform as good as the global models and some even better. Moreover, in comparison to the average models, most fused models are on average 30% more accurate. The energy cost of the fused models is decent, most of them are on par with the Average model or better and they use 2 orders of magnitude less energy than the Global model. In this case, the sink node will have to pick the 2 best models and they would be the proposed fused models. As a result, the proposed methodology in this experiment is successful.

Nonlinear Regression Models

This time, the proposed methodology will generate nonlinear local regression coefficients. As in the linear case, we cannot infer any relationship between the rewiring probability value and the fused model accuracy. The rewiring value for which the

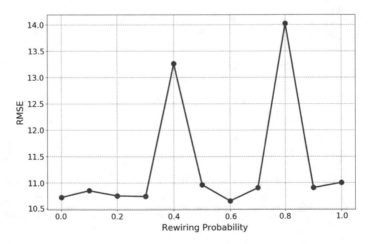

Fig. 30 Rewiring probability and median accuracy of the linear fused models

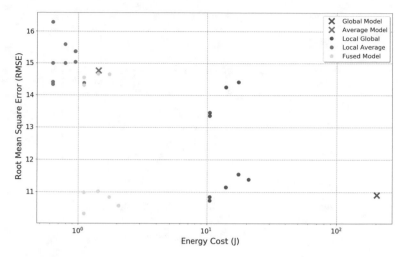

Fig. 31 Accuracy and energy consumption of each linear model in a Small World network with rewiring probability $= 0.6$

fused RMSE is minimum is 0.8 and this will be used in the experiment. As we can see the proposed methodology for the small world networks is not as effective with nonlinear regression models. This time, the Average model is more accurate than the Global model and much more energy efficient, therefore it is the point of reference for evaluating the proposed methodology in this case. The fused models' energy efficiency is on par with the Average model, but only one of them performs as good, in terms of prediction accuracy. When the sink node evaluates the top 2 performing models only one of them will be a fused model. Consequently, the experiment is unsuccessful, as the fused models do not address the limitations of the baseline solution. Therefore, when the small world network is run with nonlinear regression models, the proposed methodology is not effective (Figs. 32 and 33).

6 Conclusions and Future Work

This section summarises the chapter, the results of the performance assessment and discusses possible future improvements.

6.1 Assessment Results

Random Network Model
This network model has only one parameter - connectivity probability. Increasing the connectivity, resulted in increased fused model accuracy and energy consumption.

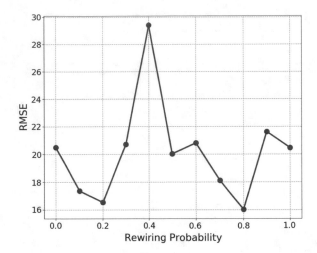

Fig. 32 Rewiring probability and median accuracy of the nonlinear fused models

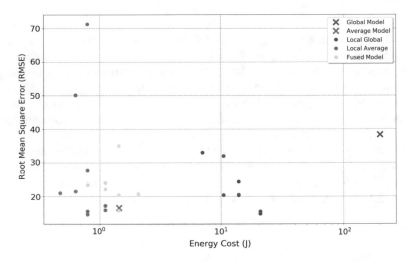

Fig. 33 Accuracy and energy consumption of each model in a Small World network with rewiring probability $= 0.8$

Next, a suitable value had to be selected, in the linear and nonlinear cases, to conduct the experiments.

When run with linear regression models the optimal connectivity probability was 0.7. In this case, the fused models offered the same accuracy as the Global model but consumed roughly 200 times less energy. Their energy consumption was on par with the Average model's but their prediction error was on average 13% less. In this case, the proposed methodology was successful.

In the nonlinear case, there was no relationship between the connectivity probability and the prediction accuracy of the fused models, but their accuracy was best at $p = 0.6$ and that was the value used for the experiment. This time, there were some outliers, which achieved the same prediction error as the 'naive' Average model. On the other hand, some of the fused models had the same accuracy as the Global model and used 2 orders of magnitude less energy. Overall, the experiment was successful. Overall, both with linear and nonlinear regression models, the fused models were outperforming the baseline solution. Therefore, we can conclude that the Random network was successful in adopting the proposed methodology of this chapter.

Small World Network Model

The parameters of this network model are the number of neighbours and the rewiring probability. Increasing the neighbours, resulted in increased fused model accuracy and energy cost, as expected. For the purposes of the experiments, the number of neighbours was set to 4 (offered best energy consumption-prediction accuracy value). When run with linear regression models, there was no evident relationship between the rewiring probability and the energy consumption or accuracy of the models. Therefore, the value, for which the fused model error was lowest, was selected for the experiment—$p = 0.6$. Again, there were some outliers, but roughly 56% of the fused models were outperforming the baseline solution, combining Global model accuracy and Average model energy efficiency.

In the other case, once more, there was no connection between the rewiring probability of the network and the accuracy of the fused models. The accuracy was best at $p = 0.8$ and this value was used for the experiment. The proposed methodology was not as effective in this case. This time, the Average model had a lower prediction error than the Global model and because of its energy efficiency, it was the point of reference for evaluating the fused models. The fused models' energy efficiency was on par with the Average model, but only one of the proposed models had higher prediction accuracy. Therefore, in this case, the proposed methodology was not effective and did not outperform the baseline solution.

Overall, the Small World network proved more unpredictable, compared to the Random network model. It was dependent on the rewiring probability (was largely responsible for the random behaviour of the network) and the positions of the nodes in the ring structure at the initialisation phase. This is why the position of the nodes in the initial ring structure was randomised at each run, to reduce bias as much as possible. The proposed methodology was efficient when the devices were generating linear regression models. It was evident that most of the fused models outperformed the baseline solution. This was not the case when run with nonlinear coefficients. At most only one fused model was performing on par with the baseline solution, all others had worse prediction accuracy. Therefore, we can conclude that the Small World network was successful in adopting the proposed methodology but only when generating linear regression coefficients.

6.2   Future Work

Should this chapter be extended in the future, this section proposes some suggestions for future work.

Larger Datasets
The chapter has experimented with datasets that have been generated from 8 and 9 sensing nodes. On the other hand, a real-life wireless sensor network can have from a few to several hundred or even thousands of sensing nodes. Therefore, it might be appropriate in the future for the experiments to use datasets that have been generated from a greater number of nodes. That will result in more realistic experiments and will test the scalability of the proposed methodology.

Network Models
The current approach experiments with only 2 network models—Random network and Small World network. Future extensions of this chapter may experiment with a larger set of network models.

Parameter Estimation Algorithms
The Particle Swarm Optimisation algorithm is used to generate the optimal regression coefficients per device. Experimenting with other regression estimation methods might be useful such as Least Squares, Gradient Descent and Regularisation. This would allow to compare the performance of the different algorithms and choose the most appropriate for the purpose of this chapter.

Real Sensing Devices
This chapter generated software that only simulates sensing devices and the communication between them. In the future, one might experiment with real IoT devices such as Raspberry Pi and see if it produces the same results.

More Nonlinear Models
Six nonlinear regression models have been experimented with, to see which one is the most appropriate for the used datasets. In the future, one might experiment with a wider range of nonlinear regression models, to produce more accurate local regression models.

6.3   Summary

This chapter has presented a lot of challenges and required a significant amount of research, especially at first. The following was achieved during the course of the chapter:

- Implemented and optimised the Particle Swarm Optimisation algorithm.
- Proved that PSO is very effective at calculating optimal regression parameters.
- Implemented two network models, in order to exploit node networking neighbourhoods.
- Proved that the proposed methodology of this chapter is effective at addressing the limitations of the baseline solution.

There are several conclusions we can draw from this chapter. Firstly, the PSO algorithm is a reliable parameter estimation approach. It exhibits a rapid convergence tendency and converged after at most 30–35 iterations, depending on the complexity of the regression model. When calculating the linear regression coefficients the results from the algorithm were identical to SciPy's lingress function [14]. Özsoy and Örkcü [23] prove in their paper that PSO is successful in estimating nonlinear regression parameters, as well.

Exploiting neighbourhoods significantly reduces the number of nodes communicating with the Data Center. It also solves the scalability issues of the baseline solution. This is because only one representative per neighbourhood sends the fused model to the sink, instead of every single node sending their local model or data. Depending on the connectivity of the network, the data center will pick the top-k fused models to use for predictive modelling. When the network is dense, the sink node will need less k models.

Overall, the proposed methodology is an efficient mechanism to tradeoff communication overhead and accuracy. It tackles the limitations of the baseline solution the expensive communication costs (Global model) and the decreased prediction accuracy, because of the naive averaging of regression models (Average model).

Acknowledgements This research is funded by the EU-H2020 GNFUV Project (#Grant 645220) and the EU-H2020 MSCA INNOVATE Project (#Grant 745829).

References

1. Anagnostopoulos, C., Hadjiefthymiades, S.: Advanced principal component-based compression schemes for wireless sensor networks. ACM Trans. Sen. Netw. **11**(1), 7:1–7:34 (2014). ISSN 1550-4859
2. Candanedo, L.M., Feldheim, V., Deramaix, D.: Data driven prediction models of energy use of appliances in a low-energy house. Energy Build. **140**, 81–97 (2017)
3. Aric Hagberg, P.S., Dan Schult. Networkx. https://networkx.github.io/ (2014). Cited 29 Oct 2018
4. Barabási, A.-L.: Network science acknowledgements random networks. Creative Commons (2014)
5. Beers, B.: What regression measures. Investopedia (2019). Cited 2 Feb 2019
6. Bhatia, R.: Why do data scientists prefer python over java? Analytics India Magazine (2018). Cited 27 Feb 2019
7. Candanedo, L.M., Feldheim, V., Deramaix, D.: Data driven prediction models of energy use of appliances in a low-energy house. Energy Build. **140**, 81–97 (2017). ISSN 0378-7788. doi:https://doi.org/10.1016/j.enbuild.2017.01.083

8. Newman, M.E.J., Watts, D., Strogatz, S.H.: Random graph models of social networks. Proc. Natl. Acad. Sci. USA **99**(1), 2566–2572 (2002). https://doi.org/10.1073/pnas.012582999
9. Engelbrecht, A.P.: Particle Swarm Optimization. Wiley (2007)
10. Harth, N., Anagnostopoulos, C.: Edge-centric efficient regression analytics. http://eprints.gla.ac.uk/160937/. April 2018
11. M. Incorporated.: Predictive modeling: The only guide you need. https://www.microstrategy.com/us/resources/introductory-guides/predictive-modeling-the-only-guide-you-need (2018). Cited 23 March 2019
12. Indu, S.D.: Wireless sensor networks: Issues and challenges. Int. J. Comput. Sci. Mob. Comput., 681–685 (20140
13. Jj and Jj. MAE and RMSE—which metric is better?—human in a machine world—medium. https://medium.com/human-in-a-machine-world/mae-and-rmse-which-metric-is-better-e60ac3bde13d (2016). Cited 1 Oct 2018
14. Jones, E., Oliphant, T., Peterson, P. et al.: SciPy: Open source scientific tools for Python. http://www.scipy.org/ (2001)
15. A. Kaveh. *Particle Swarm Optimisation*, chapter 2. Springer International Publishing, 2014
16. Keith: The history of social media: Social networking evolution! https://historycooperative.org/the-history-of-social-media/journal=HistoryCooperative (2019). Cited 21 March 2019
17. Kennedy, J., Eberhart, R.: Particle swarm optimization. In: Proceedings of ICNN'95—International Conference on Neural Networks, vol. 4, pp. 1942–1948 (1995). https://doi.org/10.1109/icnn.1995.488968
18. Lueth, K.L.: State of the IOT 2018: Number of IOT devices now at 7b—market accelerating. IoT Analytics (2018). Cited 28 Feb 2019
19. Lv, Y., Tian, Y.: Design and application of sink node for wireless sensor network. In: 2010 2nd International Conference on Industrial and Information Systems, vol. 1, pp. 487–490 (2010)
20. Math.com: Graphing equations and inequalities—slope and y-intercept—in depth. http://www.math.com/school/subject2/lessons/S2U4L2DP.html (2000–2005). Cited 24 March 2019
21. Mehl, B.: 6 IOT communication protocols for web connected devices. Kisi. https://www.getkisi.com/blog/internet-of-things-communication-protocols (2018). Cited 1 March 2019
22. Oliphant, T.: Numpy: A guide to numpy. USA: Trelgol Publishing (2006). Cited 24 March 2019
23. Özsoy, V.S., Örkcü, H.: Estimating the parameters of nonlinear regression models through particle swarm optimization. Gazi Univ. J. Sci. **29**, 187–199 (2016)
24. Pascual: Understanding regression error metrics. https://www.dataquest.io/blog/understanding-regression-error-metrics/ (2018). Cited 20 Feb 2019
25. Python, R.: Python plotting with matplotlib (guide)—real python. https://realpython.com/python-matplotlib-guide/ (2018). Cited 11 Oct 2018
26. Riordan, O., Wormald, N.: The diameter of sparse random graphs. Combinat. Prob. Comput. **19**(5–6), 835–926 (2010)
27. Srivastava, T.: 7 important model evaluation error metrics everyone should know. https://www.analyticsvidhya.com/blog/2016/02/7-important-model-evaluation-error-metrics/ (2017). Cited 24 Nov 2018
28. Swaminathan, S.: Linear regression—detailed view. https://towardsdatascience.com/linear-regression-detailed-view-ea73175f6e86 (2018). Cited 21 Oct 2018
29. V. S. Technology: What are mean squared error and root mean squared error? https://www.vernier.com/til/1014/ (2001). Cited 1 Oct 2018
30. Montoya, S.: How to calculate the Root Mean Square Error (RMSE) of an interpolated pH raster? https://www.hatarilabs.com/ih-en/how-to-calculate-the-root-mean-square-error-rmse-of-an-interpolated-ph-raster (2018). Cited 20 March 2019
31. T. P. S. University: 15.5—nonlinear regression. https://newonlinecourses.science.psu.edu/stat501/node/370/ (2018). Cited 13 Jan 2019
32. Wikipedia: Watts-strogatz model https://en.wikipedia.org/wiki/Watts-Strogatz_model (2018). Cited 3 Nov 2018

33. Wikipedia: Particle swarm optimization https://en.wikipedia.org/wiki/Particle_swarm_optimization (2018). Cited 30 Oct 2018
34. Wikipedia: Linear regression https://en.wikipedia.org/wiki/Linear_regression (2019). Cited 8 Oct 2018
35. Wikipedia: Random graph https://en.wikipedia.org/wiki/Random_graph (2019). Cited 19 Jan 2019

Machine Learning Techniques for Wireless-Powered Ambient Backscatter Communications: Enabling Intelligent IoT Networks in 6G Era

Furqan Jameel, Navuday Sharma, Muhammad Awais Khan, Imran Khan, Muhammad Mahtab Alam, George Mastorakis, and Constandinos X. Mavromoustakis

Abstract Machine learning is a rapidly evolving paradigm that has the potential to bring intelligence and automation to low-powered devices. In parallel, there have been considerable improvements in the ambient backscatter communications due to high research interest over the last few years. The combination of these two technologies is inevitable which would eventually pave the way for intelligent Internet-of-things (IoT) in 6G wireless networks. There are several use cases for machine learning-enabled ambient backscatter networks that range from healthcare network, industrial automation, and smart farming. Besides this, it would also be helpful in enabling services like ultra-reliable and low-latency communications (uRLLC), massive machine-type communications (mMTC), and enhanced mobile broadband (eMBB). Also, machine learning techniques can help backscatter communications overcome its limiting factors. The information-driven machine learning does not require the need of a tractable scientific model as the models can be prepared to deal with channel imperfections and equipment flaws in backscatter communications. Particularly, with the use of reinforcement learning approaches, the performance of

F. Jameel (✉)
Department of Communications and Networking, Aalto University, FI-02150 Espoo, Finland
e-mail: furqanjameel01@gmail.com

N. Sharma · M. M. Alam
Thomas Johann Seebeck Department of Electronics, Tallinn University of Technology,
19086 Tallinn, Estonia

M. A. Khan
Instituto de Telecomunicacoes - Aveiro, Aveiro, Portugal

I. Khan
Department of Electrical Engineering, University of Engineering & Technology, 814,
Peshawar, Pakistan

G. Mastorakis
Department of Management Science and Technology, Hellenic Mediterranean University,
72100 Agios Nikolaos, Crete, Greece

C. X. Mavromoustakis
Department of Computer Science, University of Nicosia and University of Nicosia Research
Foundation (UNRF), 46 Makedonitissas Avenue, P.O. Box 24005, 1700 Nicosia, Cyprus

© Springer Nature Switzerland AG 2020
G. Mastorakis et al. (eds.), *Convergence of Artificial Intelligence
and the Internet of Things*, Internet of Things,
https://doi.org/10.1007/978-3-030-44907-0_8

187

backscatter devices can be further improved. On-going examinations have likewise demonstrated that machine learning methodologies can be used to protect backscatter devices to enhance their ability to handle security and privacy vulnerabilities. These previously mentioned propositions alongside the ease-of-use of machine learning techniques inspire us to investigate the feasibility of machine learning-based methodologies for backscatter communications. To do such, we start this chapter by talking about the basics and different flavors of machine learning. This includes supervised learning, unsupervised learning, and reinforcement learning. We also shed light on the deep learning models like artificial neural networks (ANN) and deep Q-learning and discuss the hardware requirements of machine learning models. Then, we go on to describe some of the potential uses of machine learning in ambient backscatter communications. In the subsequent sections, we provide a detailed analysis of reinforcement learning for wireless-powered ambient backscatter devices and give some insightful results along with relevant discussion. In the end, we present some concluding remarks and highlight some future research directions.

1 Introduction to Artificial Intelligence (AI)

The so-called AI is emerging as a critical technology which has applications in myriad domains like microeconomics, biotech, and Internet-of-things (IoT) [1]. It is not just a buzz word but one of the huge breakthroughs of the modern era that aims to alter the way we interact with everyday objects. However, this immense amount of interest was not always there, especially at the beginning of the AI research. In fact, the initial attempts to describe, understand, depict, and translate the wisdom of the human being into ancient times go back to several hundred years and have a long tradition in philosophy, mathematics, psychology, neuroscience, and computer science [2]. In many cases, attempts have been made to better understand and define the concept of intelligence—that is, the cognitive capacity of human beings. By contrast, the AI traditionally refers to a branch of computer science that deals with the automation of intelligent behavior [3]. However, a precise definition is hardly possible since all directly related sciences such as psychology, biology, cognitive science, and neuroscience fail to come up with an exact definition of intelligence.

The attempts to describe and replicate intelligence can be broadly divided into four approaches that deal with the following four aspects:

- Human thinking
- Human action
- Rational thinking
- Rational action.

For example, the famous Turing test belongs to the realm of human action, since in this case an AI perfectly reproduces human action, while modern programs for image recognition and related decisions can be located more in the realm of rational action [4]. In addition to defining difficulties, part of this philosophical debate on

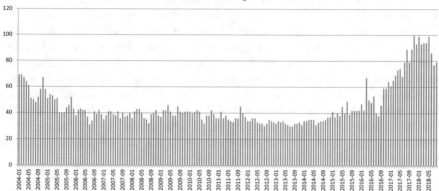

Fig. 1 An illustration of Google trends for the topic "Artificial Intelligence". The trend shows an increase in the interest in artificial intelligences and its associated domains

AI deals with the differences and consequences of, firstly, a weak or limited AI, which can intelligently solve specific problems, and second, a strong or general AI (strong/general AI).

Despite these many approaches and definitions, one central aspect can be named, which all systems designated as AI have: It is an attempt to develop a system that can independently handle complex problems. There are many ways to describe the very heterogeneous field of research of AI and its many subcategories. Some approaches deal with the problems that arise on the way to the intelligence of computer systems, others with solutions to these problems, and others with the comparisons of human intelligence [5]. These factors have recently contributed to the explosive research interest in AI, as shown in Fig. 1.

In the beginning, the development of AI often dealt with games and mathematical systems of representation of knowledge and decision making. However, since the end of the twentieth century, the technique of machine learning and, more recently, deep learning (DL) achieved great success and ultimately caused the current strong interest in AI. First approaches of the AI were based on classical principles of mathematical logic. In propositional logic, simple logical operations such as AND, OR, NOT can be combined and statements can be assigned a truth content (TRUE, FALSE), while in predicate logic arguments can be formulated and checked for their truth content [6]. Of course, logic-based AI systems are used for much more complex mathematical proofs and theorems and are still used in modern AI applications such as IBM's WATSON using logical programming languages such as PROLOG [7].

A popular field of application of AI was and is the field of human games and autonomous vehicles [8]. This approach is obvious because the abilities of the AI can be measured well and comparably on how well they play or outperform humans. The advantage of these games as a yardstick lies in their usually simple control system and easily describable possibilities of action with at the same time. As chess has very simple rules, but estimated possible moves. This huge number is beyond human imagination and, with such a large number, it is initially impossible for a program to calculate all the possibilities to develop the perfect game strategy [3]. This high number of train possibilities arises from the fact that every decision, i.e., every possible move in chess, again generates new decision alternatives and new moves, but with different starting situations and so on. These decision variants can be described as a tree or a so-called graph, in which each leaf or knot represents a possibility—in the game, this is a move—from which new and different ones result to infinity. As a tree grows, the possible moves unfold in ever new branches and ramifications to the quasi-infinite of all possible moves. Such a tree is called a decision tree, and whole areas of mathematics and computer science are concerned with the most efficient search in such branched graphs.

A very effective way of searching in decision trees are so-called heuristics. A heuristic is a process that within such to be searched for each point always determines the usefulness of a more in-depth search and thus preventing that long after the best strategy—indefinitely. In chess, this means that the possible moves are judged according to certain criteria and the possibilities that result from obviously bad moves no longer come into consideration [2]. According to this, heuristic leads to a decision tree being searched in a specific way until a satisfactory result emerges, which does not necessarily have to be the best possible result. Decision trees and the associated heuristics are a very effective process in AI for problems that can be described by a clear and unchangeable rule system.

The first successes of the AI in the field of logic and games were followed by attempts to extend the procedures to more general applications like physical layer security [9, 10], vehicular communications [11, 12], wireless social networks [13], and device-to-device communications [14]. In the 1970s, expert systems that attempt to use if-then relationships transformed a human knowledge base into computer-readable information. With the possibilities for logical reasoning and effective search-ing in these knowledge bases with the help of heuristics, the systems were able to show some success and in the 1980s raised great expectations for the possibilities of AI. A major disadvantage of these systems, however, is the immense effort involved in capturing human knowledge and transforming it into the knowledge base nec-essary for the expert system. At the beginning of the 1990s, the high expectations placed on the AI were disappointed: Many companies that had previously bought expert systems for a great deal of money abolished these. A large number of com-panies offering such systems disappeared from the market. These failures, together with a significant reduction in AI research funding, led to the first and second phases of the so-called AI Winters in the late 1970s.

2 Machine Learning Paradigm and Techniques

Despite the setbacks for research in the 1980s, the foundations for the machine learning approaches have been laid today. There are different objectives for the machine learning processes, as given in Fig. 2. However, the basic idea is simple:

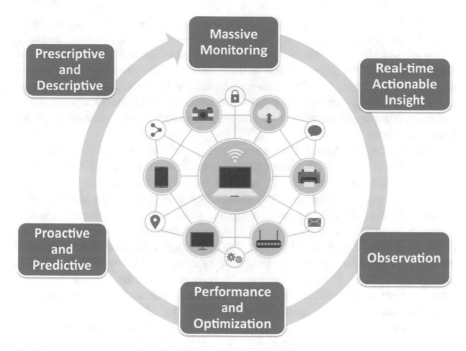

Fig. 2 Different objectives of the machine learning processes

how to get a computer program that has a specific task to learn from experience and to better fulfill the task in the future with these experiences? The difference to a static program lies in the fact that the decision rules adapt to what has been learned via feedback. The machine learning techniques are divided into three main categories of supervised machine learning, unsupervised machine learning, and reinforcement machine learning [15]. In the first system, behavior learning is initially offline, that is, separate from the application scenario. Only then will the learning would be applied and not changed. On the other hand, the online learning systems always learn and change their behavior within the application scenario and constantly adapt themselves.

2.1 Supervised Learning

In supervised machine learning, a computer program gets known sample data and is trained on the desired interpretation and the associated output. The goal is to find general rules that connect the known input data to the desired output data and then use those rules to create new output with new input data. In this sense, the computer program has something learned, and with this learned knowledge, predictions can be made about future and previously unknown input and output data. So it creates a kind of independent behavior of the computer program. The simplest procedure for such a model is that of regression, which can be explained by the following example. There is a simple linear relationship between a person's height and shoe size: the larger the human, the larger the size of the matching shoe. This relationship can be represented as a linear function, with an independent input variable (body size) and a dependent output variable (shoe size). The mathematical regression process now determines the parameters of the function and gives a model for predicting shoe sizes from body sizes.

A second important method of supervised learning is that of classification [16]. During the learning process, several values are distinguished from each other as classes, and in the later prediction individual values of a particular class are assigned. For example, one could use the left and right feet to differentiate by classifying all directions of the foot. Persons below a certain income and savings threshold would, therefore, be in the one class, namely the non-creditworthy, and above such a limit in the other class, the creditworthy one. The advantage of the classification is that it is always judged based on the interplay of several values. As a result, a person with low savings but high income in the class would be creditworthy.

Both regression and classification are predictive models that can make statements about the future [17]. They are used very effectively, for example in the area of price development, predictive maintenance, and image recognition. The difference lies in the application: Regression allows predictions about continuous values, such as the income development of a person, while classification differentiates between classes, such as creditworthiness.

2.2 Unsupervised Learning

Unsupervised machine learning works without previously known association and labeling of input data. The possible results are completely open. Therefore, the computer program can not be trained, but rather has to recognize structures in the data and transform them into interpretable information. A clear example of unsupervised learning is clustering [16].

Other less obvious methods are dimensional reduction and principal component analysis as well as density determination. Methods of unsupervised learning are used in many everyday applications. Thus, buying behavior and user behavior in online trading can be predicted and recommender systems can be created, for example for films.

2.3 Reinforcement Learning

In reinforcement machine learning, the third category of machine learning, a computer program learns directly from the experiences. It interacts with its environment and receives a reward for correct results. The program can be compared to a trained animal, for example by being rewarded in a game situation when it wins the game [17]. The goal now is for the program to remember the consequences of its action and use that knowledge to maximize its reward. The reward is accordingly the controlled variable, which is optimized in this procedure. The currently well-known example of the use of enhanced learning is AlphaGo Zero, the further development of AlphaGo. AlphaGo Zero learned the game Go so well in less than three days through increased learning without prior knowledge of the game that it played better than its predecessor and far better than the world's best human players [17]. Increased learning could prove to be an important technology in automation and robotics in particular over the next few years. For example, Fanuc's robotic arms learned to employ increased learning within just a few hours to safely grasp and move previously unknown objects.

3   Robustness of Deep (Machine) Learning Approaches

Over time, different approaches, methods and (software) technologies have been developed under the name AI. They will continue to be researched and adapted. The current AI boom is essentially based on deep learning with artificial neural networks (ANN). This is what learning is called using algorithms that simulate network structures of nerve cells. As in the term AI, a certain hint of *deep understanding* of abstract connections resonates in the everyday use of words. Although the deep learning fundamentals on the functioning of biological neural networks oriented and many media shortened only by neural networks speak, there are clear differences to the biological model [18].

The neuroscience has now developed a good understanding of how a single biological neuron. For example, a brain cell, information neurons pass on electrical impulses via chemical potentials at their synapses to a neuron. Over time, the neuron receives numerous such impulses and charges itself until a threshold potential has been reached. Then the neuron fires its impulse via its axon, which corresponds to a large data cable, at the end of which the impulse via the own synapses of the neuron is passed on again to downstream cells. This process takes place continuously in all neurons, which are processed in very different network structures [19]. An essential characteristic of biological neurons is the interconnection power or weighting with which a neuron individually transmits its electrical impulse to numerous other neurons. In addition to the network structure of the neurons, this interconnection strength or its change is an essential property for the processing of information in biological neural networks. Individual neurons have been simulated since 1952 using the Hodgkin-Huxley model, and today both simplified and more complex simulation models are in use. The simulation of entire networks can be very expensive in terms of computational complexity. Currently, in the Human Brain Project, in particular, large networks of neurons are stimulated, in terms of perspective, even in the order of magnitude of the number of biological neurons in the human brain.

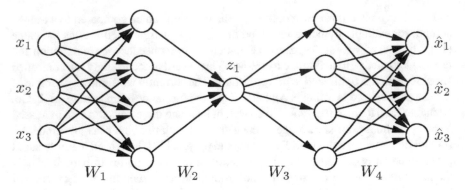

Fig. 3 An illustration of simple artificial neural network. It can be seen that inputs x_1, x_2, and x_3 are the inputs to the neural network and W_1, W_2, W_3, and W_4 act as the wights. On the other end, we have several outputs denoted as $\hat{x}_1$, $\hat{x}_2$, and $\hat{x}_3$

If we read about neural networks in the field of AI, we come across ANNs that are not aimed at an exact mapping of the biological conditions, but rather are motivated abstractly by the modeling of biological neural networks. They primarily implement the concepts of interconnecting strength or weighting and threshold information. Such ANNs fulfill their purpose in current applications. The AI boom is mainly due to the fact that the concepts of neural networks can be highly parallelized and efficiently executed on specific hardware.

The basic operation of a neural network is shown in Fig. 3. It receives input values, performs calculations and finally determines the output values. As shown in the figure, information flows in on the left side, traversing the net and flowing out on the right side. In this case, in a more complex network, the input values on the left can be the color values of the pixels of an image and the output value on the right a statement as to whether a dog is recognizable on this image. In this case, the output values can be a simple classification result, for example, a 1 (true—dog detected) or 0 (false—no dog detected). The output values can also have a more complex meaning. At each processing step, the values from the respective previous level are forwarded to the individual neurons of the next level.

The weighting of values is an essential element of the network. All incoming values are processed in the neuron in terms of their weighting and the threshold of each neuron to an output, which then passes it on to several neurons of the next layer. This process repeats itself to the last level. Between the first layer, the input neurons, and the last layer, the output neurons, are the so-called hidden neurons [20]. Due to the direction of the information flow, such a network is called a feedforward network. Naturally, more complex network structures are also possible, in which the information flows forward and sometimes also backward. For example, the processed information of a neuron layer could not only continue to flow to the next layer but also be fed back to the previous layer. Such networks are called recurrent networks.

The feedback can be a kind of *information memory* in the network and make sense depending on the application.

An empty-net must first be trained to fulfill its desired function. The weights at all points of the net must be adjusted so that the desired result is achieved. For example, a web would first have to learn whether a dog is imaged on pictures or not. This training (training) of the network is much more complex and compute-intensive than the subsequent use of the network for the recognition of patterns (inference). One method of learning is backpropagation, which is one of the supervised learning methods. Input values flow into the network and the network calculates output values. It then compares how far these calculated output values deviate from the output values, which actually ought to be correct from the input values. This deviation or error must be reduced as much as possible [21]. For this, the weights are adjusted within the network. Then the input values go through the network again and produce new output values, which in turn have a certain error. This process is repeated until the error of the output is sufficiently low. For this, the correct output values must be known for all input values. For example, the network could be trained on 10,000 images, with many images containing dogs and not the rest. After that, it can ideally be seen on new unknown pictures, whether a dog is shown or not. However, sometimes, hopefully, as rarely as possible, it will be wrong. If an ANN is trained as described above, then it is supervised learning. Yet, other approaches like deep Q-learning and deep policy gradient methods are also becoming popular due to their improved performance [21].

4 Hardware Requirements for Implementing Machine Learning Techniques

This section provides details of some of the hardware requirements for machine learning. In the context of IoT devices, it is important to understand this aspect for the practical realization of large-scale low-powered IoT networks. In addition, this section also highlights the limitations of existing hardware that hinder the realization of effective machine learning implementations.

4.1 Computational Cost: CPU Vs GPU

The machine learning has been a research topic for decades and the real breakthrough came in the past few years with the use of ANN with machine learning methods, which rudimentarily simulate processes in the nervous system. However, important drivers are not only the concepts of the ANN but above all the development of computer technology, which is used to carry out the corresponding procedures. While initially using high-performance general-purpose processors (CPUs), processors that were originally intended for graphics cards for image output have been used for some years (graphics processing unit, GPU). Currently, these are increasingly being developed into application-specific integrated circuits (ASIC) for machine learning applications [22].

To better understand the evolution of hardware for machine learning applications, it is helpful to first look at what calculations are done when using ANN with machine learning approaches. Here one still has to differentiate clearly between the training of the ANN (training) and its later use (inference), whereby the former is very computationally intensive. The hardware described in this article serves, in particular, to speed up the training. In principle, ANN consist of individual conceptual neurons arranged in specific layers. In multilayer networks, the first layer is the input layer that receives data [23]. The last layer that gives the result is the output layer. If there are additional layers (hidden neurons) between the input and output layers, the neural network becomes much more powerful and one speaks of machine learning. Between the individual layers, there are connections between neurons, which form the actual network. These compounds have different structures according to which neural networks can also be classified.

The actual *knowledge* of the network is, according to a biological neural network, in the weighting of the individual connections between the artificial neurons. This structure must first be generated, so the network is trained. A common method for this is supervised machine learning. It trains the network with known input data and output data and sets the weighting of the individual connections so that errors at the output are minimal [22]. For example, a neural network can train to distinguish dogs and cats in pictures by using images at the entrance that know which of the two species of animals are on it (output value). The training phase is completed when the neuronal network with unknown data not used for training reaches an error rate that is below a predefined value that is appropriate to the application. For instance, industry applications may require several different types of machine learning processes, as shown in Fig. 4. One can say that a neural network with more layers and more neurons, together with as many training data as possible, theoretically produces the best results, but at the same time with the number of neurons, the number of layers

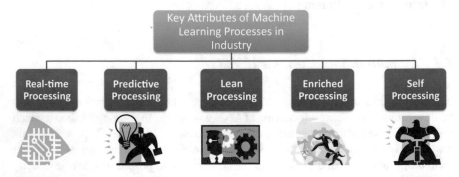

Fig. 4 Key machine learning processes in the industry

and the amount of training data, the computational effort considerably increases. These calculations can be implemented in software in different ways. However, the calculations must be usually implemented in such a way that mathematically, matrix multiplications and vector additions are performed. In the work of Chen et al., the example of matrix multiplication shows why this has a decisive influence on which hardware is particularly efficient for machine learning applications [24].

4.2 Existing Hardware Solutions

Most general-purpose processors in use today, such as the main processors in all popular computers, as well as mobile devices and servers, are fundamentally based on an architecture that John von Neumann described in 1945 and is named after him. The hallmark of this architecture is common, centralized storage for data and instruments. This is conceptually very efficient, since as powerful as possible arithmetic units are to process the programs sequentially, i.e., step by step. Such a processor is optimized for complex calculations that build on each other, but not for parallelizable tasks. This applies in principle, but today is only valid to a limited extent since the development of the CPUs in the past decades has moved a bit far from the origins. Modern CPUs have high clock rates and high processing power per clock, and command extensions allow them to perform even more complex computations in one or very few steps. Besides, these modern CPUs meanwhile can handle multiple tasks in parallel, since they contain several processor cores (currently up to 10 in smartphones, 32 in server processors) and technologies such as simultaneous multi-threading technique (SMT) that allows executing limited tasks on the same processor core [23]. Modern CPUs are very powerful, versatile and can handle complex problems quickly. However, a CPU is still unsuitable for bills that can be massively parallelled and consist of rather simple subtasks. Although the sub-steps are executed very quickly, the number of parallel tasks is limited. The large computing power of the individual cores and many optimizations of modern processors such as instruction set extensions can hardly be used or not—with the result that ultimately such a processor can not be optimally utilized with parallel computation.

In recent years, hardware has been used more and more often for such calculations, which was developed for image output. This is based on so-called GPUs. The performance of this graphics hardware has increased relatively recently, especially when compared to CPUs. GPUs consist of similar single components like CPUs, but they differ significantly in the overall architecture. To calculate individual pixels, GPUs used to use small calculation cores, so-called shaders, which were optimized for certain functions and could only execute them. There were specialized shaders, for example, to calculate the color, transparency or geometry of individual pixels or image areas [25]. Whether the individual functions were used, however, depended heavily on the software. In order to better utilize the hardware in general, modern GPUs are based on universal shaders, so-called unified shader architectures. These generalized shaders are capable of performing any of the desired functions as needed.

Condition is that each shader can be programmed directly, which makes it a small universal processor.

This capability now allows such GPUs to no longer be used only for image rendering, but also allows them to perform other calculations, making them general-purpose computation on the graphics processing unit (GPGPU). When used as a GPGPU, each shader can now be considered as a kind of general-purpose kernel [26]. Although such a kernel is considerably weaker and clocked much lower in comparison to a CPU core, modern GPUs have thousands of corresponding shaders, two orders of magnitude more than one CPU. Another difference to the CPU is that the memory of a graphics card is connected by about a factor of ten faster, which is particularly advantageous for large amounts of data. A third possibility to perform calculations is the use of application-specific integrated circuits (ASIC). Compared to CPUs and GPUs, these are not universal processors, which in principle are capable of performing almost any computation. ASICs are specially designed circuits for a specific task only. The boundary at which a modified or supplemented general-purpose processor stops and an ASIC begins is quite fluid, but not essential for the selection of machine learning hardware.

The current implementations of ANN are based on the fact that essentially a large number of matrix operations are executed. As shown by the example of matrix multiplication, such tasks are inherently parallelizable, so they can be broken down into many fairly simple computations, most of which can occur simultaneously [25]. However, such an application can hardly benefit from the optimizations of modern, even powerful CPUs with their still limited ability to calculate in parallel. Rather, GPUs, originally developed for graphics hardware or accelerator cards, can fully exploit their potential here. This is also the main reason that many machine learning applications made their breakthrough only with the use of GPUs [26]. Previously, only very expensive mainframes were able to perform appropriate calculations in a timely manner. Great potential for the future also has ASICs specializing in matrix operations, as they are currently being used gradually.

5   Ambient Backscatter Communications and Machine Learning

There is little doubt that IoT is presently one of the most rapidly developing technologies in wireless networking domain [27]. At the moment, it has the capability to totally reform how we interact and communicate with day-to-day items. Above all, it can change how we see our general surroundings ranging from everyday errands to other specific fields like sensors, vehicular technology, banking, and healthcare services [28–30]. In any case, before we receive the rewards of IoT in regular day-to-day existence, we would need to discover approaches to guarantee self-supportable and reliable communication techniques [1]. As a key enabler of these IoT technologies, ultra-low power backscatter communication has recently been proposed that can also enable ultra-reliable and low-latency communications (uRLLC), massive machine-type communications (mMTC), and enhanced mobile broadband (eMBB) services [31]. The fundamentals of backscatter communication are provided in the following subsection.

5.1 Basics of Backscatter Communications

The future achievement and quick adoption of IoT depends significantly on the capacity of devices to convey consistently, without reducing the quality of services. To address the previously mentioned issues, a few arrangements have been proposed in the form of device-to-device (D2D) communications, Zigbee, and Insteon. However, none offers the pervasiveness, the simple of ease of use, and the scope of utility guaranteed by backscatter communications [33]. This is indicated by myriad applications of backscatter communications in entertainment, healthcare, and sports. Backscatter communications are fundamentally valuable in circumstances where network capacities extend to smaller than short-range devices that exchange the information on low power [34]. A standard backscatter framework comprises of a backscatter transmitter, a receiver, and a carrier wave producing RF source.

Of late, ambient backscatter correspondences is gaining significant research interest since it enables exploiting already existing RF signals. Through this exploitation, the ambient backscatter communication reduces the dependability on any specific RF source, thus, providing more freedom to devices. Also, the ambient backscatter setup does not require many changes in the hardware setup and does not need any expensive hardware to operate [35]. These factors greatly improve the utility of these devices in areas like industrial sensing, healthcare, and home automation. Some of the more specific applications are provided in the Fig. 5.

5.2 Applications of Machine Learning in Ambient Backscatter Communications

Due to the immense potential of combining backscatter communications with emerging machine learning techniques, we anticipate that great breakthroughs can be achieved in future wireless networks. Additionally, it would provide a feasible solution for enabling intelligence into the low-powered IoT networks in 6G era To this end, the following explanation features a set of the key uses of machine learning techniques in backscatter communications.

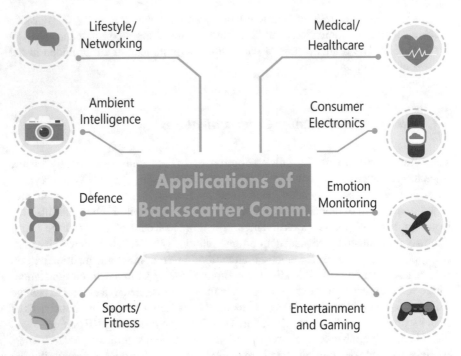

Fig. 5 Applications of backscatter communications. These applications range from the connected consumer electronics like smart watch and wearables [32] to ambient intelligence and defense systems and security. Other applications may include the entertainment and gaming along with sports and fitness for vital measurements of the players

5.2.1 Impedance Matching

Impedance matching is a critical issue for backscatter devices. Whenever left unaddressed, the backscattered signals make impedance matching difficult to tackle. Of late, machine learning methodologies have demonstrated promising results for impedance matching for other wireless technologies. Reinforcement learning strategies, like Q-learning, are considered to be most important in this context. Because of their simplicity and lightweight implementation, these learning-based procedures can effectively improve the backscatter communications, especially for less dynamic indoor conditions.

5.2.2 Improved Energy Harvesting

Reliable information sharing is one of the key foundations of static and mobile IoT nodes [8]. Because of this reason, we note a quick increment in research towards energy harvesting techniques for backscatter communications. Notwithstanding, a significant part of wireless power transmission is finding the ideal splitting conditions between the power used for harvesting and information processing. To take care of this issue, machine learning models can be utilized to discover the optimal points for splitting. In such a manner, ANN systems have just demonstrated an extraordinary guarantee by beating conventional methodologies [36]. With the assistance of machine learning, these machine learning techniques can turn out to be less power consuming for a variety of communication environments.

5.3 Selection of Reflection Coefficient

The reflection coefficient is one of the critical parameters of a backscatter device. It determines how much power needs to be reflected while communicating to the receiver. This is also helpful in minimizing the interference to other backscatter devices since they may use the same resources for backscattering the signal. Machine learning can be extremely beneficial in selecting the appropriate reflection coefficient. More specifically, time series analysis based recurrent neural networks can be used to predict the reflection coefficient based on the service requirements of the network.

5.3.1 Channel Access Schemes

The enormous scale of backscatter devices makes it difficult to exploit conventional multiple access techniques. As of late, machine learning techniques have been considered for non-orthogonal multiple access (NOMA) [37] and have demonstrated significant improvements. A regular long short-term memory (LSTM) system can be fused into a NOMA framework for backscatter communications to reduce the interference and improve the spectral efficiency of the system.

5.3.2 Determination of RF Source

A basic part of the manageability of IoT devices is the use of existing wireless resources. Thus, backscatter communication is considered as one of the critical technologies to empower IoT devices in future networks. As a key standard, the backscatter transmitter needs to utilize existing RF signals for transmitting the detected information to a receiver. For indoor situations, the Wi-Fi and Bluetooth are fitting RF hotspots for ambient backscatter communications. On the other hand, for open conditions, the RF waves generated from the base station (BS) and FM/ TV towers are appropriate RF sources. Truth be told, a comparative analysis of different machine learning techniques can be the way to tackle the issue of RF source choice in ambient backscatter communication frameworks.

5.3.3 Channel Estimation

Absence of an appropriate channel estimation scheme is one of the key restricting elements of backscatter communications. This constraint is increasingly felt for backscatter communications where the backscatter transmitter needs to tweak and reflect ambient RF signals. Besides, since channel estimation is performed regularly, a lot of power is devoured by the backscatter device during every cyclic estimation. In this regard, machine learning-based channel estimation can be very helpful. For example, carefully prepared neural networks can be utilized to anticipate the channel imperfections. Also, an augmentation of the deep neural network can be successfully utilized for predicting channel response, thereby, reducing the high energy loss of the devices.

6 System Model

We consider a backscatter communication network having a source and multiple backscatter transmitters and receivers, as shown in the Fig. 6. The backscatter devices receive the RF carrier from the nearby RF source and backscatter their data to the receivers in the vicinity. The backscattering regions are divided into the small cluster, whereby each cluster has a single backscattering transmitter and multiple receivers.

We consider the omnidirectional reflection of the signal, as given in Fig. 6. We denote the channel gain between RF source and the backscatter tag as h_a and that between the backscatter tag and the gateway h_b. Without loss of generality, we consider that all the channel gains follow the independent and identical distributed (i.i.d) Rayleigh distribution.

The receivers in a particular cluster are scattered as homogeneous Poisson point process (PPP) in R^2. It is evident that the channel resources are also used by the other clusters. Therefore, the signals from other clusters are also received at any particular receiver. The number of receivers per unit area is represented as λ in the process Φ. Let us define Φ as the process with a stationary number of users (N) in an area W from the Poisson distribution with mean

$$\lambda W = \lambda L^2. \tag{1}$$

The state of a cluster can be considered normal or it can undergo some faulty states, as given in Table 1.

These faults can be caused by channel or link failure and are stored in a memory (for applying Q-learning as discussed in coming sections). The received signal at the i-th receiver can be written as

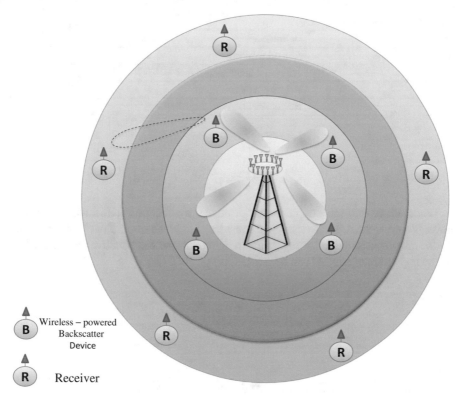

Fig. 6 An illustration of backscatter network setup. The network consists of an RF source with multiple backscatter devices placed in a circular region. The backscatter devices are considered to transfer their collected data to the receivers

$$y_{a,i} = \sqrt{\frac{\beta P}{P_{l,a}}} h_{a,i} s_{a,i} + n_a,$$
(2)

where $y_{a,i}$ denotes the received signal, $s_{a,i}$ is the normalized signal, P, and $P_{l,a}$ represents the transmit power and the path loss experienced by the backscatter device, respectively. Furthermore, β is the reflection coefficient of the backscatter devices and n_a is the zero-mean additive white Gaussian noise with zero mean and N_0 variance. Assuming there are M clusters in the vicinity of RF source, the received signal-to-interference-and noise ratio (SINR) at the gateway can be written as

$$\gamma_i = \frac{\beta \Omega_{i,b} |h_{b,i}|^2}{N_0 + \sum_{j:o_j \in M \setminus o_0} \beta_j \Omega_{j,b} |h_{b,j}|^2},$$
(3)

where o_j is the coordinate of the j-th backscattering cluster, o_0 is the main cluster which is communicating in the same transmit time interval, and

Table 1 Definition of faulty states

Fault	Definition	Rate
0	Normal working	f_0
1	Lossy and attenuated signal	f_1
2	Neighbor clusters silent	f_2
3	Neighbor clusters transmitting	f_3

$$\Omega_{i,b} = \frac{\Omega_{i,a}}{P_{l,b}}. \tag{4}$$

Our main task in this chapter is to optimize the effective SINR through efficient power control. The effective SINR is considered to be an important metric to incorporate uncompensated reflecting paths. In light of above, the effective received SINR of the main cluster (in dB) can be written as

$$\bar{\gamma}_i \equiv 10\log(\frac{1}{N}\sum_{i=1}^{N}\gamma_i). \tag{5}$$

7 Power Control in Ambient Backscatter Communications

In this section, we provide the details of our reinforcement learning-based algorithm for power control.

As illustrated in Fig. 7, we aim to design a tabular Q-learning based power control algorithm. To do so, we aim to meet the requirements of the effective SINR through under channel fluctuations and noise variations. This is done by tracking a fault register with an agent. The agent carries out a set of actions $a_t \in A$, and the set of states $s_t \in S$ at time t. The agent is rewarded positively r_t for correct action and negatively for an incorrect action. Different states and actions are defined in Table 2. Here, η is the power control command in reaction to channel variations due to network faults in Table 1.

Consider the state-action value function as $Q(s, a)$ denoted by selecting an action a under a state s. Based on the state-action table, we can write the expected discount reward function as

$$Q(s, a) = (1 - \alpha)Q(s, a) + \alpha[r + \zeta \max_{\hat{a} \in A} Q(\hat{s}, \hat{a})], \tag{6}$$

where $\hat{a}$ represents the next action, $\hat{s}$ denotes the next state, ζ is the discount factor, and α is the learning rate. When the $\zeta \to 0$, the agent mainly considers the immediate next reward, whereas, for $\zeta \to 1$ focuses on the future reward.

Fig. 7 A reinforcement learning setup with set of actions, states, and rewards. In general, an agent explores the environment based on a set of states and actions. Subsequently, a positive/negative or no reward is given to the agent

Table 2 Power control actions and states at t

Action a_t	Definition
0	$\eta = 0$
1	$\eta = -1$
2	$\eta = -1$
State s_t	**Definition**
0	Power is unchanged
1	Power increased
2	Power decreased

We consider each episode of the algorithm as a time when the agent interacts with the environment. This period, also known as transmit time interval, is represented as τ. During each iteration, the agent tries to maximize the impact of its decided action, which also introduces the well-known exploitation-exploration tradeoff in the system. Particularly, the agents adopt a greedy strategy and select a random action a with a probability ω, where $0 < \omega < 1$. This exploration rate decreases every iteration until it reaches a minimum threshold. After completion of every action, the effective SINR (gain/ loss) is calculated as

$$\Delta \bar{\gamma} = \bar{\gamma}_t - \bar{\gamma}_{t-1}. \tag{7}$$

The Algorithm 1 shows the detailed steps of the power control scheme.

Algorithm 1 Reinforcement learning algorithm for ambient bakcscatter communications.

1: **procedure** *Q-Learning Process*
2: **Input:** Target effective SINR $\bar{\gamma}_0$, Initially computed effective SINR from the environment $\bar{\gamma}_{init}$.
3: **Start:**
        Define power control states S.
        Define power control actions A.
        Define exploration rate ω.
        Define decay rate d.
        Define the minimum exploration threshold ω_{min}
        Initialize $\bar{\gamma} := \bar{\gamma}_{init}$, $s := 0$, $a := 0$, and $t := 0$
4:    **for** each episode **do**
        Start simulator and send transmit the backscattered message
5:       **while** each time slot **do**
        $\bar{\gamma} \leq \bar{\gamma}_0$ or $t \leq \tau$
        Select the exploration rate $\max(\omega.d, \omega_{\min})$
        Select r randomly from the uniform distribution.
        If $r \leq \omega$ then
        Select an action randomly such that $a \in A$.
        else
      · Select an action that maximize the reward, such that $a = \arg\max_{\hat{a} \in A} Q(s, \hat{a})$.
        end
        Obtain the reward r_t by performing the action a_t.
        Check the impact on the next state.
        Update the table entry as per $Q(s, a) = (1 - \alpha)Q(s, a) + \alpha[r + \zeta \max_{\hat{a} \in A} Q(\hat{s}, \hat{a})]$
        Change the state $s := \hat{s}$.
6:       **end while**
7:    **end for**
8: **Output:** Optimized sequence of power control commands for the target SINR against different network actions.
9: **end procedure**

---

8 Simulation Results and Discussion

This section provides the results and relevant discussion based on the algorithm proposed in the previous section. Here, we provide a comparison of the proposed technique with the conventional power allocation scheme. Unless mentioned otherwise, we consider that the backscatter tags have been distributed in the area of 200 m × 200 m. The distribution of backscatter devices follows a Poisson point process with varying densities as shown in the forthcoming plots. We have taken into account three performance metrics for evaluation, i.e., effective capacity, amount of interference generated by the backscatter devices, and the mean energy per bit consumed for data transmission. The Fig. 8 shows that the amount of interference increases with an increase in density of backscatter devices. Initially, the amount of interference for reinforcement learning is higher than the equal power allocation technique. But, as the number of backscatter devices increases in the network, the reinforcement

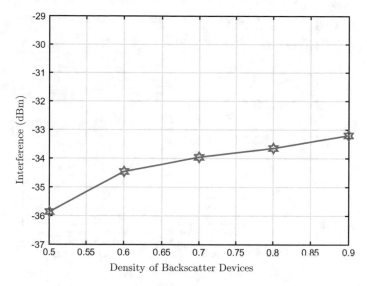

Fig. 8 Interference level versus the density of backscatter communications

learning starts to outperform the equal power allocation technique but effectively managing the interference.

The Fig. 9 shows that for both reinforcement learning and equal power allocation, the capacity increases with an increase in the density of backscatter tags. However,

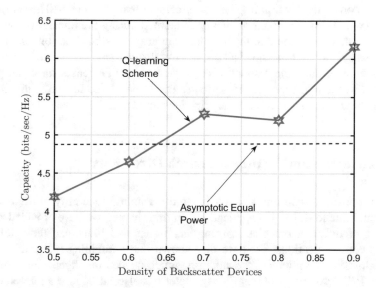

Fig. 9 Effective capacity of the system against the density of backscatter communications

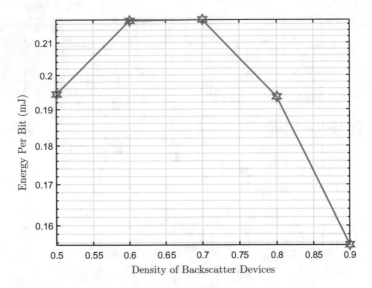

Fig. 10 Energy per bit against the density of backscatter communications

the equal power allocation approaches a ceiling at the higher number of backscatter devices. Moreover, it is also clear that the reinforcement learning technique outperforms the conventional equal power allocation method.

In the Fig. 10, we have demonstrated the utility of reinforcement learning from the perspective of energy consumption. Note that we do not incorporate the amount of energy consumed for training the reinforcement learning model. The energy per bit only considers the amount of energy consumed during the transfer of data bits. In this figure, we can see that the amount of energy consumed per bit declines for equal power allocation and since all backscatter devices are assumed to transmit the same amount of data, the amount of consumed energy decreases for equal power allocation as the number of devices grows. The same is true for the reinforcement learning, however, in this case, the energy consumed per bit is relatively low.

9   Conclusion and Future Research Directions

Ambient backscatter communication is rapidly gaining popularity in the research community due to its myriad applications. In this chapter, we have provided a brief overview of different machine learning techniques and illustrated their utility in wireless communications. We have also shown that combining machine learning with backscatter communications can bring much-needed intelligence into the low-powered IoT networks in 6G era. This chapter has also introduced a reinforcement learning framework to enable power allocation in backscatter communications. More specifically, a multi-cluster backscatter interference minimizing problem was formulated. Using the reinforcement learning-based algorithm, power allocation commands were generated which have shown that our proposed machine learning scheme improves the performance of backscatter networks

Now, we identify some possible issues which need to be addressed. These issues act as the main motivation for future research in machine learning-based wireless-powered ambient backscatter systems. The one possible issue in backscatter systems exploiting machine learning techniques is the multiple connections of ultra-low powered and short-range tags on the same bandwidth resource. The use of multiple antennas in machine learning-based wireless powered backscatter systems may drastically consume its small energy reservoirs which leads the issue of energy/coverage trade-off. The second possible issue is the mobility and interference management in machine learning-based wireless powered backscatter systems. There are no solutions existed in the literature on mobility management, and due to this, the problem of mobility management for the devices using machine learning needs to be investigated. Moreover, the management of interference is difficult in ultra-low powered and short-range backscatter systems because the transmitting device is unable to communicate to the neighbor devices which leads to simultaneous transmission of all tags.

References

1. Alsharif, M.H., Kelechi, A.H., Kim, S., Khan, I., Kim, J., Kim, J.H.: Enabling hardware green internet of things: a review of substantial issues. IEEE Access 1–1 (2019). https://doi.org/10.1109/ACCESS.2019.2926800
2. Rondeau, T.W., Bostian, C.W.: Artificial Intelligence in Wireless Communications. Artech House (2009)
3. Rondeau, T.W.: Application of artificial intelligence to wireless communications. Ph.D. thesis, Virginia Tech (2007)
4. Montoya, A., Restrepo, D.C., Ovalle, D.A.: Artificial intelligence for wireless sensor networks enhancement. Smart Wirel. Sens. Netw. 73–81 (2010)
5. Sharma, P., Liu, H., Wang, H., Zhang, S.: Securing wireless communications of connected vehicles with artificial intelligence. In: IEEE International Symposium on Technologies for Homeland Security (HST). IEEE , pp. 1–7 (2017)
6. Li, R., Zhao, Z., Zhou, X., Ding, G., Chen, Y., Wang, Z., Zhang, H.: Intelligent 5G: when cellular networks meet artificial intelligence. IEEE Wirel. Commun. 24(5), 175–183 (2017)
7. Lally, A., Fodor, P.: Natural language processing with prolog in the IBM WATSON system. Assoc. Logic Program. (ALP) Newsl
8. Jameel, F., Chang, Z., Huang, J., Ristaniemi, T.: Internet of autonomous vehicles: architecture, features, and socio-technological challenges. IEEE Wirel. Commun. 26(4), 21–29 (2019)
9. Jameel, F., Khan, F., Haider, M.A.A., Haq, A.U.: Secrecy analysis of relay assisted device-to-device systems under channel uncertainty. In: International Conference on Frontiers of Information Technology (FIT), pp. 345–349 (2017)
10. Jameel, F., Haider, M.A.A., Butt, A.A.: Physical layer security under Rayleigh/Weibull and Hoyt/Weibull fading. In: 2017 13th International Conference on Emerging Technologies (ICET), pp. 1–5 (2017)

11. Jameel, F., Hamid, Z., Jabeen, F., Javed, M.A.: Impact of co-channel interference on the performance of VANETs under—fading. AEU—Int. J. Electron. Commun. **83**, 263–269 (2018). http://www.sciencedirect.com/science/article/pii/S1434841117315546

12. Jameel, F., Wyne, S., Nawaz, S.J., Chang, Z.: Propagation channels for mmwave vehicular communications: state-of-the-art and future research directions. IEEE Wirel. Commun. **26**(1), 144–150 (2019)

13. Jameel, F., Wyne, S., Jayakody, D.N.K., Kaddoum, G., OKennedy, R.: Wireless Social networks: a survey of recent advances, applications and challenges. IEEE Access **6**, 59589–59617 (2018)

14. Jameel, F., Kumar, S., Chang, Z., Hamalainan, T., Ristaniemi, T.: Operator revenue analysis for device-to-device communications overlaying cellular network. In: IEEE Conference on Standards for Communications and Networking (CSCN), pp. 1–6 (2018)

15. Jiang, C., Zhang, H., Ren, Y., Han, Z., Chen, K.-C., Hanzo, L.: Machine learning paradigms for next-generation wireless networks. IEEE Wirel. Commun. **24**(2), 98–105 (2016)

16. Alsheikh, M.A., Lin, S., Niyato, D., Tan, H.-P.: Machine learning in wireless sensor networks: algorithms, strategies, and applications. IEEE Commun. Surv. Tutor. **16**(4), 1996–2018 (2014)

17. Silver, D., Hubert, T., Schrittwieser, J., Antonoglou, I., Lai, M., Guez, A., Lanctot, M., Sifre, L., Kumaran, D., Graepel, T., et al.: A general reinforcement learning algorithm that masters chess, shogi, and Go through self-play. Science **362**(6419), 1140–1144 (2018)

18. Wang, T., Wen, C.-K., Wang, H., Gao, F., Jiang, T., Jin, S.: Deep learning for wireless physical layer: Opportunities and challenges. China Commun. **14**(11), 92–111 (2017)

19. Wang, X., Gao, L., Mao, S., Pandey, S.: DeepFi: deep learning for indoor fingerprinting using channel state information. In: IEEE Wireless Communications and Networking Conference (WCNC). IEEE, 1666–1671 (2015)

20. Wen, C.-K., Shih, W.-T., Jin, S.: Deep learning for massive MIMO CSI feedback. IEEE Wirel. Commun. Lett. **7**(5), 748–751 (2018)

21. Rajendran, S., Meert, W., Giustiniano, D., Lenders, V., Pollin, S.: Deep learning models for wireless signal classification with distributed low-cost spectrum sensors. IEEE Trans. Cogn. Commun. Netw. **4**(3), 433–445 (2018)

22. Li, B., Najafi, M.H., Lilja, D.J.: Using stochastic computing to reduce the hardware requirements for a restricted Boltzmann machine classifier. In: Proceedings of the 2016 ACM/SIGDA International Symposium on Field-Programmable Gate Arrays, ACM, pp. 36–41 (2016)

23. Sze, V., Chen, Y.-H., Emer, J., Suleiman, A., Zhang, Z.: Hardware for machine learning: challenges and opportunities. In: IEEE Custom Integrated Circuits Conference (CICC). IEEE, pp. 1–8 (2017)

24. Chen, Y., Chen, T., Xu, Z., Sun, N., Temam, O.: DianNao family: energy-efficient hardware accelerators for machine learning. Commun. ACM **59**(11), 105–112 (2016)

25. Lew, J., Shah, D.A., Pati, S., Cattell, S., Zhang, M., Sandhupatla, A., Ng, C., Goli, N., Sinclair, M.D., Rogers, T.G.: Analyzing machine learning workloads using a detailed GPU simulator. In: IEEE International Symposium on Performance Analysis of Systems and Software (ISPASS). IEEE, pp. 151–152 (2019)

26. Choi, W, Duraisamy, K., Kim, R.G., Doppa, J.R., Pande, P.P., Marculescu, R., Marculescu, D.: Hybrid network-on-chip architectures for accelerating deep learning kernels on heterogeneous manycore platforms. In: Proceedings of the International Conference on Compilers, Architectures and Synthesis for Embedded Systems, ACM, p. 13 (2016)

27. Jameel, F., Javed, M.A., Jayakody, D.N., Hassan, S.A.: On secrecy performance of industrial Internet of things. Internet Technol. Lett. **1**(2), e32 (2018)

28. Jameel, F., Javed, M.A., Ngo, D.T.: Performance analysis of cooperative V2V and V2I communications under correlated fading. IEEE Trans. Intell. Trans. Syst. 1–9 (2019)

29. Assimonis, S.D., Daskalakis, S.-N., Bletsas, A.: Sensitive and efficient RF harvesting supply for batteryless backscatter sensor networks. IEEE Trans. Microw. Theory Tech. **64**(4), 1327–1338 (2016)

30. Jameel, F., Duan, R., Chang, Z., Liljemark, A., Ristaniemi, T., Jantti, R.: Applications of backscatter communications for healthcare networks. IEEE Netw. **33**(6), 50–57 (2019)

31. Jabeen, T., Ali, Z., Khan, W.U., Jameel, F., Khan, I., Sidhu, G.A.S., Choi, B.J.: Joint power allocation and link selection for multi-carrier buffer aided relay network. Electronics **8**(6), 686 (2019)
32. Awais, M., Raza, M., Ali, K., Ali, Z., Irfan, M., Chughtai, O., Khan, I., Kim, S., Ur Rehman, M.: An Internet of Things based bed-egress alerting paradigm using wearable sensors in elderly care environment. Sensors **19**(11), 2498 (2019)
33. Jameel, F., Ristaniemi, T., Khan, I., Lee, B.M.: Simultaneous harvest-and-transmit ambient backscatter communications under Rayleigh fading. EURASIP J. Wirel. Commun. Netw. **2019**(1), 166 (2019)
34. Lu, X., Niyato, D., Jiang, H., Kim, D.I., Xiao, Y., Han, Z.: Ambient backscatter assisted wireless powered communications. IEEE Wirel. Commun. **25**(2), 170–177 (2018)
35. Van Huynh, N., Hoang, D.T., Lu, X., Niyato, D., Wang, P., Kim, D.I.: Ambient backscatter communications: a contemporary survey. IEEE Commun. Surv. & Tutor. **20**(4), 2889–2922 (2018)
36. Min, M., Xiao, L., Chen, Y., Cheng, P., Wu, D., Zhuang, W.: Learning-based computation offloading for IoT devices with energy harvesting. IEEE Trans. Veh. Technol. **68**(2), 1930–1941 (2019)
37. Gui, G., Huang, H., Song, Y., Sari, H.: Deep learning for an effective non-orthogonal multiple access scheme. IEEE Trans. Veh. Technol. **67**(9), 8440–8450 (2018)

Processing Systems for Deep Learning Inference on Edge Devices

Mário Véstias

Abstract Deep learning models are taking place at many artificial intelligence tasks. These models are achieving better results but need more computing power and memory. Therefore, training and inference of deep learning models are made at cloud centers with high-performance platforms. In many applications, it is more beneficial or required to have the inference at the edge near the source of data or action requests avoiding the need to transmit the data to a cloud service and wait for the answer. In many scenarios, transmission of data to the cloud is not reliable or even impossible, or has a high latency with uncertainty about the round-trip delay of the communication, which is not acceptable for applications sensitive to latency with real-time decisions. Other factors like security and privacy of data force the data to stay in the edge. With all these disadvantages, inference is migrating partial or totally to the edge. The problem, is that deep learning models are quite hungry in terms of computation, memory and energy which are not available in today's edge computing devices. Therefore, artificial intelligence devices are being deployed by different companies with different markets in mind targeting edge computing. In this chapter we describe the actual state of algorithms and models for deep learning and analyze the state of the art of computing devices and platforms to deploy deep learning on edge. We describe the existing computing devices for deep learning and analyze them in terms of different metrics, like performance, power and flexibility. Different technologies are to be considered including GPU, CPU, FPGA and ASIC. We will explain the trends in computing devices for deep learning on edge, what has been researched, and what should we expect from future devices.

1 Introduction

The amazing evolution of semiconductor technologies made possible the deployment of computing devices near of almost any source of data or processing request in a myriad of applications ranging from industrial, IoT (Internet-of-Things), automotive,

M. Véstias (✉)
INESC-ID/ISEL/Instituto Politécnico de Lisboa, Lisbon, Portugal
e-mail: mvestias@deetc.isel.ipl.pt

© Springer Nature Switzerland AG 2020
G. Mastorakis et al. (eds.), *Convergence of Artificial Intelligence
and the Internet of Things*, Internet of Things,
https://doi.org/10.1007/978-3-030-44907-0_9

213

surveillance and security cameras, drones, airplanes, satellites, medical and many others.

Edge devices collect data from their attached sensors, which could be imaging, audio, or environment data. Billions of edge devices generate enormous amounts of data that have to be processed based on which some measures have to be taken by the same or other edge devices. To reduce the complexity of this devices and energy necessary for their processing, the practice has been to send all data to cloud to be processed by high-performance computing environments. However, in a vast set of applications, it is advantageous or necessary to have the decision task in the edge near the source of data so that important decisions are made at the edge device instead of sending the information to the cloud and wait for a reply. Transmitting data to the cloud requires a reliable communication channel, which even with high communication bandwidth has a high and non-deterministic round-trip delay. Many edge applications are sensitive to latency (e.g., autonomous driving) and non real-time processing is unacceptable. Security and privacy of data is another aspect that is called into question by cloud computing, not only along the communication medium but also at the cloud. In some applications, knowing the type of sensors and data permits optimizations at the edge that cannot be deployed if data is processed by a general algorithm at the cloud.

Considering all these aspects, data processing and analysis is migrating partially or totally into the edge depending on the requirements of the application in terms of criticality, security, latency, power and communication bandwidth. For some critical applications or when communication bandwidth is unreliable or insufficient, the edge systems are autonomous and must analyze data and take decisions without the help of an external system.

Devices and sensors at the edge are becoming more complex and in a vast set of applications a large volume of data is collected and should be analyzed and processed locally. Traditional algorithms for data analysis of complex information are not scalable, that is, are unable to keep with the constant increase of data volume. This is where machine learning enters the scenario with new and scalable approaches with great results in a set of intelligent tasks, like image classification and language translation.

Machine learning is a subset of artificial intelligence (AI), a technology whose objective is to take decisions and actions like humans based on the interpretation of external data. It is now applied in many domains, including medical, finance, transportation and many other activities that involve taking decisions based on data.

Well-known machine learning (ML) algorithms (e.g., reinforcement learning, fuzzy systems and deep learning) are being extensively studied and applied for data analysis at the edge. This chapter focus on deep learning in edge devices, a machine learning technique with very good results in applications like image classification, speech recognition, biomedicine, medical image analysis, board games and others.

Deep learning (DL) models, like deep neural networks (DNN), require large amounts of memory and computations. They are executed in devices and platforms with high performance computing and high memory availability. The problem is that these high-performance platforms cannot be used in edge devices which have restrictions on size, cost and power.

Running complex tasks, in particular machine learning models, with high computing requirements in low cost and low energy platforms is a big challenge for hardware and software developers. One of these challenges is associated with the constant evolution of machine learning models. Computing platforms of edge devices must be programmable and adaptable to new tasks and running models and at the same time keep a high computing power with high energy efficiency. These requires hardware/software architectures with some hardware flexibility to keep the high computing power of new models. Latency is also an aspect to be considered in edge computing platforms. Many applications running on edge must provide an output in real-time and/or within a latency constraint. This is a major reason for the high computing requirements over many edge computing platforms. Edge computing avoids or reduces communication latencies to transfer data to cloud servers but still has to keep the computing requirements. Energy and power requirements is another major issue to be considered in the design of the edge platform [28, 30, 31]. Computing efficiency is a must follow metric that guarantees the highest computing rate at the lowest energy consumption. Dedicated hardware improves power efficiency but somehow reduces flexibility and programmability of the platform. Cost is an important aspect for edge devices and is usually traded off for performance. Therefore edge devices are designed not for the best performance but for the best performance efficiency, that is, the highest performance at the lowest cost. Besides all these design aspects of edge computing platforms, communication of input at output data between the edge platforms and the cloud or other edges must also be guaranteed. The edge computing architecture must integrate intelligent communications where devices collaborate with each other in the share and processing of data.

All these requirements for edge computing platforms brought new design players offering computing solutions for edge devices.

This chapter focus on the processing solutions for deep learning on edge. It gives an overview and characterizes the deep learning domain, the requirements over edge devices to run deep learning, the technologies available to design these devices and commercial solutions and their applicability.

2 Deep Learning

Machine learning algorithms are a subject of intense research and development for some decades. Well known traditional algorithms are K-means, decision trees, support vector machine (SVM), Markov models, artificial neural networks, among others. From this set of algorithms, this chapter focus on ANNs.

An artificial neural network (ANN) consists of input, output and hidden layers, which are all layers between the input and the output layers. Each layer contains nodes or neurons that map all its inputs into an output to be used by the nodes of the next layer. The input layer generates the inputs of the first hidden layer and the output layer receives the inputs from the last hidden layer and produces the classification result. A neural network with more than three hidden layers is known as deep neural network.

Considering a neuron with n inputs from the outputs of n neurons of the previous layer, $x_1, x_2, x_3, \ldots, x_n$. Each connection between neurons has a weight, $w_1, w_2, w_3, \ldots, w_n$. The output of a neuron is computed as inner product between the inputs and the weights plus a bias value.

Deep neural networks are ANNs with many layers. The concept is not new but at the time it was proposed was very difficult to apply. In the 1990s, in spite of the success of the first deep neural network for hand written classification [25], the community still preferred traditional algorithms because they were easier to implement and train and also less computationally demanding. An obstacle to the design of neural networks with many layers was its training. ANNs with many layers were very difficult to train. Also, at the time, the computing power was not enough to train deep neural networks in acceptable times. This forced the reduction of input training vectors and therefore the network models were overfitted.

Later, with the publication of deep belief networks [19], deep neural networks have gained popularity for its very good results on several AI tasks and seen by many as a superset of the traditional machine learning algorithms.

The convergence of several algorithmic and architectural aspects have made deep learning a reality. New training algorithms extended with dropout methods [20], new activation functions [10], residual neural networks [17], etc., permitted to work with larger networks, that is, with more parameters and more layers. The advances in GPUs (Graphics Processing Unit) have increased the computational capacity for training and at the same time opened deep learning to a vast community due to the availability of low cost GPUs. A third factor was the availability of larger training sets. From the 1990s until today there was a large boost in all these factors.

The computing performance of processors has increased five orders of magnitude mainly driven by IC technology improvements. The utilization of the massive parallelism of GPUs for general purpose computing (GPGPU) made available computing platforms with ten times better performance when compared to multi-core CPUs. This is the major factor behind the feasibility of training of such large networks. Even though, it may take several days to train them.

The training data size has also increased by four orders of magnitude due to the easy access to data from the Internet. While traditional AI algorithms do not keep improving accuracy with the size of the training set, deep learning improves accuracy with more training instances. This permitted to improve the accuracy of the algorithm using more training instances.

With more processing power available and more training instances, deep neural networks have increased in size by almost four orders of magnitude. More parameters means more expressiveness of the network being able to identify more features.

Using a hierarchy of multiple hidden layers it is possible to extract complex features from input data through correlations between features of previous hidden layers [24]. For example, when a deep neural network is applied to classify an image, the initial layers detect simple features, like lines, circles, etc. The next layers identify

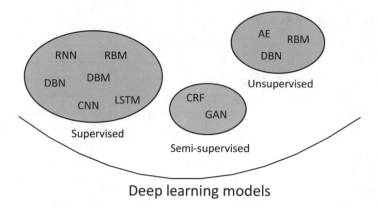

Fig. 1 Deep learning models

more complex features composed of the lines and circles identified in the first layer. Determining the best network for a specific learning problem is still an open issue. How many layers and neurons, how should they be connected, different types of layers, etc. is still an empirical process mainly based on a trial en error process.

Although empirical, deep neural models follow different structures for different classes of problems. Each deep neural model adopt one learning model that can be supervised, unsupervised or semi-supervised (see Fig. 1).

Many deep learning models have been proposed in the last decades: convolutional neural network (CNN), deep belief network (DBN), recurrent neural network (RNN), long short-term memory (LSTM), autoencoder (AE), restricted Boltzman machine (RBM), deep Boltzman machine (DBM), conditional random field (CRF), generative adversarial network (GAN) and new models are proposed every year. From these, the most common and more used model is the CNN applied to image analysis.

A traditional neural network can be applied to images. However, even with images of medium size, the number of input neurons is high and consequently there would be a high number of weights from the input neurons to the first hidden layer, which is very difficult to train with good results. A better approach is to take into consideration the type of input data. Convolutional neural networks consider the spatial information between pixels, that is, the output of a neuron of the input layer is the result of the convolution between a small subset of the image and a kernel of weights. CNNs have introduced new types of layers, like convolutional, pooling and fully connected.

The convolutional layer receives a set of input feature maps (IFM) and generates a set of output feature maps (OFM). A feature map is a 2D matrix of neurons and several feature maps form a 3D volume of feature maps. A 3D block of weights is convolved with the input feature maps and computes the dot product between the kernel and the feature maps generating an output feature map. Each different kernel produces one OFM. The pooling layer reduces a pooling windows (e.g. 2 × 2) to a single output to subsample the IFMs and achieve translation invariance. The fully connected layer is the layer of a traditional neural network. An activation function

Table 1 Complexity of known CNN models

CNN model	MAC operations (G-ops)	Parameters (millions)
AlexNet	724	61
ENet [33]	2	0.4
ResNet-18 [17]	1.83	11.7
ResNet-50 [17]	3.9	25.5
ResNet-101 [17]		44.5
ResNet-152 [17]	11.3	60.2
GoogleNet [43]	0.85	4
Inception-v1	1.43	7
Inception-v3 [44]	5	23
Inception-v4 [42]	12.3	42.7
VGG-16 [38]	15.5	138
MobileNet-v2 [36]	0.3	3.4

is used to decide the output of all neurons. A recent and frequently used activation function is ReLU (Rectified Linear Unit). The function is 0 when the input is negative and is equal to the input when the input is positive.

The first large CNN was published in 2012 [23] and won the ImageNet Large Scale Visual Recognition Competition (ILSVRC) in the same year. Since then, many CNN have been proposed for image classification with a variable number of parameters, layers and operations (see Table 1).

A common feature of convolutional neural networks is the high number of parameters and the high computational complexity. Running these models without any optimizations in edge devices is clearly infeasible. Recently, there has been some network proposals with much less parameters and operations. MobileNet-v2 [36] is a recent example of a reduced network, not only in the number of parameters, but also in the number of operations.

3   Complexity Reduction of Deep Learning Models

The main objective of the first generation of deep neural networks was to improve accuracy. This lead to very large models with a huge number of computations and weights. However, to keep increasing the size of networks it is important at the same time to reduce the complexity of the network. This is even more evident for networks to be deployed in edge devices. The second generation of DNNs addresses not only accuracy but also throughput, cost and energy. Towards this objective, a set of architectural and model optimizations have been proposed to improve the implementation of the model while minimizing the effects over accuracy.

Two types of optimizations can be identified among all proposals: (1) reduction of the bit width of weights and activations (quantization) and (2) reduction of the number of weights and operations.

Quantizationn refers to the process of reducing the number of bits used to represent DNN data. It can be fixed, when the same representation is used for all data types of the model, or variable (hybrid quantization) when different quantization is used for weights and activations at different layers. The first works focused on weight quantization, but recently both weights and activations are considered for quantization. If we consider both training and inference, only a few works started to look at training with reduced bit-width [12].

The most used type of quantization is to convert from floating-point to fixed-point. In [15] it was shown for some known CNN models that both activations and weights can be represented with 8-bit fixed-point. The quantization has a high impact over the energy and area of the multiply-accumulate units, as well as the energy and cost associated with the memory. Some commercial chips already explore data quantization. The Tensor Processing Unit [18], for example, suppports 8-bit arithmetic. The same quantization was introduced by NVIDIA with its PASCAL GPU that also includes 8-bit arithmetic. Bit-width quantization was further refined with weights and activations represented with a different number of bits across different layers with a very small impact in accuracy (less than 1%) [22]. In [49] an hybrid quantization schema was proposed and implemented in a ZYNQ XC7Z045, where data in first and the last layers use 8-bit, while data in hidden layers use 2 and/or 1 bit.

Quantization was taken to the limits using only one bit to represent weights $\subset \{-1, 1\}$ [6]. With binary weights the multiplications become additions and subtractions. Binarized Neural Networks (BNN) is the extreme data representation where both weights and activations are represented with a single bit [5] reducing multiplications to a XNOR. BNNs have some accuracy degradation that can be reduced using some modifications in the processing flow, like output scale to maintain the dynamic range of the weights, and keeping the first and last layers with maximum precision.

Accuracy can be improved using 2-bit activations [53]. If we allow two bits for weight representation $\in \{-1, 0, 1\}$ then the accuracy loss is reduced and the zero weights can be removed to avoid computations with zero and reduce storage size [26].

Reducing the size of the model is another class of techniques that can be used alone or integrated with quantization methods. Usually, networks are over-parameterized to simplify the training, which means that many weights are redundant and can be removed without changing the accuracy of the model—pruning. Pruning is one method that removes zero-valued weights or weights with a magnitude close to zero. A threshold value is used to remove weights above this value. A higher threshold increases the weight pruning but also increases accuracy error. Weight pruning can be applied during training [16, 32] or during inference—dynamic pruning [1]. The method is quite efficient in fully connected layers. For example, 89 % of the weights of AlexNet can be removed without loss in accuracy [16].

Pruning introduces sparsity in the kernels of weights and so a custom dataflow hardware is required. Some works have reduced the hardware complexity of pruned dataflows by using structured pruning that consists of pruning blocks of weights instead of single weights. The size of blocks can be adapted to he size of the data-parallel architecture of the computing device [34, 52].

Some methods also take advantage of the zeros produced by the non-linear activation function ReLU. The output feature map of a layer has several zeros sparsely distributed. Considering the zero activations, the feature map can be compressed reducing the size of required external memory as well as the energy associated with memory accesses. Compression was adopted in [4] where run-length enconder/decoders are used to compress feature maps. In terms of computations, multiplication with zero-valued activations can be skipped. The operation can be accomplished by just gating the read of the activation and the multiplication which has no effect in the execution time of the model but reduces the energy cost [4]. Additionally, the execution time can be reduced by skipping the MAC (Multiply-ACcumulate) cycle [1]. These methods have no effect over the accuracy, but its effectiveness is limited by the existence of zero values in the model.

The eep learning model itself can be changed to reduce complexity without accuracy degradation or with negligible accuracy reduction. Models using convolution windows with different sizes requires a more general hardware structure. Besides, the concatenation of several smaller filters produce the same result of a larger filter with less operations. Some works, like MobileNet [36] replace 3D convolutions by 2D convolutions followed by 1×1 convolutions with a considerable reduction in the number of operations.

All these optimizations reduce the complexity of the network and, therefore the computing platform requirements. However, the computing requirements are still high and so new and dedicated architectures are being researched and developed.

4 Deep Learning Computation

Deep learning has two computing phases with different computing requirements: training and inference. Training can be done offline or online at computing centers while inference is usually online and can be done at the cloud or at the edge. Training determines the accuracy of the network and requires a data representation format that reduces computing errors, requires high memory size and bandwidth to deal with large sets of training data and large number of network parameters. Inference, on the other side, can be executed with the same platform used to train the network or in other device that only has to implement the inference. When implemented in edge devices, metrics like energy, area, cost and latency are important requirements (see Fig. 2).

Cloud devices are used for both training and inference. Training requires high-performance, high precision and high flexibility to support many deep learning models. The same platforms are used for inference, but in this case, since it is a shared

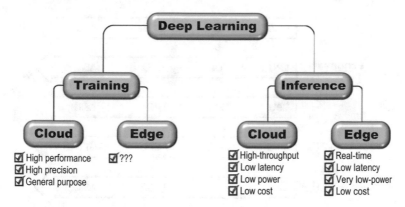

Fig. 2 Hardware requirements for deep learning: training and inference

service requested on demand it is important to guarantee the service with high-throughput, low latency, low power and low cost. At the edge, training is still an open area of research, but in a near future it is expected to have edge devices training themselves for its tasks. A solution is to run the first train in the cloud and then incrementally train the network at the edge using spare processing time when the device is not doing inference. At the moment, inference is the only process at the edge. Inference requirements depend on the application but in general real-time and low latency inference is important to guarantee on-time decisions, very low power to increase the life-time of the device that it is usually battery supplied and low cost to turn the device commercially viable.

Deployment of machine learning algorithms is now possible because of the advances in integrated circuit technology. New faster, cheaper and easier to design devices lead the machine learning design and development to a new design dimension. Developing machine learning is now more than just software optimization but also hardware redesign and algorithm optimization to obtain a more efficient hardware implementation. This increases the complexity of the problem not only because the design space increases but also because hardware and software must be designed together to get the best device for a particular machine learning application.

While deep learning algorithms are still in its infancy with many open issues about the algorithm itself and some optimizations, the market of chipsets of deep learning is expected to grow more than $40\times$ in the next six years [45]. Since the introduction of AlexNet [23] in 2012 that dozens of companies have announced to the market their new chipsets for deep learning. However, still there are no standards associated with AI chips design and therefore most of the devices cover specific deep learning models or implement particular models and optimizations. While some chips are flexible enough to implement any deep learning model, others are model specific and can not be used to run different models.

The device technologies currently available to run deep learning are CPU (Central Processing Unit), GPU (Graphics Processing Unit), FPGA (Field-Programmable

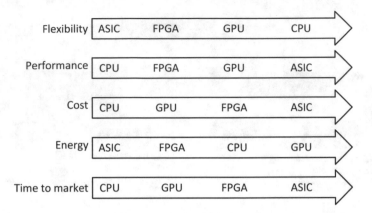

Fig. 3 Device technologies for deep learning

Gate Array), ASIC (Applications Specific Integrated Circuit) and SoC (System-on-Chip). Each of these technologies offers different levels of flexibility, cost, performance, power and time-to-market (see Fig. 3).

The flexibility of the device is the level of how easy is to change its functionality. CPU is the most flexible, but exhibits the lowest performance and energy efficiency. ASICs are the less flexible because the function is hardwired in the silicon fabric. FPGAs are more flexible than ASICs because the hardware can be reconfigured for a specific function but less flexible than CPU or GPU since software is easier to modify. GPUs offer the best performance among the flexible devices and are cheaper than FPGAs. This is why they are predominantly used for deep learning on the cloud. However, in spite of their high energy efficiency, they have high energy consumption because of its massive parallel processing. To keep the high performance computing but with a reduced energy consumption companies are deploying ASIC at the cost of loosing the flexibility. This is a risk since deep learning algorithms are still evolving and looking for the best algorithmic solutions.

GPU are widely used for training in the cloud because they are flexible and its many core architecture and high memory bandwidth offers computation powers over 100 TFLOPs. Recent GPUs include 16-bit floating-point operations together with single and double precision floating-point. While the last can be used for training, 16-bit floating-point units are enough for inference, which improves the inference efficiency. ASIC circuits for specific deep learning models increase performance and power efficiency at the cost of reduced flexibility. The latest Tensor Processing Unit (TPU) from Google is an example of a chip for both training and inference with 420 TFLOPs and 2.4 TBytes of memory bandwidth at very high power efficiency.

GPUs and ASICs are good platforms for cloud computing but the high energy consumption of GPUs and fixed architecture of ASIC make it unfeasible for edge devices with tight energy, cost and flexibility constraints. This barrier pushed edge devices to microcontrollers, CPUs and DSP (Digital Signal Processor). The problem is that these programmable devices cannot guarantee the required performance of the new machine learning applications, in particular deep learning. A tradeoff between flexibility and specificity leaves designers between better performance and lower device cost [46]. This leaves space for FPGAs as devices for edge computing.

Field programmable gate arrays can be reprogrammed to implement a specific deep learning architecture improving the performance while being flexible. At this development stage of deep learning algorithms, flexible and high-performance devices with low power requirements, like FPGAs, are the appropriate target device. Recently, with the introduction of SoC (System-on-Chip) FPGAs, hardware/software systems have reduced design times while increasing the programmability with a processor and the performance with integrated reconfigurable logic.

In the following sections, an overview of devices applied to deep learning at the edge is given.

5 Computing Platforms for Edge Devices

Edge devices cover a wide range of applications, many of these include or will include some form of machine learning. Therefore, there has a been a shift towards the design and implementation of artificial intelligence (AI) chips with enough computing power to infer deep learning networks at the edge with high energy efficiency. Most of the recent AI chips are ASICs with a specific functionality for different targets, like surveillance cameras, drones and IoT. However, ASIC architectures are hardwired and cannot be modified to run different models. Hardware flexibility at moderate performance can be achieved with reconfigurable computing. Modified or new network models can be design in hardware with fast design times with good performance and energy efficiency. Considering these aspects, several coarse-grained reconfigurable devices and FPGAs have been proposed to implement deep neural networks.

GPUs are easy to use, have very high performances and are programmable. So, they can run any model and layer. The downside is that they are general-purpose processors and, therefore, not power efficient. The power requirements of GPU are not feasible for an edge device with limited power availability. Embedded GPUs, like Jetson TX2, are the only alternative to run inference at the edge, but still requiring from 7 to 15 W of power.

In the following two sections we describe the state-of-the-art reconfigurable architectures and ASICs for deep learning on the edge. GPUs are not considered since today's GPUs are more appropriate for cloud training and inference.

5.1 Reconfigurable Platforms for Edge Computing

Reconfigurable computing technology offers hardware flexibility which permits the adaptation of the hardware to a specific DNN model, network configuration and specific network and architecture optimizations.

Eyeriss [4] is an example of a reconfigurable accelerator for convolutional neural networks. The architecture uses an array of 168 processing elements with a four-level memory hierarchy (see Fig. 4).

PEs are interconnected with a network-on-chip (NoC). The NoC is a scalable interconnection network quite efficient to implement the communication layer of a many-core architecture. The NoC can be adapted according to the dataflow of data among the PE dictated by each specific layer of the neural network model. To improve energy efficiency, the architecture has a run-length compression decoder and encoder that takes advantage of zeros introduced in the feature maps by the ReLU activation function that reduces data movement to and from external memory. Each PE has a local memory, the fastest access to weights and neurons. PEs can communicate with neighbor PEs, with global memory and with external memory.

One of the objectives of Eyeriss is to improve energy efficiency through data movement reduction and data reduction. To reduce data movement, the dataflow is optimized for each CNN to reduce data movement to all kind of memories. Since accesses to external memory and global memory are costly in terms of energy consumption, the architecture maximizes data reuse, that is, weights and input activations are cached in on-chip memories to be reused without requiring new external memory accesses. Data reduction is implemented by doing zeros compression with run-length encoders which reduces the external memory accesses.

The accelerator executes a CNN layer by layer. For each layer, it loads the configuration bits serially to reconfigure the accelerator in about 100 us. Therefore, the architecture is optimized for the execution of each layer and their filters, including

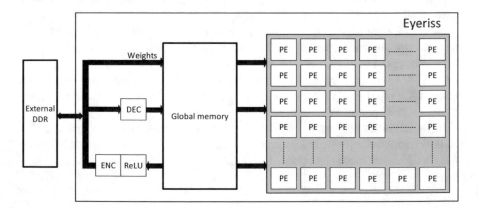

Fig. 4 The reconfigurable architecture of Eyeriss

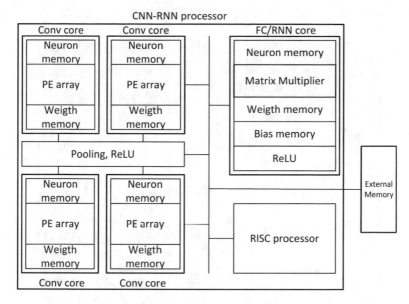

Fig. 5 Reconfigurable architecture of the hybrid CNN-RNN

the maping of PEs and the NoC configuration. After the initial configuration, the accelerator loads bocks of the input feature maps from external memory, computes the output feature maps and stores them to be available for the next layer.

The architecture was tested with AlexNet at 200 MHz and data represented with 16-bits, achieving a performance of 46.2 GOPs (multiplications and additions), an energy efficiency of 166 GOPs/W and a measured power of 278 mW.

An hybrid reconfigurable processor was proposed in [37] for convolutional neural networks and recurrent neural networks. The problem associated with the design of a chip for both network models is that convolutional layers require a high number of computations with a small number of filters and weights. On the other side, fully connected layers and RNN layers require a massive amount of weights for a relative small number of computations. So, chips dedicated to one model are inefficient when running the other model.

The proposed architecture has two main modules: one for convolutional layers and the other for fully connected and RN layers (see Fig. 5).

The convolutional module has four convolution clusters. Each convolution cluster has four cores with 48 PEs each. Each convolution cluster performs convolutions using four convolution cores and sends the result to an accumulation core. PEs mainly run multiply-accumulations to calculate convolutions and inner products. The results of convolutions are sent to a central module to run pooling and the activation function (ReLU function, in this case). The fully connected-RL module performs inner products with a matrix multiplication core. Matrix multiplication is implemented with a quantization table, that removes duplicated multiplications, and eight 16-bit

DSP	DSP	DSP	DSP	DSP	DSP	DSP	DSP
MLUT	MLUT	MLUT	MLUT	MLUT	MLUT	MLUT	MLUT
MLUT	MLUT	MLUT	MLUT	MLUT	MLUT	MLUT	MLUT
DSP	DSP	DSP	DSP	DSP	DSP	DSP	DSP
MLUT	MLUT	MLUT	MLUT	MLUT	MLUT	MLUT	MLUT
MLUT	MLUT	MLUT	MLUT	MLUT	MLUT	MLUT	MLUT
DSP	DSP	DSP	DSP	DSP	DSP	DSP	DSP
MLUT	MLUT	MLUT	MLUT	MLUT	MLUT	MLUT	MLUT
MLUT	MLUT	MLUT	MLUT	MLUT	MLUT	MLUT	MLUT
DSP	DSP	DSP	DSP	DSP	DSP	DSP	DSP
MLUT	MLUT	MLUT	MLUT	MLUT	MLUT	MLUT	MLUT
MLUT	MLUT	MLUT	MLUT	MLUT	MLUT	MLUT	MLUT
DSP	DSP	DSP	DSP	DSP	DSP	DSP	DSP
MLUT	MLUT	MLUT	MLUT	MLUT	MLUT	MLUT	MLUT
MLUT	MLUT	MLUT	MLUT	MLUT	MLUT	MLUT	MLUT
DSP	DSP	DSP	DSP	DSP	DSP	DSP	DSP
MLUT	MLUT	MLUT	MLUT	MLUT	MLUT	MLUT	MLUT
MLUT	MLUT	MLUT	MLUT	MLUT	MLUT	MLUT	MLUT
DSP	DSP	DSP	DSP	DSP	DSP	DSP	DSP
MLUT	MLUT	MLUT	MLUT	MLUT	MLUT	MLUT	MLUT
MLUT	MLUT	MLUT	MLUT	MLUT	MLUT	MLUT	MLUT

Fig. 6 eFPGA array of FlexLogic for AI

fixed-point multipliers to determine the products used to update the table. The multipliers are reconfigurable, that is, a single 16-bit multiplier implements four 4-bit multiplications, 2 8-bit multiplications or a single 16-bit multiplication. The reconfigurability of the multipliers permits to take advantage of weights quantizations, that is, representation of weights with different sizes. The smaller the weights, the more multiplications can be run in parallel, since a single 16-bit multiplication can be configured to implement four 4-bit multiplications, which improves performance.

With 4-bit weights the architecture has a performance of 1.2 TOPs, an average power of 279 mW and an energy efficiency of 3.9 TOPs/W.

Flex Logic a supplier of IPs for embedded FPGA (eFPGA) has recently deployed a new member of their eFPGA family for artificial intelligence—EFLX4K [8].

Neural networks exhibit a high parallelism of multiply-accumulation (MAC) operations. Therefore, one way to improve the performance of neural network hardware architectures is to increase the number of available MACs in a scalable and flexible way. This design approach was followed in the EFLX4K AI core that has increased the number of multiply-accumulate units (MAC) per square millimeter of FPGA silicon in their array (see Fig. 6).

The new member of the family increased the number of MACs relative to the number of distributed memory and LUT blocks (MLUT). The architecture of the MAC has also changed to an hybrid structure whose data size can be configured as

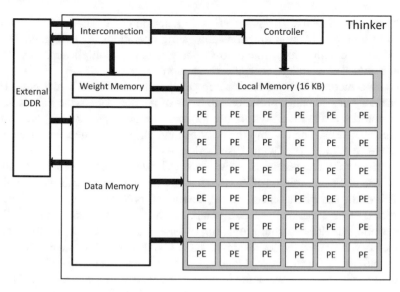

Fig. 7 Architecture of the Thinker core

8×8, 16×16, 8×16 or 16×8. This flexibility permits to adapt the MAC units to the size of activations and weights of neural networks. A single core with 441 MACs configured with 8-bit running at 1 GHz (depends on the IC technology) delivers 441 GMACs per second. The core can be tiled up to a 7×7 array with a peak performance of 22 TMACs per second.

The Flex Logic core was utilized in a research project from Harvard (presented in Hot Chips 2018) to design an accelerator to run deep learning in IoT. Their 25 mm^2 chip (SMIV) fabricated in 16-nm technology uses an eFPGA with 80 MACs and 44 Kbits of RAM. The results have shown improvements in power and area efficiency.

The Thinker chip [51] is a reconfigurable chip for artificial intelligence developed in the Microeletronics Institute of Tsinghua University that uses reconfigurability to increase the number of supported deep learning models, like convolutional neural networks, recursive neural networks and fully connected networks. The chip introduces reconfigurability at three different levels: at the core array of computing units, at the distribution of external memory bandwidth and at the lowest level where data can be represented with 8 and 16 bits (see architecture in Fig. 7).

The architecture has two 16×16 arrays of processing elements (PE). PEs are organized in clusters that can be configured to implement different functions. Neurons and activations are stored on-chip in two multi-bank data memories with 144 KB each. Each array has a local memory to store weights read from an on-chip shared memory of weights.

The computing unit of each PE can be configured to execute one 16×16 multiplication or two 8×16 multiplications. There are two types of PEs, depending on the functionality. The general PE supports the execution of MAC operations for the three types of DNN. The super PE is an extension of the general PE that also executes pooling, activation functions and gating operations for RNN models. PEs include some optimizations, like zero-skipping to avoid multiplications by 0 and, consequently, a reduction in energy consumption.

The data memory is used to store output data from a layer to the next. Each data memory has 48 banks providing 96 bytes of parallel data to the PE array. The local memories of the PE arrays are used to store weights and have 16 ports to feed 16 columns of PEs. The on-chip memory bandwidth is configured for different AI algorithms to improve bandwidth distribution.

To run convolution operations, activations are loaded from the on-chip data memory and sent to the first column of PEs that forward the activations to the next neighbor PEs. PEs of the same column compute different points (with multiply-accumulation operations) of the same output feature map. The final result is sent to a super PE to apply pooling and the activation function, and the final activation is sent back to the data memory to be used by the next layer. Fully connected layers does not exhibit the same parallelism of convolutional layers, namely the parallel computation of different points of the same output feature map. To reuse weights of fully connected layers, multiple batches of images can be computed in parallel, with one batch associated with each row of PEs. This approach permits to keep full usage of all PEs in both types of layers.

The architecture was tested with different networks covering all supported DNN models. The networks were trained with fixed-point representation. At 200 MHz the chip has a peak performance of 410 GOPs. On average, they obtained a performance of 181 GOPs, a power of 335 mW and an energy efficiency of 580 GOPs/W with 16-bit configuration. The energy efficiency and the performance doubles for the 8-bit configuration.

In [9] a dynamically reconfigurable processor (DRP) was developed to accelerate embedded AI applications. DRP is a reconfigurable hardware core with an array of coarse-grained processing elements the can be dynamically reconfigured (see Fig. 8).

PEs handle 16-bit fixed-point and floating-point mixed precision. PEs have access to small memories (4 KB and 64 KB) in the array. MACs are organized in 16×64 channels. Each group of four MACs shares 64-bit input and output FIFOs. A single MAC has three modes of operation according to the size of weights and activations: (1) 16-bit mode where data is represented with 16-bit mixed precision, (2) binary-weight where weights are represented with a single bit (processing throughput increases $16\times$), and (3) binary data where both weights and activations are represented with a single bit.

The DRP processor deals with large models with dynamic reconfiguration. When the reconfiguration stream is in the local memories of the PE array, the reconfiguration takes about 1 ns, but when the reconfiguration is in external memory, then reconfiguration takes 1 ms. A full reconfiguration is used to reconfigure the array for a specific layer, while the fast reconfiguration is used to partially process a single layer.

Initially, DMAs transfer weights from external memory to the input FIFOs and from these to the on-chip memory on the array of PEs. Processing cores multiply

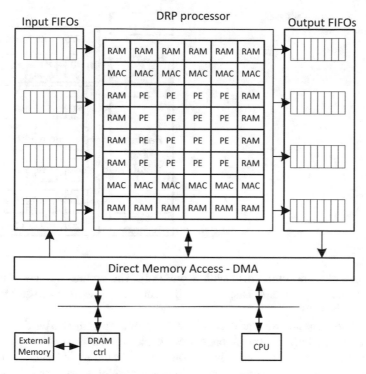

Fig. 8 Architecture of DRP, a dynamically reconfigurable processor for AI

the activations by the weights. The results are sent back in order for normalization, activation, and pooling operations. The final activations are sent back to memory. Several output activations are calculated in parallel by the same number of parallel MACs.

The chip has a measured performance of 960 GOPs in 16-bit mode, but there is no reference to the energy efficiency.

FPGAs are increasingly being used for deep learning inference, in particular, convolutional neural networks, since they can be fine-grain reconfigured to adapt to each CNN model offering good performance and energy efficiency [11, 41].

Some approaches consider a general hardware core for convolution [35, 39], while others proposed a pipeline architecture where all layers are implemented in a pipelined fashion [2, 29]. Initial implementations of DNN with FPGA considered high-density FPGAs. With the migration of deep learning to the edge low density FPGAs are now also considered as target devices. In [47] small CNNs are implemented in a ZYNQ XC7Z020 with a performance of 13 GOPs with 16 bit fixed-point data. In [13] the same FPGA is used to implemented big CNN models, like VGG16, with data represented with 8 bits achieving performances of 84 GOPs.

Recently, a configurable architecture was proposed to implement large CNN in low density FPGAs, achieving a peak performance close to 400 GOPs in a ZYNQ

Fig. 9 Block diagram of the
LiteCNN architecture

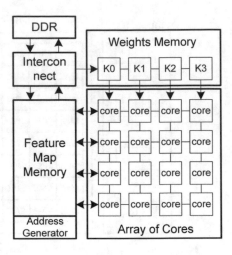

XC7Z020 with data represented with 8-bits [48]. The architecture consists of an
array of processing cores to calculate dot-products and a memory to store feature
maps (see Fig. 9).

3D convolutions are calculated as a linear dot-product identical to the dot-products
of fully connected layers. The architecture executes one layer at a time and is configured with the features of each layer. The image or the input feature maps of a
layer are stored in a multi-port on-chip memory and weights are stored also on-chip
with each kernel of weights stored in a different memory. Multiple activations are
broadcasted to the cores that calculate dot-products between a vector of activations
and a vector of weights. The result is sent back to the multi-port memory. Before
being stored in this memory, the activation function (ReLU) is applied to the result
of the convolution to produce an activation. The process repeats until finishing the
convolution between the image and the kernels. After that, the next kernels are loaded
from memory and the process repeats until running all kernels of a layer.

The array of cores contains an array of identical units to calculate the dot-product
between activations and weights. Each line of cores receive a different activation
and each column of cores receive the same kernel. So, each core determines one
activation of the output feature maps. Cores in the same column generate activations
of the same feature map, while cores in different columns generate activations of
different feature maps.

The architecture explores inter-output parallelism (multiple output feature maps
are calculated in parallel), intra-output parallelism (multiple activations of a feature
map are calculated in parallel) and dot-product parallelism (multiple activations and
weights from each kernel are read and processed in parallel).

The architecture was tested for the inference of AlexNet in a ZYNQ XC7Z020
and achieved an average performance of 133 GOPs. Assuming an average power of
4W, the energy efficiency is 33 GOPs/W.

Table 2 Main features of reconfigurable architectures for deep neural networks

	Eyeriss [4]	DNPU [37]	SMIV [8]	Thinker [51]	DRP [9]	LiteCNN [48]
Year	2017	2017	2018	2018	2018	2018
Tech. (nm)	65	65	—	65	28	28
Freq. (MHZ)	250	200	—	200	500	200
Data	16b fix	4b/16b fix	8b/16b fix	8b/16b fix	1b/16b fix	8b/16b fix
Model	CNN	CNN/RNN	DNN	CNN/RNN	DNN	CNN
GOPs/W	246	3900/1000	—	1000 (8b)	—	33
Peak GOPs	67	1200/300	—	410	1000	410

Table 2 resumes the main features of reconfigurable architectures for deep neural networks.

All architectures are coarse-grain reconfigurable, except LiteCNN that targets fine-grained FPGAs. As can be observed from the table, all reconfigurable architectures adopted fixed-point representation with 8, 16 and even 1 bit to represent data (weights and activations). Since it is known from the literature that inference can be executed with small datawidth, compared to that needed for network training, without negligible accuracy loss, all architectures have adopted smaller representations that improves performance and energy efficiency, and reduces memory size and bandwidth requirements.

While some architectures consider a particular neural network model (e.g., CNN), others support two models (e.g., CNN and RNN) and others support general DNN models by implementing in hardware the basics of neural network processing common to many deep neural network models (DNN) and moving all other specific computations to a general-purpose processing unit.

The performance varies among architectures since it depends on the number of cores of the architecture. The energy efficiency is related to the flexibility of the architecture. The more configurability the less the efficiency since the fabric logic and interconnection associated with the reconfiguration of the device consumes power, area and reduces performance.

5.2 ASICs for Edge Computing

Fine and coarse-grain reconfigurable devices increases flexibility but reduces performance and energy efficiency. ASICs offer the best performance with the best energy efficiency but reduced or none flexibility. Some companies started to provide IP cores for DNN processing, while others provide full SoC solutions. In the following, several dedicated architectures for deep neural networks processing are described.

Fig. 10 Block diagram of
the NeuPro architecture

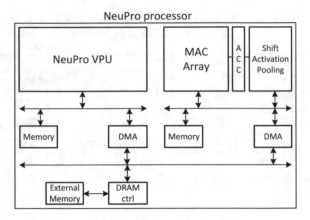

Since the architectures are proprietary and commercial, the information available is
somehow more restricted and, therefore, only a general description is provided for
each along with a few characterization parameters.

In [14] a hardware accelerator for audio and video processing for different scenarios, including edge, was announced to accelerate CNN processing. The architecture
uses a matrix processing engine (MPE) with 168×168 cores, where each core has
local memories, for data and weights, and a MAC unit. The chip delivers 16.8 TOPs
at 300 MHz with only 700 mW, an energy efficiency of 24 TOPs/W. The chip has
support for large network models: ResNet, MobileNet and ShiftNet.

NeuPro [27] is a neural network processor IP for general-purpose machine learning with a set of embedded target applications, like advanced driver-assistance system, drones, surveillance, among others. NeuPro has two main cores: (1) the NeuPro
Vector Processing Unit (VPU) and the NeuPro Engine (see Fig. 10).

The NeuPro VPU is a vector processor to control the NeuPro Engine and to execute
other layer functions not included in the processing engine. The NeuPro Engine is
the core processing of the NeuPro and executes all layers of a CNN with dedicated
hardware.

Four different configurations of NeuPro are available with different number of
MACs (512, 1024, 2048, 4096). Each MAC is programmable and can execute an
8×8 MAC in a single cycle, one 16×8 MAC in two cycles or one 16×16 MAC
in four cycles. Accumulations are implemented with 32-bit accumulators. There is a
dedicated module to calculate the activation function (it supports different activation
functions: ReLU, sigmoid, tangent hyperbolic, etc.). Pooling is also implemented
with a dedicated hardware module with support for average and maximum pooling
over 2×2 and 3×3 windows. The NeuPro Engine implements a stream pipeline
including the convolutional, the activation and the pooling layer without local memory access, reducing the required memory bandwidth to external memory.

The smallest configuration of NeuPro has a performance of 2 TOPs, while the
configuration with 4096 MACs has a performance of 12.5 TOPs running at 1.53
MHz.

Fig. 11 Block diagram of
the DNA processor

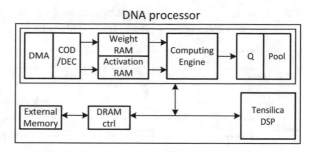

DesignWare EV6x IP processors from Synopsys [40] are used for embedded vision processing. The IP processor has a 32-bit scalar core, a vector DSP core and a convolutional neural network accelerator. The CNN accelerator is optional. The DSP core integrates a scalar 32-bit core and a 512-bit vector DSP. The EV6x processor can be configured with one core (EV61), two cores (EV62) or four cores (EV64).

The DSP core is able to execute one scalar operation and three vector operations per cycle. The vector DSP is configurable with up to 512 bits and different MAC bitwidths. The CNN core supports any CNN, including AlexNet, VGG16, GoogLeNet, Yolo, Faster R-CNN, SqueezeNet, and ResNet, and considers data quantization to 12 bits and 8 bits, to reduce memory bandwidth and power requirements. The configuration with the best performance delivers 4.5 TMACs and a power efficiency of 2 TMACs/W.

Tensilica DNA [3] is a processor from Cadence for deep neural network acceleration for IoT, smart home, drones, surveillance, ADAS and autonomous vehicles. The DNA processor integrates a Tensilica DSP and one hardware accelerator (see Fig. 11).

The engine only computes non-zero elements with a sparse compute engine. So, the architecture supports pruned networks. A compression/decompression unit is used to reduce weight data volume. The MAC engine can be configured with 256, 512 or 1024 MACs. Each MAC operates over 8-bit integers at maximum throughput. However, it also supports 16-bit integers at half rate and 16-bit float at one quarter rate. One DNA processor can have up to four engines with a total of 4096 MACs.

The configuration with 4096 MACs at 1 GHz and network pruning can achieve an average performance of 12 TMACs with an energy efficiency of 3.4 TMACs/W and a total power usage of 850 mW.

The Tensilica processor is used to control the accelerator and to execute all new layers not supported by the DNN hardware accelerator. The DNA accelerator can execute multiple types of network layers, including convolution, fully connected, LSTM, LRN and pooling.

Intel also entered the market of processors for deep learning at the edge with the Movidius Myriad X Vision Processing Unit (VPU) [21]. It is also a SoC with a 128-bit VLIW vector processor and a dedicated computing engine for deep neural network inference at the edge. The chip has 2.5 MBytes of internal memory with a memory bandwidth of 450 GBytes/s. The DNN computing engine achieves up to

Table 3 Main features of dedicated architectures for deep neural networks

	EV6x [40]	DNA [3]	Myriad X [21]	LightSpeeur [14]	NeuPro [27]
Year	2017	2017	2017	2018	2018
Tech. (nm)	16	16	16	22	16
Freq. (GHZ)	—	1	—	0.3	1.53
Data	8b/12b fix	8b/16b fix	8b/16b fix	—	8b/16b fix
Model	DNN	DNN	DNN	CNN	DNN
TOPs/W	4	6.8	—	24	—
Peak TOPs	9	24	4	16.8	12.5

1 TOPs and the whole processor has an overall peak performance of 4 TOPs. The MAC units support 16-bit floating-point and 8-bit and 16-bit fixed-point.

Table 3 resumes the main features of dedicated architectures for deep neural networks.

Like in the reconfigurable architectures, dedicated architectures have also adopted fixed-point representation with 8, 12 or 16 to represent data. In spite of being dedicated architectures, all of them implement flexible MAC units that can be splited in smaller MACs to increase the number of parallel MACs. Compared to the reconfigurable architectures, the fabrication technologies are better and therefore achieve higher performances and energy efficiencies.

Since dedicated architectures cannot be changed after fabrication, all try to support a vast set of neural network models. So, typically they have a module with massive parallelism to run MACs (the basic operations of most neural network models) and a separate generic processing unit to run all other specific functions.

The performance also varies among architectures since it depends on the number of cores of the architecture. The energy efficiency depends on the technology, data size and area efficiency of the architecture. These different features justify the differences in peak performance among the architectures.

Neural networks try to mimic the behavior of human brain. However, knowing that our brain has around 10^{11} neurons each with 10,000 connections to other neurons, the conclusion is that all DNN accelerators are quite far from the complexity of the brain network. Using the brain processing structure could be a better model to design neural networks with potentially more performance and energy efficiency. The subject is known as Neuromorphic Computing.

Different models were considered for neuromorphic computation, including the network models discussed above and a new type of neural network called Spiking Neural Network (SNN). In this model, the output of neurons is a sequence of pulses encoded in space and time. Many differential dynamic equations has been proposed to model the neuron behavior, but require complex implementations in hardware. A simplified algorithm that models the neuron behavior with good results and is easy to implement in hardware is the Leaky Integrate-and-Fire (LIF). In this algorithm pulses received by a neuron are weighted and summed to obtain the potential of a neuron, which is then added to the previous potential of the neuron. A pulse is generated at the output of the neuron if the potential exceeds a given threshold.

In a deep neural network neurons connect only to a specific subset of neurons. SNNs have no restrictions over the interconnection among neurons. Any neuron can connect to an arbitrary number of neurons. So, the complexity of a SNN hardware implementation resides in the interconnection structure. Usually, neurons are organized as hierarchical clusters of neurons interconnected with a network-on-chip (NoC) or crossbars.

In 2014 IBM produced the TrueNorth, a neuromorphic processor with 4096 cores, each simulating 256 neurons for a total of more than one million neurons. Each neuron is connected to other 256 neurons. a total of 268 millions of synapse-like connections with an average power of 70 mW.

Intel also announced in 2017 a neuromorphic chip for AI [7] that implements an asynchronous spiking neural network—Loihi. The many-core chip has 128 neuromorphic cores, each implements 1024 spiking neurons, and 130,000 synaptic connections. Loihi does not train because it can learn in real-time.

Neuromorphic computing has a great potential but results are needed to prove the advantages of this model. In terms of hardware, the main challenge now is how to design neuromorphic chips with more neurons and connections.

6 Inference at the Edge: Present and Future

Machine learning and in particular deep learning are still in its infancy with a large space of options, open issues and uncertainties. Some neural networks already achieve superior accuracy compared to humans for specific tasks. However, there are still many cognitive problems without useful or efficient models. Knowing the size of the models that solve only a few AI tasks, it is obvious that the next generations of DNN chips will have to be orders of magnitude higher than today.

This uncertainty has dictated the research and development of chips for edge inference. Since it is still not clear which model to follow, if there will be a general DNN model or new and more appropriate models, processing devices for DL include some form of programmability.

Coarse-grain reconfigurable technology was adopted by many to design DNN accelerators. There is some commitment with general deep neural network features, like MAC units and specific layer models, but the reconfigurability gives the architecture some flexibility to adapt to the DNN structure to optimize performance and energy efficiency. Configurable MACs with different weight and activation sizes, configurable datapaths to best adapt to a DNN model, are common to all coarse-grained accelerators.

While flexible, coarse-grained accelerators have a fixed structure optimized for a class of DNN. Many optimization techniques applied to reduce the complexity of DNN require specific modules or specific datapaths to implement it. These can be considered during the design of the coarse-grain structure but create area and throughput inefficiencies when not used to run a particular model. These limiting factors are solved with fine-grained reconfigurable computing, mostly FPGAs. An FPGA can be reconfigured to implement any hardware structure, limited only by the available resources of the device. This is the ideal structure to implement DNN models and associated optimization techniques and a fast platform to deploy new algorithms and techniques. The drawback is that fine-grain reconfigurable structures reduce operating frequencies, increase design area and increase energy consumption.

ASIC devices offer the best performance, area and power when compared to reconfigurable solutions and GPUs with a fixed hardware structure. Since there is no general neural model, companies delivering ASICs for deep learning have to stick to a model or a set of models, which limits its applicability. Two different approaches have been followed in the design of ASICs for AI tasks: (1) IP core for AI and (2) SoC chips for AI.

IP core accelerators are deployed to be integrated with digital signal processing systems. The core is dedicated for deep learning acceleration and can be configured with different number of processing units. SoC chips for AI integrate a dedicated unit for DNN acceleration and a high-performance processor to run all other DSP tasks and any layers not supported by the DNN accelerator. Some forms of flexibility are introduced in these ASIC solutions to deal with the evolution of deep learning models and its optimizations. At the core of the DNN processing, MAC units support different sizes of weights and activations. At the algorithmic level, different types of layers are implemented to increase the target DNN models that can be executed in the device.

In SoC solutions, a vector digital signal processor is usually used to run other digital signal processing algorithms and/or to run other DNN models or layers not supported by the dedicated engine.

The tasks and functionalities of edge devices are increasing faster than the technology improvement of computing devices. We are expecting more functionality at a lower cost and lower power. CPUs and GPUs are good proof-of-concept models but unable to fulfill the expected power requirements. Even FPGAs with better energy efficiency than CPU and GPUs at low energy consumption are unable to accomplish these energy goals in part because ASICs can do much better even knowing they have low adaptability to deep learning models.

SoC solutions with a dedicated DNN accelerator integrated with a high-performance processor is now the most appropriate architecture for deep learning inference on edge devices. Different edge applications require different DL applications with diverse complexity. SoC devices that can execute with highest performance and energy efficiency a broad range of DL applications becomes expensive when used in edge devices not requiring all the functionality of the chip. On the other side, deploying different edge devices for different DL applications reduces the volume production of each device, which increases its cost. The midterm seems to be the best practice case, where SoC edge devices are designed to cope with a set of related applications. However, since DL research is still in its infancy, new SoC devices are expected in the near future to cope with new DL models.

Following this line of thought, Xilinx has recently announced the Versal Adaptive Compute Acceleration Platform (ACAP) [50] that delivers three types of computing engines in the same SoC chip. It includes scalar software programmable engines, adaptable fine-grained reconfigurable engines and intelligent engines for domain specific tasks, like AI and DSP tasks, with computing and energy efficiency. It is a generalization of SoC ASIC solutions with an extra fine-grained reconfigurable engine to implement new accelerators not implemented as an intelligent dedicated engine. This bring FPGAs again into the edge computing arena and so it is expected that edge devices will be equipped with FPGAs and ASICs.

Cloud to edge migration of DL services is now restricted to inference. Chip designers of edge devices have concentrated their efforts on chips and IPs for inference. However, it is expected that in a near future, some form of incremental training will be necessary in edge devices also. Training requires other type of data representations, other kind of operations and it is more demanding in terms of computational effort. These requires other DL engines, more performing devices and maybe different training methods that facilitates the implementation of chips that integrate both inference and training.

7 Conclusions

Deep learning algorithms are drawing the attention of researchers and chip vendors since they are expected to be every where in our day lives. The first approaches for deep learning inference were based on cloud computing, but for some deep learning applications this task is migrating partially or totally to the edge.

Edge inference has brought new challenges, open issues and vendor opportunities that require new network models, network optimizations and new computing devices with high performance and energy efficiency.

In this chapter we reviewed the deep learning concept and what to expect from edge devices to run their inferences. Edge computing large DL models can be greatly improved using algorithmic and architectural optimizations. A brief overview of the most used optimizations was also described in this chapter.

We have also extensively described computing architectures, IP cores, processors and SoC devices for edge computing of DL models. Two main types of processing approaches were identified: reconfigurable computing with reconfigurable devices and dedicated computing with ASIC devices. Both present different tradeoffs in terms of performance, cost, energy and programmability. Both solutions are expected to appear in future edge devices.

As a final remark we have identified some design challenges and research directions in the long path of migration of DL inference to edge devices.

References

1. Albericio, J., Judd, P., Hetherington, T., Aamodt, T., Jerger, N.E., Moshovos, A.: Cnvlutin: ineffectual-neuron-free deep neural network computing. In: 2016 ACM/IEEE 43rd Annual International Symposium on Computer Architecture (ISCA), pp. 1–13 (2016). https://doi.org/10.1109/ISCA.2016.11
2. Aydonat, U., O'Connell, S., Capalija, D., Ling, A.C., Chiu, G.R.: An opencl™deep learning accelerator on arria 10. In: Proceedings of the 2017 ACM/SIGDA International Symposium on Field-Programmable Gate Arrays, pp. 55–64. FPGA'17, ACM, New York, NY, USA (2017). https://doi.org/10.1145/3020078.3021738
3. Cadence: Tensilica DNA Processor IP For AI Inference (Oct 2017). https://ip.cadence.com/uploads/datasheets/TIP_PB_AI_Processor_FINAL.pdf
4. Chen, Y., Krishna, T., Emer, J.S., Sze, V.: Eyeriss: An energy-efficient reconfigurable accelerator for deep convolutional neural networks. IEEE J. Solid-State Circuits **52**(1), 127–138 (2017). https://doi.org/10.1109/JSSC.2016.2616357
5. Courbariaux, M., Bengio, Y.: Binarynet: training deep neural networks with weights and activations constrained to +1 or −1. CoRR **abs/1602.02830** (2016). http://arxiv.org/abs/1602.02830
6. Courbariaux, M., Bengio, Y., David, J.: Binaryconnect: training deep neural networks with binary weights during propagations. CoRR **abs/1511.00363** (2015). http://arxiv.org/abs/1511.00363
7. Davies, M., Srinivasa, N., Lin, T., Chinya, G., Cao, Y., Choday, S.H., Dimou, G., Joshi, P., Imam, N., Jain, S., Liao, Y., Lin, C., Lines, A., Liu, R., Mathaikutty, D., McCoy, S., Paul, A., Tse, J., Venkataramanan, G., Weng, Y., Wild, A., Yang, Y., Wang, H.: Loihi: a neuromorphic manycore processor with on-chip learning. IEEE Micro **38**(1), 82–99 (2018). https://doi.org/10.1109/MM.2018.112130359
8. Flex Logic Technologies, Inc.: Flex Logic Improves Deep Learning Performance by 10X with new EFLX4K AI eFPGA Core (June 2018)
9. Fujii, T., Toi, T., Tanaka, T., Togawa, K., Kitaoka, T., Nishino, K., Nakamura, N., Nakahara, H., Motomura, M.: New generation dynamically reconfigurable processor technology for accelerating embedded AI applications. In: 2018 IEEE Symposium on VLSI Circuits, pp. 41–42 (June 2018). https://doi.org/10.1109/VLSIC.2018.8502438
10. Glorot, X., Bordes, A., Bengio, Y.: Deep sparse rectifier neural networks. In: Gordon, G., Dunson, D., Dudík, M. (eds.) Proceedings of the Fourteenth International Conference on Artificial Intelligence and Statistics. Proceedings of Machine Learning Research, vol. 15, pp. 315–323. PMLR, Fort Lauderdale, FL, USA (11–13 Apr 2011). http://proceedings.mlr.press/v15/glorot11a.html
11. Guo, K., Zeng, S., Yu, J., Wang, Y., Yang, H.: A survey of FPGA based neural network accelerator. CoRR **abs/1712.08934** (2017). http://arxiv.org/abs/1712.08934
12. Guo, S., Wang, L., Chen, B., Dou, Q., Tang, Y., Li, Z.: Fixcaffe: Training cnn with low precision arithmetic operations by fixed point caffe. In: APPT (2017)
13. Guo, K., Sui, L., Qiu, J., Yu, J., Wang, J., Yao, S., Han, S., Wang, Y., Yang, H.: Angel-eye: a complete design flow for mapping cnn onto embedded fpga. IEEE Trans. Comput. Aided Des. Integr. Circ. Syst. **37**(1), 35–47 (2018). https://doi.org/10.1109/TCAD.2017.2705069
14. Gyrfalcon Technology: Lightspeeur 2803S Neural Accelerator (Jan 2018)
15. Gysel, P., Motamedi, M., Ghiasi, S.: Hardware-oriented approximation of convolutional neural networks. In: Proceedings of the 4th International Conference on Learning Representations (2016)
16. Han, S., Mao, H., Dally, W.J.: Deep compression: compressing deep neural network with pruning, trained quantization and Huffman coding. In: 4th International Conference on Learning Representations, ICLR 2016, San Juan, Puerto Rico, 2–4 May 2016, Conference Track Proceedings (2016)
17. He, K., Zhang, X., Ren, S., Sun, J.: Deep residual learning for image recognition. CoRR **abs/1512.03385** (2015). http://arxiv.org/abs/1512.03385

18. Higginbotham, S.: Google Takes Unconventional Route with Homegrown Machine Learning Chips (May 2016)
19. Hinton, G.E., Salakhutdinov, R.R.: Reducing the dimensionality of data with neural networks. Science **313**(5786), 504–507 (2006). https://doi.org/10.1126/science.1127647
20. Hinton, G.E., Srivastava, N., Krizhevsky, A., Sutskever, I., Salakhutdinov, R.: Improving neural networks by preventing co-adaptation of feature detectors. CoRR **abs/1207.0580** (2012)
21. Intel: Intel Movidius Myriad X VPU (Aug 2017)
22. Judd, P., Albericio, J., Hetherington, T., Aamodt, T.M., Moshovos, A.: Stripes: Bit-serial deep neural network computing. In: 2016 49th Annual IEEE/ACM International Symposium on Microarchitecture (MICRO), pp. 1–12 (Oct 2016). https://doi.org/10.1109/MICRO.2016.7783722
23. Krizhevsky, A., Sutskever, I., Hinton, G.E.: Imagenet classification with deep convolutional neural networks. In: Proceedings of the 25th International Conference on Neural Information Processing Systems, vol. 1, pp. 1097–1105. NIPS'12, Curran Associates Inc., USA (2012)
24. LeCun, Y., Bengio, Y., Hinton, G.E.: Deep learning. Nature **521**(7553), 436–444 (2015). https://doi.org/10.1038/nature14539
25. LeCun, Y., Boser, B.E., Denker, J.S., Henderson, D., Howard, R.E., Hubbard, W.E., Jackel, L.D.: Handwritten digit recognition with a back-propagation network. In: Touretzky, D.S. (ed.) Advances in Neural Information Processing Systems 2, pp. 396–404. Morgan-Kaufmann (1990)
26. Li, F., Liu, B.: Ternary weight networks. CoRR **abs/1605.04711** (2016). http://arxiv.org/abs/1605.04711
27. Linley Group: Ceva NeuPro Accelerates Neural Nets (Jan 2018)
28. Liu, Y., Yang, C., Jiang, L., Xie, S., Zhang, Y.: Intelligent edge computing for iot-based energy management in smart cities. IEEE Netw. **33**(2), 111–117 (2019). https://doi.org/10.1109/MNET.2019.1800254. March
29. Liu, Z., Dou, Y., Jiang, J., Xu, J., Li, S., Zhou, Y., Xu, Y.: Throughput-optimized FPGA accelerator for deep convolutional neural networks. ACM Trans. Reconfig. Technol. Syst. **10**(3), 17:1–17:23 (Jul 2017). https://doi.org/10.1145/3079758
30. Markakis, E.K., Karras, K., Zotos, N., Sideris, A., Moysiadis, T., Corsaro, A., Alexiou, G., Skianis, C., Mastorakis, G., Mavromoustakis, C.X., Pallis, E.: Exegesis: Extreme edge resource harvesting for a virtualized fog environment. IEEE Communications Magazine **55**(7), 173–179 (July 2017). https://doi.org/10.1109/MCOM.2017.1600730
31. Mavromoustakis, C.X., Batalla, J.M., Mastorakis, G., Markakis, E., Pallis, E.: Socially oriented edge computing for energy awareness in iot architectures. IEEE Commun. Mag. **56**(7), 139–145 (2018). https://doi.org/10.1109/MCOM.2018.1700600. July
32. Nurvitadhi, E., Venkatesh, G., Sim, J., Marr, D., Huang, R., Ong Gee Hock, J., Liew, Y.T., Srivatsan, K., Moss, D., Subhaschandra, S., Boudoukh, G.: Can FPGAs beat GPUs in accelerating next-generation deep neural networks? In: Proceedings of the 2017 ACM/SIGDA International Symposium on Field-Programmable Gate Arrays, pp. 5–14. FPGA '17, ACM, New York, NY, USA (2017). https://doi.org/10.1145/3020078.3021740
33. Paszke, A., Chaurasia, A., Kim, S., Culurciello, E.: Enet: A deep neural network architecture for real-time semantic segmentation. CoRR **abs/1606.02147** (2016). http://arxiv.org/abs/1606.02147
34. Peres, T., Gonalves, A., Véstias, M.: Faster convolutional neural networks in low density FPGAs using block pruning. In: 15th Annual International Symposium on Applied Reconfigurable Computing (April 2019)
35. Qiao, Y., Shen, J., Xiao, T., Yang, Q., Wen, M., Zhang, C.: Fpga-accelerated deep convolutional neural networks for high throughput and energy efficiency. Concu. Comput. Pract. Exp. **29**(20), e3850–n/a (2017). https://doi.org/10.1002/cpe.3850,e3850cpe.3850
36. Sandler, M., Howard, A.G., Zhu, M., Zhmoginov, A., Chen, L.: Inverted residuals and linear bottlenecks: mobile networks for classification, detection and segmentation. CoRR **abs/1801.04381** (2018). http://arxiv.org/abs/1801.04381

37. Shin, D., Lee, J., Lee, J., Yoo, H.: 14.2 DNPU: An 8.1tops/w reconfigurable CNN-RNN processor for general-purpose deep neural networks. In: 2017 IEEE International Solid-State Circuits Conference (ISSCC), pp. 240–241 (Feb 2017). https://doi.org/10.1109/ISSCC.2017.7870350

38. Simonyan, K., Zisserman, A.: Very deep convolutional networks for large-scale image recognition. CoRR **abs/1409.1556** (2014). http://arxiv.org/abs/1409.1556

39. Suda, N., Chandra, V., Dasika, G., Mohanty, A., Ma, Y., Vrudhula, S., Seo, J.s., Cao, Y.: Throughput-optimized opencl-based fpga accelerator for large-scale convolutional neural networks. In: Proceedings of the 2016 ACM/SIGDA International Symposium on Field-Programmable Gate Arrays, pp. 16–25. FPGA '16, ACM, New York, NY, USA (2016). https://doi.org/10.1145/2847263.2847276

40. Synopsys: DesignWare EV6x Vision Processors (Oct 2017). https://www.synopsys.com/dw/ipdir.php?ds=ev6x-vision-processors

41. Sze, V., Chen, Y., Yang, T., Emer, J.S.: Efficient processing of deep neural networks: a tutorial and survey. Proc. of the IEEE **105**(12), 2295–2329 (2017). https://doi.org/10.1109/JPROC.2017.2761740. Dec

42. Szegedy, C., Ioffe, S., Vanhoucke, V.: Inception-v4, inception-resnet and the impact of residual connections on learning. CoRR **abs/1602.07261** (2016). http://arxiv.org/abs/1602.07261

43. Szegedy, C., Liu, W., Jia, Y., Sermanet, P., Reed, S.E., Anguelov, D., Erhan, D., Vanhoucke, V., Rabinovich, A.: Going deeper with convolutions. CoRR **abs/1409.4842** (2014). http://arxiv.org/abs/1409.4842

44. Szegedy, C., Vanhoucke, V., Ioffe, S., Shlens, J., Wojna, Z.: Rethinking the inception architecture for computer vision. CoRR **abs/1512.00567** (2015). http://arxiv.org/abs/1512.00567

45. Tractica: Deep Learning Chipsets (Jan 2019). https://www.tractica.com/research/deep-learning-chipsets/

46. Valdes Pena, M.D., Rodriguez-Andina, J.J., Manic, M.: The internet of things: the role of reconfigurable platforms. IEEE Ind. Electron. Mag. **11**(3), 6–19 (2017). https://doi.org/10.1109/MIE.2017.2724579. Sep

47. Venieris, S.I., Bouganis, C.: fpgaconvnet: Mapping regular and irregular convolutional neural networks on FPGAs. IEEE Trans. Neural Netw. Learn. Syst. 1–17 (2018). https://doi.org/10.1109/TNNLS.2018.2844093

48. Véstias, M., Duarte, R.P., Sousa, J.T.d., Neto, H.: Lite-CNN: a high-performance architecture to execute CNNs in low density FPGAs. In: Proceedings of the 28th International Conference on Field Programmable Logic and Applications (2018)

49. Wang, J., Lou, Q., Zhang, X., Zhu, C., Lin, Y., Chen., D.: Design flow of accelerating hybrid extremely low bit-width neural network in embedded FPGA. In: 28th International Conference on Field-Programmable Logic and Applications (2018)

50. Xilinx: Versal: the first adaptive compute acceleration platform (acap) (Oct 2018), https://www.xilinx.com/support/documentation/white_papers/wp505-versal-acap.pdf

51. Yin, S., Ouyang, P., Tang, S., Tu, F., Li, X., Zheng, S., Lu, T., Gu, J., Liu, L., Wei, S.: A high energy efficient reconfigurable hybrid neural network processor for deep learning applications. IEEE J. Solid-State Circ. **53**(4), 968–982 (2018). https://doi.org/10.1109/JSSC.2017.2778281. April

52. Yu, J., Lukefahr, A., Palframan, D., Dasika, G., Das, R., Mahlke, S.: Scalpel: Customizing DNN pruning to the underlying hardware parallelism. In: 2017 ACM/IEEE 44th Annual International Symposium on Computer Architecture (ISCA), pp. 548–560 (June 2017). https://doi.org/10.1145/3079856.3080215

53. Zhou, S., Ni, Z., Zhou, X., Wen, H., Wu, Y., Zou, Y.: Dorefa-net: training low bitwidth convolutional neural networks with low bitwidth gradients. CoRR **abs/1606.06160** (2016). http://arxiv.org/abs/1606.06160

Power Domain Based Multiple Access for IoT Deployment: Two-Way Transmission Mode and Performance Analysis

Tu-Trinh Thi Nguyen, Dinh-Thuan Do, Imran Khan, George Mastorakis, and Constandinos X. Mavromoustakis

Abstract As one of the promising radio access techniques in two-way relaying network, we consider Power Domain based Multiple Access (PDMA) and such PDMA is effective way to deploy in 5G communication networks. In this paper, a cooperative PDMA two-way scheme with two-hop transmission is proposed to enhance the outage performance under consideration on how exact successive interference cancellation (SIC) performs at each receiver. In the proposed scheme, performance gap of two NOMA users are examined, and power allocation factors are main impairment in such performance evaluation. In order to reveal the benefits of the proposed scheme, we choose full-duplex at relay to improve bandwidth efficiency. As important result, its achieved outage probability is mathematical analyzed with imperfect SIC taken into account. Our examination shows that the proposed scheme can significantly outperforms existing schemes in terms of achieving a acceptable outage probability, i.e. a lower outage probability given.

T.-T. Thi Nguyen
Industrial University of Ho Chi Minh City, Ho Chi Minh City, Vietnam
e-mail: nguyenthitutrinh@iuh.edu.vn

D.-T. Do (✉)
Wireless Communications Research Group, Faculty of Electrical and Electronics Engineering, Ton Duc Thang University, Ho Chi Minh City, Vietnam
e-mail: dodinhthuan@tdtu.edu.vn

I. Khan
Department of Electrical Engineering, University of Engineering & Technology Peshawar, P.O.B 814, Peshawar, Pakistan
e-mail: imran_khan@uetpeshawar.edu.pk

G. Mastorakis
Department of Management Science and Technology, Hellenic Mediterranean University, Agios Nikolaos, 72100 Crete, Greece
e-mail: gmastorakis@staff.teicrete.gr

C. X. Mavromoustakis
Department of Computer Science, University of Nicosia and University of Nicosia Research Foundation (UNRF), 46 Makedonitissas Avenue, P.O. Box 24005, 1700 Nicosia, Cyprus
e-mail: mavromoustakis.c@unic.ac.cy

© Springer Nature Switzerland AG 2020
G. Mastorakis et al. (eds.), *Convergence of Artificial Intelligence and the Internet of Things*, Internet of Things,
https://doi.org/10.1007/978-3-030-44907-0_10

241

1 Introduction

Power Domain based multiple access (PDMA) is considered new paradigm to share same source in non-orthogoanl manner. Such PDMA is proposed as one of the essential component for the fifth generation (5G) networks for improving system performance including the spectrum efficiency and providing massive connectivity [1, 2]. We recall two famous schemes, namely frequency division multiple access (FDMA) and time division multiple access (TDMA) in conventional orthogonal multiple access (OMA) schemes, the devices or users having poor channel conditions provide the low spectrum efficiency since the bandwidth resources are occupied by these users [3]. However, resource of network including the same frequency, time and code with different levels of power are charaterizations of communication in PDMA systems [4, 5]. According to [6], if the power allocation factors in PDMA can choose in reasonable manner, the system throughput and user fairness can be guaranteed. More particularly, removing the interferences of weak users is required to employ at users having strong channel conditions by using successive interference cancellation (SIC) techniques [7]. Therefore, the PDMA systems with the guaranteed fairness of users consequently outperform orthogoanl multiple access (OMA) in spectral efficiency and system throughput [8]. Various system models of cooperative relay [9–13] are proposed to combine with PDMA. New concept of cooperative PDMA has recently attracted growing interests to further enhance the performance of NOMA systems [14–16]. The first advantage of PDMA is to allocate more power to the users with low channel qualities, however, the performance improvement brought by PDMA is relatively constrained. In cooperative communications, with the assistance of several relays the source transmits information to the desired destinations. Such copperative aims to further improve the system performance. In [17], half-duplex (HD) and full-duplex (FD) relay are examined. In particular, the outage and throughput performance of cooperative PDMA systems was studied, where user near to the source works as using decode-and-forward (DF) strategy to enhance the performance of the far user [17]. In orther work, a cooperative communication scheme with NOMA was proposed in [18]. In such model, the users who are assigned high channel qualities can serve as relays to enhance the communication between the users who have poor channel conditions and the base station (BS). Furthermore, a cooperative NOMA transmission scheme with a dedicated relay was introduced in order to improve the communication reliability for users with poor channel conditions as recent work in [19]. Liu et al. [20], three users were served by BS, and the user nearest to the BS was operated as a relay to perform the transmissio between BS and the other two users and they introduced a collaborative NOMA assisted relaying system. To improve the secure performance of two NOMA users, Dinh-Thuan et al. [21] proposed relay selection schemes, where best link was proposed to decide whether the relaying transmission is necessary to serve the far user.

   In this paper, we focus on impact of imperfect SIC as performance analysis in a cooperative PDMA two-way relaying system with full-duplex scheme deployed at the relay. Different from the existing result in work [20], where the relay can serve to simultaneously transmit and receive symbols to far PDMA users by using power allocation coefficients in our proposed scheme.

The remainder paper is structured as follows: Sect. 2 presents the system model of our proposed PDMA scheme and related SNR calculation. Section 3 analytically characterizes the outage probability to show performance gap among two NOMA users of our proposed scheme. Then numerical results are illustrated in Sect. 4 to exhibit the theoretical expressions in Sect. 3. Finally, Sect. 5 concludes this work and introduces future works.

2   System Model

This study considers a scenario as in Fig. 1, in which two sources transmit signal to two destinations via intermediate node. Here, the base station (BS) operates as relay to assist a group of transmitters distributed in a place intends to communicate with their corresponding destinations distributed in another place. In this case, the poor channel conditions and/or physical obstructions are main reasons of nonexistence of the direct transmission between sources and destinations. In such a situation, perfect DF mode is required to decode signal exactly. The first group contains U_1, U_3 while U_2, U_4 belong to the second group and these users employ NOMA scheme. In this scenario, signal processing procedure happens in two hop of the relaying network cor-

Fig. 1 System model of Two-way transmission mode in Power Domain based Multiple Access

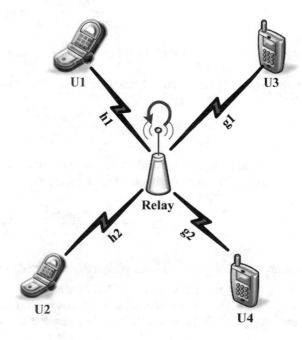

responding two time slots. To handle uplink transmission for such NOMA network, both two source nodes U_1, U_2 transmit signal x_1, x_2 simultaneously, respectively in the first hop and it is performed during the considered first time slot. We call h_1, h_2 as two channels which serve uplink NOMA. While transmission from the relay to destination in the second hop is considered as downlink NOMA and these channels are denoted by g_1, g_2. In FD scenario, to represent residual self-interference, we call f as self-interference channel among transmit/receive antenna pair equipped at relay. Regarding on imperfect SIC, interference channels are denoted as k_1, k_2. In the first phase, the power allocation factors are $\alpha_1 P$, $\alpha_2 P$ to distinguish power levels for two symbols (x_1, x_2) corresponding separated services for the first phase and $\alpha_3 P$, $\alpha_4 P$ are re-assigned for two symbols in the second transmission phase and P, P_r in this consideration are the total transmit power at the transmitters, relay respectively. Concurrently, the superimposed composite signal $\sqrt{P_r \alpha_3} x_1 + \sqrt{P_r \alpha_4} x_2$ is re-transmitted by the relay to U_3, U_4.

It is worth noting that such relay can operate in FD mode and two-hop transmission which leads to processing delay with small time epoch. The constraint of power allocation fractions in NOMA are $\alpha_1 + \alpha_2 = 1$, $\alpha_3 + \alpha_4 = 1$ and their assignment give impacts on system performance significantly. As important duty in NOMA, SIC is performed at both relay and destination to detect remaining signal and two cases need be considered (imperfect SIC and perfect SIC).

Therefore, the signal to interference plus noise ratio (SINR) to detect x_1 at relay can be formulated by

$$\gamma_{r_1} = \frac{\alpha_1 \rho_s |h_1|^2}{\alpha_2 \rho_s |h_2|^2 + \rho_r |f|^2 + 1}, \tag{1}$$

where $\rho_s = P/\sigma_0^2$ is transmit SNR of source. σ_0^2 is assumed as noise variance for all noise terms.

In perfect SIC situation, the signal to noise ratio (SNR) at relay node to decode x_2 is determined in FD mode by

$$\gamma_{r_2}^{pSIC} = \frac{\alpha_2 \rho_s |h_2|^2}{\rho_r |f|^2 + 1}, \tag{2}$$

In worse situation, imperfect SIC happens at FD/HD-assisted relay, these expressions in term of SNR can be recomputed respectively as

$$\gamma_{r_2} = \frac{\alpha_2 \rho_s |h_2|^2}{\alpha_1 \rho_s |k_1|^2 + \rho_r |f|^2 + 1}, \tag{3}$$

where $k_1 \sim \mathcal{CN}\left(0, \kappa \lambda_{h_1}\right)$ is interference term related to imperfect SIC; $\rho_s \triangleq P_s/\sigma_0^2$ and $\rho_r \triangleq P_r/\sigma_0^2$ are transmission SNR of source and relay node, respectively. Regarding to second hop transmission, the destination intends to decode its own data and hence the SINR can be achieved at destination U_3 by

$$\gamma_{U_3} = \frac{\alpha_3 \rho_r |g_1|^2}{\alpha_4 \rho_r |g_1|^2 + 1}. \tag{4}$$

Additionally, destination U_4 only recovers its data after successfully decompositing destination U_3 data and applying SIC. Then, the SINR and SNR at second user U_4 to detect the first user symbol for SIC purpose and to discover its own data are respectively given by

$$\gamma_{U_4 \leftarrow 1} = \frac{\alpha_3 \rho_r |g_2|^2}{\alpha_4 \rho_r |g_2|^2 + 1}, \tag{5}$$

The SNR and SINR at second destination U_4, in two cases including perfect SIC and imperfect SIC are respectively as

$$\gamma_{U_4}^{pSIC} = \alpha_4 \rho_r |g_2|^2, \tag{6}$$

and

$$\gamma_{U_4} = \frac{\alpha_4 \rho_r |g_2|^2}{\alpha_3 \rho_r |k_2|^2 + 1}. \tag{7}$$

3 Performance Evaluation: Outage Performance Analysis

3.1 The Outage Probability of the First User Pair

The first user pair is evaluated through outage probability. In particular, the exact outage probability of the first user pair is given by

$$OP_1 = \Pr\left(\gamma_{r_1} < \gamma_0^1 \cup \gamma_{U_1} < \gamma_0^1\right)$$
$$= 1 - \Pr\left(\frac{\alpha_1 \rho_s |h_1|^2}{\alpha_2 \rho_s |h_2|^2 + \rho_r |f|^2 + 1} \geq \gamma_0^1, \frac{\alpha_3 \rho_r |g_1|^2}{\alpha_4 \rho_r |g_1|^2 + 1} \geq \gamma_0^1\right). \tag{8}$$

Proposition 1 *The closed-form expression of the first user pair in term of outage probability can be given by*

$$OP_1 = 1 - \Pr\left(\frac{\alpha_1 \rho_s |h_1|^2}{\alpha_2 \rho_s |h_2|^2 + \rho_r |f|^2 + 1} \geq \gamma_0^1\right) \Pr\left(\frac{\alpha_3 \rho_r |g_1|^2}{\alpha_4 \rho_r |g_1|^2 + 1} \geq \gamma_0^1\right)$$
$$= 1 - \frac{1}{\lambda_f \lambda_{h_2}} \exp\left(-\frac{\gamma_0^1}{\alpha_1 \rho_s \lambda_{h_1}}\right) \left(\frac{\gamma_0^1 \rho_r}{\alpha_1 \rho_s \lambda_{h_1}} + \frac{1}{\lambda_f}\right)^{-1} \left(\frac{\gamma_0^1 \alpha_2}{\alpha_1 \lambda_{h_1}} + \frac{1}{\lambda_{h_2}}\right)^{-1}$$
$$\times \exp\left(-\frac{\gamma_0^1}{(\alpha_3 - \alpha_4 \gamma_0^1) \rho_r \lambda_{g_1}}\right). \tag{9}$$

Proof See in Appendix 5.

3.1.1 The Outage Probability of the Second User Pair with Imperfect SIC

In this case, γ_0^2 is assumed to be the predefined SNR threshold of destination U_4. In similar way, by consideration on outage condition, the second user pair meet outage case as follow

$$
O P_2^{ipSIC} = \Pr\left(\gamma_{r_1} < \gamma_0^1 \cup \gamma_{r_2} < \gamma_0^2 \cup \gamma_{U_4 \leftarrow 1} < \gamma_0^1 \cup \gamma_{U_4} < \gamma_0^2\right)
$$

$$
= 1 - \underbrace{\Pr\left(\frac{\alpha_1 \rho_s |h_1|^2}{\alpha_2 \rho_s |h_2|^2 + \rho_r |f|^2 + 1} \geq \gamma_0^1, \frac{\alpha_2 \rho_s |h_2|^2}{\alpha_1 \rho_s |k_1|^2 + \rho_r |f|^2 + 1} \geq \gamma_0^2\right)}_{\triangleq \mathcal{O}_1}
$$

$$
\times \underbrace{\Pr\left(\frac{\alpha_3 \rho_r |g_2|^2}{\alpha_4 \rho_r |g_2|^2 + 1} \geq \gamma_0^1, \frac{\alpha_4 \rho_r |g_2|^2}{\alpha_3 \rho_r |k_2|^2 + 1} \geq \gamma_0^2\right)}_{\triangleq \mathcal{O}_2}.
$$

(10)

Proposition 2 *The closed-form expression of the second user pair in term of outage probability can be computed by*

$$
O P_2^{ipSIC} = 1 - \mathcal{O}_1 \mathcal{O}_2,
$$

(11)

In which

$$
\mathcal{O}_1 = \frac{1}{\lambda_{h_2}} \left(\frac{\alpha_2 \gamma_0^1}{\alpha_1 \lambda_{h_1}} + \frac{1}{\lambda_{h_2}}\right)^{-1} \exp\left(-\frac{\gamma_0^1}{\alpha_1 \rho_s \lambda_{h_1}} - \frac{\gamma_0^2}{\alpha_2 \rho_s}\left(\frac{\alpha_2 \gamma_0^1}{\alpha_1 \lambda_{h_1}} + \frac{1}{\lambda_{h_2}}\right)\right)
$$

$$
\times \frac{1}{\lambda_f}\left(\frac{\rho_r \gamma_0^1}{\alpha_1 \rho_s \lambda_{h_1}} + \frac{\rho_r \gamma_0^2}{\alpha_2 \rho_s}\left(\frac{\alpha_2 \gamma_0^1}{\alpha_1 \lambda_{h_1}} + \frac{1}{\lambda_{h_2}}\right) + \frac{1}{\lambda_f}\right)^{-1}
$$

$$
\times \frac{1}{\lambda_{k_r}}\left(\frac{\alpha_1 \gamma_0^2}{\alpha_2}\left(\frac{\alpha_2 \gamma_0^1}{\alpha_1 \lambda_{h_1}} + \frac{1}{\lambda_{h_2}}\right) + \frac{1}{\lambda_{k_r}}\right)^{-1},
$$

(12)

and

$$
\mathcal{O}_2 = \frac{1}{\lambda_{k_2}}\left(\frac{\alpha_3 \gamma_0^2}{\alpha_4 \lambda_{g_2}} + \frac{1}{\lambda_{k_2}}\right)^{-1} \exp\left(-\frac{\gamma_0^2}{\alpha_4 \rho_r \lambda_{g_2}} - \Phi\left(\frac{\alpha_3 \gamma_0^2}{\alpha_4 \lambda_{g_2}} + \frac{1}{\lambda_{k_2}}\right)\right)
$$

$$
+ \exp\left(-\frac{\gamma_0^1}{(\upsilon_3 - \gamma_0^1 \upsilon_4)\rho_r \lambda_{g_2}}\right)
$$

$$
\times \left(1 - \exp\left(-\frac{1}{\lambda_{k_2}} \max\left(0, \frac{1}{\alpha_3 \rho_r}\left(\frac{\alpha_4 \gamma_0^1}{(\alpha_3 - \gamma_0^1 \alpha_4)\gamma_0^2} - 1\right)\right)\right)\right).
$$

(13)

Proof See in Appendix 5.

3.1.2 The Outage Probability of the Second User Pair with Perfect SIC

In this case, γ_0^2 is assumed to be the predefined SNR threshold of destination U_4. In similar way, by consideration on outage condition, the second user pair meets outage case as follow

$$
\begin{aligned}
OP_2^{pSIC} &= \Pr\left(\gamma_{r_1} < \gamma_0^1 \cup \gamma_{r_2}^{pSIC} < \gamma_0^2 \cup \gamma_{U_4 \leftarrow 1} < \gamma_0^1 \cup \gamma_{U_4}^{pSIC} < \gamma_0^2\right) \\
&= 1 - \underbrace{\Pr\left(\frac{\alpha_1 \rho_s |h_1|^2}{\alpha_2 \rho_s |h_2|^2 + \rho_r |f|^2 + 1} \geq \gamma_0^1, \frac{\alpha_2 \rho_s |h_2|^2}{\rho_r |f|^2 + 1} \geq \gamma_0^2\right)}_{\triangleq \zeta_1} \\
&\quad \times \underbrace{\Pr\left(\frac{\alpha_3 \rho_r |g_2|^2}{\alpha_4 \rho_r |g_2|^2 + 1} \geq \gamma_0^1, \alpha_4 \rho_r |g_2|^2 \geq \gamma_0^2\right)}_{\triangleq \zeta_2}.
\end{aligned}
\tag{14}
$$

Proposition 3 *The closed-form expression of the second user pair in term of outage probability can be computed by*

$$
OP_2^{pSIC} = 1 - \zeta_1 \zeta_2
\tag{15}
$$

In which

$$
\begin{aligned}
\zeta_1 =&\; \frac{1}{\lambda_{h_2}} \left(\frac{\alpha_2 \gamma_0^1}{\alpha_1 \lambda_{h_1}} + \frac{1}{\lambda_{h_2}}\right)^{-1} \exp\left(-\frac{\gamma_0^1}{\alpha_1 \rho_s \lambda_{h_1}} - \frac{\gamma_0^2}{\alpha_2 \rho_s}\left(\frac{\alpha_2 \gamma_0^1}{\alpha_1 \lambda_{h_1}} + \frac{1}{\lambda_{h_2}}\right)\right) \\
&\times \frac{1}{\lambda_f}\left(\frac{\rho_r \gamma_0^1}{\alpha_1 \rho_s \lambda_{h_1}} + \frac{\rho_r \gamma_0^2}{\alpha_2 \rho_s}\left(\frac{\alpha_2 \gamma_0^1}{\alpha_1 \lambda_{h_1}} + \frac{1}{\lambda_{h_2}}\right) + \frac{1}{\lambda_f}\right)^{-1}
\end{aligned}
\tag{16}
$$

and

$$
\zeta_2 = \exp\left(-\frac{1}{\lambda_{g_2}} \max\left(\frac{\gamma_0^1}{(\alpha_3 - \gamma_0^1 \alpha_4)\rho_r}, \frac{\gamma_0^2}{\alpha_4 \rho_r}\right)\right)
\tag{17}
$$

Proof See in Appendix 5.

4 Numerical Results

In this section, numerical results and Monte Carlo simulations are provided to validate the analytical results presented in this paper. Particularly, the parameters used in the simulations are set as follow. The Monte Carlo simulation results are averaged over 10^5 independent trials.

Figure 2 examines impact of transmit SNR at source on outage performance. It can be seen clearly that outage performance of the first user pair is lower than that of the second user pair. However, performance gap of the second user pair for ipSIC and pSIC is small. It shows that outage performance will be remain constant at high ρ as such ρ is greater than 35 (dB).

Figure 3 illustrates how target rate R_2 make performance of the second user pair change. We considers two cases of transmit SNR $\rho = 10, 40$ (dB) at source related to outage performance evaluation. It can be seen clearly that outage event happens at $R_2 = 5$. Besides, the simulations results match with analytical result very well. It indicated exactness of derived expressions. In this case, performance gap of the second user pair for ipSIC and pSIC is large at lower target rate R_2.

We explore impact of interference of FD on outage performance as in Fig. 4. It confirms how transmit SNR at source ρ make fluctuations in term of outage performance at the second user pair change. Four cases of transmit SNR $\rho = 10, 20, 30, 40$ (dB) are compared in ipSIC situation. It can be seen clearly that outage event happens

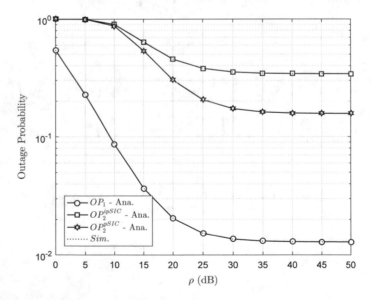

Fig. 2 Outage probability versus SNR with $\alpha_1 = 0.9$, $\alpha_2 = 0.1$, $\alpha_3 = 0.9$, $\alpha_4 = 0.1$, $\kappa = 0.001$, $\lambda_f = 0.01$, $R_1 = 1$ BPCU and $R_2 = 2$

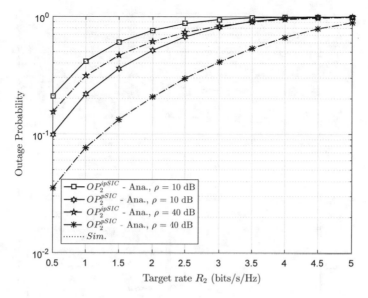

Fig. 3 Outage probability with $\alpha_1 = 0.8$, $\alpha_2 = 0.2$, $\alpha_3 = 0.8$, $\alpha_4 = 0.2$, $\kappa = 0.003$, $\lambda_f = 0.01$ and $R_1 = 0.5$ BPCU

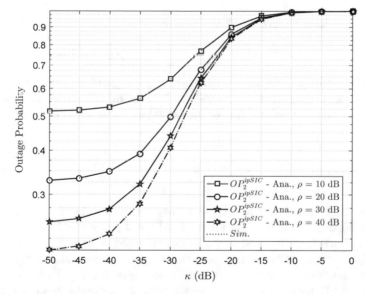

Fig. 4 Outage probability with $\alpha_1 = 0.8$, $\alpha_2 = 0.2$, $\alpha_3 = 0.8$, $\alpha_4 = 0.2$, $R_1 = 0.5$ BPCU, $R_2 = 2$ BPCU and $\lambda_f = 0.01$

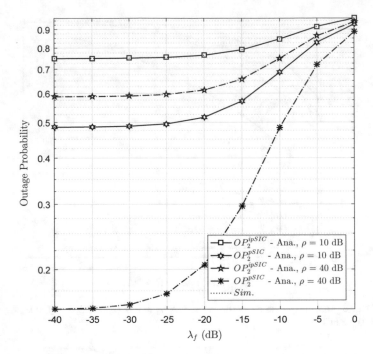

Fig. 5 Outage probability with $\alpha_1 = 0.8$, $\alpha_2 = 0.2$, $\alpha_3 = 0.8$, $\alpha_4 = 0.2$, $R_1 = 0.5$ BPCU, $R_2 = 2$ BPCU and $\kappa = 0.003$

at $\kappa = -10 to 0$ (dB). In this experiment, performance gap of the second user pair for ipSIC is large at lower value of κ. Similarly, how strong of interference channel λ_f make impact on outage probability of the second user pair as in Fig. 5.

5 Conclusion

This paper analyzed a two-way NOMA network with hardware impairments. With regard to impairment of SIC, i.e. imperfect SIC and perfect SIC were used in the receiver to provide comparison study. To examine system performance, the outage probability was considered in this paper under varying various parameters. Specially, we derived the closed-form expressions for the outage probability. In order to evaluate the impact of FD impairment on the system at high SNRs, the simulation results were also further evaluated. It can be confirmed that the worse outage performance can be raised when the larger impairments FD level was observed. In future work, more users and more pair of group are need be investigated.

Appendix

Proof of Proposition 1

OP_1 can be obtained

$$OP_1 = 1 - \underbrace{\Pr\left(|h_1|^2 \geq \frac{\gamma_0^1}{\alpha_1 \rho_s}\left(\alpha_2 \rho_s |h_2|^2 + \rho_r |f|^2 + 1\right)\right)}_{\triangleq \mathcal{O}_3}$$

$$\times \underbrace{\left(1 - F_{|g_1|^2}\left(\frac{\gamma_0^1}{(\alpha_3 - \alpha_4 \gamma_0^1)\rho_r}\right)\right)}_{\triangleq \mathcal{O}_4} \tag{18}$$

It need be computed these outage expressions as below

$$\mathcal{O}_3 = \Pr\left(|h_1|^2 \geq \frac{\gamma_0^1}{\alpha_1 \rho_s}\left(\alpha_2 \rho_s |h_2|^2 + \rho_r |f|^2 + 1\right)\right)$$

$$= \Pr\left(1 - F_{|h_1|^2}\left(\frac{\gamma_0^1}{\alpha_1 \rho_s}\left(\alpha_2 \rho_s |h_2|^2 + \rho_r |f|^2 + 1\right)\right)\right) \tag{19}$$

After the implementation of the calculation, we have

$$\mathcal{O}_3(x, y) = E_{|h_1|^2}\left\{\Pr\left(|h_1|^2 \geq \frac{\gamma_0^1}{v_1 \rho_s}\left(\alpha_2 \rho_s |h_2|^2 + \rho_r |f|^2 + 1\right)\right)\right\}$$

$$= \int_0^\infty f_{|f|^2}(x)\, dx \int_0^\infty \exp\left(-\frac{\gamma_0^1}{\alpha_1 \rho_s \lambda_{h_1}}\left(\alpha_2 \rho_s y + \rho_r x + 1\right)\right) f_{|h_2|^2}(y)\, dy$$

$$= \frac{1}{\lambda_f \lambda_{h_2}} \exp\left(-\frac{\gamma_0^1}{\alpha_1 \rho_s \lambda_{h_1}}\right) \int_0^\infty \exp\left(-\left(\frac{\gamma_0^1 \rho_r}{\alpha_1 \rho_s \lambda_{h_1}} + \frac{1}{\lambda_f}\right) x\right) dx$$

$$\times \int_0^\infty \exp\left(-\left(\frac{\gamma_0^1 \alpha_2}{\alpha_1 \lambda_{h_1}} + \frac{1}{\lambda_{h_2}}\right) y\right) dy$$

$$= \frac{1}{\lambda_f \lambda_{h_2}} \exp\left(-\frac{\gamma_0^1}{\alpha_1 \rho_s \lambda_{h_1}}\right)\left(\frac{\gamma_0^1 \rho_r}{\alpha_1 \rho_s \lambda_{h_1}} + \frac{1}{\lambda_f}\right)^{-1}\left(\frac{\gamma_0^1 \alpha_2}{\alpha_1 \lambda_{h_1}} + \frac{1}{\lambda_{h_2}}\right)^{-1} \tag{20}$$

It is noted that, we have following results

$$\int_0^\infty \exp\left(-\left(\frac{\gamma_0^1 \rho_r}{\alpha_1 \rho_s \lambda_{h_1}} + \frac{1}{\lambda_f}\right) x\right) dx = \left(\frac{\gamma_0^1 \rho_r}{\alpha_1 \rho_s \lambda_{h_1}} + \frac{1}{\lambda_f}\right)^{-1}, \qquad (21)$$

$$\int_0^\infty \exp\left(-\left(\frac{\gamma_0^1 \alpha_2}{\alpha_1 \lambda_{h_1}} + \frac{1}{\lambda_{h_2}}\right) y\right) dy = \left(\frac{\gamma_0^1 \alpha_2}{\alpha_1 \lambda_{h_1}} + \frac{1}{\lambda_{h_2}}\right)^{-1}, \qquad (22)$$

$$f_{|h_2|^2}(y) = \frac{1}{\lambda_{h_2}} \exp\left(-\frac{y}{\lambda_{h_2}}\right), \qquad (23)$$

$$f_{|f|^2}(x) = \frac{1}{\lambda_f} \exp\left(-\frac{x}{\lambda_f}\right). \qquad (24)$$

To further computation, we have

$$\mathcal{O}_3 = \frac{1}{\lambda_f \lambda_{h_2}} \exp\left(-\frac{\gamma_0^1}{\alpha_1 \rho_s \lambda_{h_1}}\right) \left(\frac{\gamma_0^1 \rho_r}{\alpha_1 \rho_s \lambda_{h_1}} + \frac{1}{\lambda_f}\right)^{-1} \left(\frac{\gamma_0^1 \alpha_2}{\alpha_1 \lambda_{h_1}} + \frac{1}{\lambda_{h_2}}\right)^{-1} \qquad (25)$$

and

$$\mathcal{O}_4 = 1 - F_{|g_1|^2}\left(\frac{\gamma_0^1}{(\alpha_3 - \alpha_4 \gamma_0^1) \rho_r}\right) = \exp\left(-\frac{\gamma_0^1}{(\alpha_3 - \alpha_4 \gamma_0^1) \rho_r \lambda_{g_1}}\right) \qquad (26)$$

Proof of Proposition 2

We have following equation as

$$\begin{aligned}
\mathcal{O}_1 &= \Pr\left(\frac{\alpha_1 \rho_s |h_1|^2}{\alpha_2 \rho_s |h_2|^2 + \rho_r |f|^2 + 1} \geq \gamma_0^1, \frac{\alpha_2 \rho_s |h_2|^2}{\alpha_1 \rho_s |k_1|^2 + \rho_r |f|^2 + 1} \geq \gamma_0^2\right) \\
&= \Pr\left(|h_1|^2 \geq \frac{\gamma_0^1}{\alpha_1 \rho_s}\left(\alpha_2 \rho_s |h_2|^2 + \rho_r |f|^2 + 1\right), |h_2|^2 \geq \frac{\gamma_0^2}{\alpha_2 \rho_s}\left(\alpha_1 \rho_s |k_1|^2 + \rho_r |f|^2 + 1\right)\right)
\end{aligned}$$

$$(27)$$

After placing the following variables $x = |f|^2$, $y = |h_2|^2$, $z = |k_1|^2$ and performing calculations, we have

$$
\mathcal{O}_1 = \exp\left(-\frac{\gamma_0^1}{\alpha_1 \rho_s \lambda_{h_1}}\right) \int_0^\infty \exp\left(-\frac{\rho_r \gamma_0^1}{\alpha_1 \rho_s \lambda_{h_1}} x\right) f_{|f|^2}(x) dx
$$

$$
\times \int_{\frac{\gamma_0^2}{\alpha_2 \rho_s}(\alpha_1 \rho_s z + \rho_r x + 1)}^\infty \exp\left(-\frac{\alpha_2 \gamma_0^1}{\alpha_1 \lambda_{h_1}} y\right) f_{|h_2|^2}(y) dy \int_0^\infty f_{|k_1|^2}(z) dz
$$

$$
= \frac{1}{\lambda_{h_2}}\left(\frac{\alpha_2 \gamma_0^1}{\alpha_1 \lambda_{h_1}} + \frac{1}{\lambda_{h_2}}\right)^{-1} \exp\left(-\frac{\gamma_0^1}{\alpha_1 \rho_s \lambda_{h_1}} - \frac{\gamma_0^2}{\alpha_2 \rho_s}\left(\frac{\alpha_2 \gamma_0^1}{\alpha_1 \lambda_{h_1}} + \frac{1}{\lambda_{h_2}}\right)\right)
$$

$$
\times \int_0^\infty \exp\left(-\left(\frac{\rho_r \gamma_0^1}{\alpha_1 \rho_s \lambda_{h_1}} + \frac{\rho_r \gamma_0^2}{\alpha_2 \rho_s}\left(\frac{\alpha_2 \gamma_0^1}{\alpha_1 \lambda_{h_1}} + \frac{1}{\lambda_{h_2}}\right)\right) x\right) f_{|f|^2}(x) dx \tag{28}
$$

$$
\times \int_0^\infty \exp\left(-\frac{\alpha_1 \gamma_0^2}{\alpha_2}\left(\frac{\alpha_2 \gamma_0^1}{\alpha_1 \lambda_{h_1}} + \frac{1}{\lambda_{h_2}}\right) z\right) f_{|k_1|^2}(z) dz
$$

$$
= \frac{1}{\lambda_{h_2}}\left(\frac{\alpha_2 \gamma_0^1}{\alpha_1 \lambda_{h_1}} + \frac{1}{\lambda_{h_2}}\right)^{-1} \exp\left(-\frac{\gamma_0^1}{\alpha_1 \rho_s \lambda_{h_1}} - \frac{\gamma_0^2}{\alpha_2 \rho_s}\left(\frac{\alpha_2 \gamma_0^1}{\alpha_1 \lambda_{h_1}} + \frac{1}{\lambda_{h_2}}\right)\right)
$$

$$
\times \frac{1}{\lambda_f}\left(\frac{\rho_r \gamma_0^1}{\alpha_1 \rho_s \lambda_{h_1}} + \frac{\rho_r \gamma_0^2}{\alpha_2 \rho_s}\left(\frac{\alpha_2 \gamma_0^1}{\alpha_1 \lambda_{h_1}} + \frac{1}{\lambda_{h_2}}\right) + \frac{1}{\lambda_f}\right)^{-1}
$$

$$
\times \frac{1}{\lambda_{k_1}}\left(\frac{\alpha_1 \gamma_0^2}{\alpha_2}\left(\frac{\alpha_2 \gamma_0^1}{\alpha_1 \lambda_{h_1}} + \frac{1}{\lambda_{h_2}}\right) + \frac{1}{\lambda_{k_1}}\right)^{-1}
$$

The following results can be given as

$$
\int_{\frac{\gamma_0^2}{\alpha_2 \rho_s}(\alpha_1 \rho_s z + \rho_r x + 1)}^\infty \exp\left(-\frac{\alpha_2 \gamma_0^1}{\alpha_1 \lambda_{h_1}} y\right) f_{|h_2|^2}(y) dy = \frac{1}{\lambda_{h_2}}\left(\frac{\alpha_2 \gamma_0^1}{\alpha_1 \lambda_{h_1}} + \frac{1}{\lambda_{h_2}}\right)^{-1}
$$

$$
\times \exp\left(-\frac{\gamma_0^2}{\alpha_2 \rho_s}\left(\frac{\alpha_2 \gamma_0^1}{\alpha_1 \lambda_{h_1}} + \frac{1}{\lambda_{h_2}}\right)\right) \tag{29}
$$

$$
\times \exp\left(-\frac{\alpha_1 \gamma_0^2}{\alpha_2}\left(\frac{\alpha_2 \gamma_0^1}{\alpha_1 \lambda_{h_1}} + \frac{1}{\lambda_{h_2}}\right) z\right)
$$

$$
\times \exp\left(-\frac{\rho_r \gamma_0^2}{\alpha_2 \rho_s}\left(\frac{\alpha_2 \gamma_0^1}{\alpha_1 \lambda_{h_1}} + \frac{1}{\lambda_{h_2}}\right) x\right),
$$

$$\int_0^\infty \exp\left(-\left(\frac{\rho_r \gamma_0^1}{\alpha_1 \rho_s \lambda_{h_1}} + \frac{\rho_r \gamma_0^2}{\alpha_2 \rho_s}\left(\frac{\alpha_2 \gamma_0^1}{\alpha_1 \lambda_{h_1}} + \frac{1}{\lambda_{h_2}}\right)\right)x\right) f_{|f|^2}(x)dx$$

$$= \frac{1}{\lambda_f}\left(\frac{\rho_r \gamma_0^1}{\alpha_1 \rho_s \lambda_{h_1}} + \frac{\rho_r \gamma_0^2}{\alpha_2 \rho_s}\left(\frac{\alpha_2 \gamma_0^1}{\alpha_1 \lambda_{h_1}} + \frac{1}{\lambda_{h_2}}\right) + \frac{1}{\lambda_f}\right)^{-1}, \tag{30}$$

$$\int_0^\infty \exp\left(-\frac{\alpha_1 \gamma_0^2}{\alpha_2}\left(\frac{\alpha_2 \gamma_0^1}{\alpha_1} + \frac{1}{\lambda_{h_2}}\right)z\right) f_{|k_1|^2}(z)dz$$

$$= \frac{1}{\lambda_{k_1}}\left(\frac{\alpha_1 \gamma_0^2}{\alpha_2}\left(\frac{\alpha_2 \gamma_0^1}{\alpha_1} + \frac{1}{\lambda_{h_2}}\right) + \frac{1}{\lambda_{k_1}}\right)^{-1}. \tag{31}$$

To further compute $\mathcal{O}_2$, we have

$$\mathcal{O}_2 = \Pr\left(\frac{\alpha_3 \rho_r |g_2|^2}{\alpha_4 \rho_r |g_2|^2 + 1} \geq \gamma_0^1, \frac{\alpha_4 \rho_r |g_2|^2}{\alpha_3 \rho_r |k_2|^2 + 1} \geq \gamma_0^2\right)$$

$$= \Pr\left(|g_2|^2 \geq \frac{\gamma_0^1}{(\alpha_3 - \gamma_0^1 \alpha_4)\rho_r}, |g_2|^2 \geq \frac{\gamma_0^2}{\alpha_4 \rho_r}\left(\alpha_3 \rho_r |k_2|^2 + 1\right)\right)$$

$$= \Pr\left(|g_2|^2 \geq \max\left(\frac{\gamma_0^1}{(\alpha_3 - \gamma_0^1 \alpha_4)\rho_r}, \frac{\gamma_0^2}{\alpha_4 \rho_r}\left(\alpha_3 \rho_r |k_2|^2 + 1\right)\right)\right)$$

$$= \Pr\left(|g_2|^2 \geq \frac{\gamma_0^2}{\alpha_4 \rho_r}\left(\alpha_3 \rho_r |k_2|^2 + 1\right), |k_2|^2 > \frac{1}{\alpha_3 \rho_r}\left(\frac{\alpha_4 \gamma_0^1}{(v_3 - \gamma_0^1 \alpha_4)\gamma_0^2} - 1\right)\right)$$

$$+ \Pr\left(|g_2|^2 \geq \frac{\gamma_0^1}{(\alpha_3 - \gamma_0^1 \alpha_4)\rho_r}, |k_2|^2 < \frac{1}{\alpha_3 \rho_r}\left(\frac{\alpha_4 \gamma_0^1}{(\alpha_3 - \gamma_0^1 \alpha_4)\gamma_0^2} - 1\right)\right)$$

$$= \underbrace{\Pr\left(|g_2|^2 \geq \frac{\gamma_0^2}{v_4 \rho_r}\left(v_3 \rho_r |k_2|^2 + 1\right), |k_2|^2 > \max\left(0, \frac{1}{v_3 \rho_r}\left(\frac{v_4 \gamma_0^1}{(v_3 - \gamma_0^1 v_4)\gamma_0^2} - 1\right)\right)\right)}_{\triangleq \Delta_1}$$

$$+ \underbrace{\Pr\left(|g_2|^2 \geq \frac{\gamma_0^1}{(v_3 - \gamma_0^1 v_4)\rho_r}, |k_2|^2 < \max\left(0, \frac{1}{v_3 \rho_r}\left(\frac{v_4 \gamma_0^1}{(v_3 - \gamma_0^1 v_4)\gamma_0^2} - 1\right)\right)\right)}_{\triangleq \Delta_2}$$

$$\tag{32}$$

If $|k_2|^2 > \frac{1}{v_3 \rho_r}\left(\frac{v_4 \gamma_0^1}{(v_3 - \gamma_0^1 v_4)\gamma_0^2} - 1\right) \Rightarrow \max\left(\frac{\gamma_0^1}{(v_3 - \gamma_0^1 v_4)\rho_r}, \frac{\gamma_0^2}{v_4 \rho_r}\left(v_3 \rho_r |k_2|^2 + 1\right)\right) = \frac{\gamma_0^2}{v_4 \rho_r}\left(v_3 \rho_r |k_2|^2 + 1\right)$. Else $|k_2|^2 < \frac{1}{v_3 \rho_r}\left(\frac{v_4 \gamma_0^1}{(v_3 - \gamma_0^1 v_4)\gamma_0^2} - 1\right)$

$\Rightarrow \max\left(\frac{\gamma_0^1}{(v_3 - \gamma_0^1 v_4)\rho_r}, \frac{\gamma_0^2}{v_4 \rho_r}\left(v_3 \rho_r |k_2|^2 + 1\right)\right) = \frac{\gamma_0^1}{(v_3 - \gamma_0^1 v_4)\rho_r}$.

It worth noting that, some important results can be achieved as

$$\Delta_1 = \Pr\left(|g_2|^2 \geq \frac{\gamma_0^2}{\alpha_4\rho_r}\left(\alpha_3\rho_r|k_2|^2 + 1\right), |k_2|^2 > \underbrace{\max\left(0, \frac{1}{\alpha_3\rho_r}\left(\frac{\alpha_4\gamma_0^1}{\left(\alpha_3 - \gamma_0^1\alpha_4\right)\gamma_0^2} - 1\right)\right)}_{\triangleq \Phi}\right)$$

$$= \int_{\Phi}^{\infty}\left(1 - F_{|g_2|^2}\left(\frac{\gamma_0^2}{\alpha_4\rho_r}\left(\alpha_3\rho_r x + 1\right)\right)\right) f_{|k_2|^2}(x)\, dx$$

$$= \frac{1}{\lambda_{k_2}}\exp\left(-\frac{\gamma_0^2}{\alpha_4\rho_r\lambda_{g_2}}\right)\int_{\Phi}^{\infty}\exp\left(-\left(\frac{\alpha_3\gamma_0^2}{\alpha_4\lambda_{g_2}} + \frac{1}{\lambda_{k_2}}\right)x\right)dx$$

$$= \frac{1}{\lambda_{k_2}}\left(\frac{\alpha_3\gamma_0^2}{\alpha_4\lambda_{g_2}} + \frac{1}{\lambda_{k_2}}\right)^{-1}\exp\left(-\frac{\gamma_0^2}{\alpha_4\rho_r\lambda_{g_2}} - \Phi\left(\frac{\alpha_3\gamma_0^2}{\alpha_4\lambda_{g_2}} + \frac{1}{\lambda_{k_2}}\right)\right)$$

(33)

and

$$\Delta_2 = \Pr\left(|g_2|^2 \geq \frac{\gamma_0^1}{\left(\alpha_3 - \gamma_0^1\alpha_4\right)\rho_r}, |k_2|^2 < \max\left(0, \frac{1}{\alpha_3\rho_r}\left(\frac{\alpha_4\gamma_0^1}{\left(\alpha_3 - \gamma_0^1\alpha_4\right)\gamma_0^2} - 1\right)\right)\right)$$

$$= \Pr\left(|g_2|^2 \geq \frac{\gamma_0^1}{\left(\alpha_3 - \gamma_0^1\alpha_4\right)\rho_r}\right)\Pr\left(|k_2|^2 < \max\left(0, \frac{1}{\alpha_3\rho_r}\left(\frac{\alpha_4\gamma_0^1}{\left(\alpha_3 - \gamma_0^1\alpha_4\right)\gamma_0^2} - 1\right)\right)\right)$$

$$= \left(1 - F_{|g_2|^2}\left(\frac{\gamma_0^1}{\left(v_3 - \gamma_0^1 v_4\right)\rho_r}\right)\right)\left(F_{|k_2|^2}\left(\max\left(0, \frac{1}{\alpha_3\rho_r}\left(\frac{\alpha_4\gamma_0^1}{\left(\alpha_3 - \gamma_0^1\alpha_4\right)\gamma_0^2} - 1\right)\right)\right)\right)$$

$$= \exp\left(-\frac{\gamma_0^1}{\left(v_3 - \gamma_0^1 v_4\right)\rho_r\lambda_{g_2}}\right)\left(1 - \exp\left(-\frac{1}{\lambda_{k_2}}\max\left(0, \frac{1}{\alpha_3\rho_r}\left(\frac{\alpha_4\gamma_0^1}{\left(\alpha_3 - \gamma_0^1\alpha_4\right)\gamma_0^2} - 1\right)\right)\right)\right)$$

(34)

Substituting (33) and (34) into (32), we obtain final formula of

$$\mathcal{O}_2 = \Delta_1 + \Delta_2$$

$$= \frac{1}{\lambda_{k_2}}\left(\frac{\alpha_3\gamma_0^2}{\alpha_4\lambda_{g_2}} + \frac{1}{\lambda_{k_2}}\right)^{-1}\exp\left(-\frac{\gamma_0^2}{\alpha_4\rho_r\lambda_{g_2}} - \Phi\left(\frac{\alpha_3\gamma_0^2}{\alpha_4\lambda_{g_2}} + \frac{1}{\lambda_{k_2}}\right)\right)$$

$$+ \exp\left(-\frac{\gamma_0^1}{\left(v_3 - \gamma_0^1 v_4\right)\rho_r\lambda_{g_2}}\right)$$

$$\left(1 - \exp\left(-\frac{1}{\lambda_{k_2}}\max\left(0, \frac{1}{\alpha_3\rho_r}\left(\frac{\alpha_4\gamma_0^1}{\left(\alpha_3 - \gamma_0^1\alpha_4\right)\gamma_0^2} - 1\right)\right)\right)\right)$$

(35)

It completes Proposition 2.

Proof of Proposition 3

We have following equation as

$$\zeta_1 = \Pr\left(\frac{\alpha_1\rho_s|h_1|^2}{\alpha_2\rho_s|h_2|^2 + \rho_r|f|^2 + 1} \geq \gamma_0^1, \frac{\alpha_2\rho_s|h_2|^2}{\rho_r|f|^2 + 1} \geq \gamma_0^2\right)$$

$$= \Pr\left(|h_1|^2 \geq \frac{\gamma_0^1}{\alpha_1\rho_s}\left(\alpha_2\rho_s|h_2|^2 + \rho_r|f|^2 + 1\right), |h_2|^2 \geq \frac{\gamma_0^2}{\alpha_2\rho_s}\left(\rho_r|f|^2 + 1\right)\right)$$

$$\tag{36}$$

After placing the following variables $x = |f|^2$, $y = |h_2|^2$ and performing calculations, we have

$$\zeta_1 = \exp\left(-\frac{\gamma_0^1}{\alpha_1\rho_s\lambda_{h_1}}\right)\int_0^\infty \exp\left(-\frac{\rho_r\gamma_0^1}{\alpha_1\rho_s\lambda_{h_1}}x\right)f_{|f|^2}(x)dx$$

$$\times \int_{\frac{\gamma_0^2}{\alpha_2\rho_s}(\rho_r x+1)}^\infty \exp\left(-\frac{\alpha_2\gamma_0^1}{\alpha_1\lambda_{h_1}}y\right)f_{|h_2|^2}(y)dy$$

$$= \frac{1}{\lambda_{h_2}}\left(\frac{\alpha_2\gamma_0^1}{\alpha_1\lambda_{h_1}} + \frac{1}{\lambda_{h_2}}\right)^{-1}\exp\left(-\frac{\gamma_0^1}{\alpha_1\rho_s\lambda_{h_1}} - \frac{\gamma_0^2}{\alpha_2\rho_s}\left(\frac{\alpha_2\gamma_0^1}{\alpha_1\lambda_{h_1}} + \frac{1}{\lambda_{h_2}}\right)\right) \tag{37}$$

$$\times \int_0^\infty \exp\left(-\left(\frac{\rho_r\gamma_0^1}{\alpha_1\rho_s\lambda_{h_1}} + \frac{\rho_r\gamma_0^2}{\alpha_2\rho_s}\left(\frac{\alpha_2\gamma_0^1}{\alpha_1\lambda_{h_1}} + \frac{1}{\lambda_{h_2}}\right)\right)x\right)f_{|f|^2}(x)dx$$

$$= \frac{1}{\lambda_{h_2}}\left(\frac{\alpha_2\gamma_0^1}{\alpha_1\lambda_{h_1}} + \frac{1}{\lambda_{h_2}}\right)^{-1}\exp\left(-\frac{\gamma_0^1}{\alpha_1\rho_s\lambda_{h_1}} - \frac{\gamma_0^2}{\alpha_2\rho_s}\left(\frac{\alpha_2\gamma_0^1}{\alpha_1\lambda_{h_1}} + \frac{1}{\lambda_{h_2}}\right)\right)$$

$$\times \frac{1}{\lambda_f}\left(\frac{\rho_r\gamma_0^1}{\alpha_1\rho_s\lambda_{h_1}} + \frac{\rho_r\gamma_0^2}{\alpha_2\rho_s}\left(\frac{\alpha_2\gamma_0^1}{\alpha_1\lambda_{h_1}} + \frac{1}{\lambda_{h_2}}\right) + \frac{1}{\lambda_f}\right)^{-1}$$

And to further examine ζ_2, it can be given by

$$\zeta_2 = \Pr\left(\frac{\alpha_3\rho_r|g_2|^2}{\alpha_4\rho_r|g_2|^2 + 1} \geq \gamma_0^1, \alpha_4\rho_r|g_2|^2 \geq \gamma_0^2\right)$$

$$= \Pr\left(|g_2|^2 \geq \frac{\gamma_0^1}{(\alpha_3 - \gamma_0^1\alpha_4)\rho_r}, |g_2|^2 \geq \frac{\gamma_0^2}{\alpha_4\rho_r}\right)$$

$$= \Pr\left(|g_2|^2 \geq \max\left(\frac{\gamma_0^1}{(\alpha_3 - \gamma_0^1\alpha_4)\rho_r}, \frac{\gamma_0^2}{\alpha_4\rho_r}\right)\right)$$

$$= \exp\left(-\frac{1}{\lambda_{g_2}}\max\left(\frac{\gamma_0^1}{(\alpha_3 - \gamma_0^1\alpha_4)\,\rho_r}, \frac{\gamma_0^2}{\alpha_4\rho_r}\right)\right)$$

(38)

It completes Proposition 3.

References

1. Ding, Z., Liu, Y., Choi, J., Sun, Q., Elkashlan, M., Chih-Lin, I., Poor, H.V.: Application of non-orthogonal multiple access in LTE and 5G networks. IEEE Commun. Mag. **55**, 185–191 (2017)
2. Liu, Y., Qin, Z., Elkashlan, M., Ding, Z., Nallanathan, A., Hanzo, L.: Nonorthogonal multiple access for 5G and beyond. Proc. IEEE **105**, 2347–2381 (2017)
3. Ding, Z., Lei, X., Karagiannidis, G.K., Schober, R., Yuan, J., Bhargava, V.: A survey on non-orthogonal multiple access for 5G networks: research challenges and future trends. IEEE J. Sel. Areas Commun. **35**, 2181–2195 (2017)
4. Liu, Y., Qin, Z., Elkashlan, M., Gao, Y., Hanzo, L.: Enhancing the physical layer security of non-orthogonal multiple access in large-scale networks. IEEE Trans. Wireless Commun. **16**, 1656–1672 (2017)
5. Liu, Y., Ding, Z., Elkashlan, M., Poor, H.V.: Cooperative nonorthogonal multiple access with simultaneous wireless information and power transfer. IEEE J. Select. Areas Commun. **34**, 938–953 (2016)
6. Ding, Z., Yang, Z., Fan, P., Poor, H.V.: On the performance of nonorthogonal multiple access in 5G systems with randomly deployed users. IEEE Signal Process. Lett. **21**, 1501–1505 (2014)
7. Saito, Y., Benjebbour, A., Kishiyama, Y., Nakamura, T.: Systemlevel performance of downlink non-orthogonal multiple access (NOMA) under various environments. IEEE 81st Vehicular Technology Conference (VTC Spring), pp. 1–5 (2015)
8. Ding, Z., Fan, P., Poor, V.: Impact of user pairing on 5G nonorthogonal multiple access downlink transmissions. IEEE Trans. Veh. Technol. **65**, 6010–6023 (2015)
9. Do, D.-T., Nguyen, H.-S., Voznak, M., Nguyen, T.-S.: Interference cancellation at receivers in cache-enabled wireless networks. Radioengineering **26**, 869–877 (2017)
10. Nguyen, X.-X., Do, D.-T.: Optimal power allocation and throughput performance of full-duplex DF relaying networks with wireless power transfer-aware channel. EURASIP J. Wireless Commun. Netw. **152** (2017)
11. Nguyen, X.-X., Do, D.-T.: Maximum harvested energy policy in full-duplex relaying networks with SWIPT. Int. J. Commun. Syst. (Wiley) **30** (2017)
12. Nguyen, T.L., Do, D.T.: A new look at AF two-way relaying networks: energy harvesting architecture and impact of co-channel interference. Ann. Telecommun. **72**, 669–678 (2017)
13. Nguyen, K.-T., Do, D.-T., Nguyen, X.-X., Nguyen, N.-T., Ha, D.-H.: Wireless information and power transfer for full duplex relaying networks: performance analysis. Radioengineering, 53–62 (2015)
14. Do, D.-T., Van Nguyen, M.-S.: NOMA in downlink SDMA with limited feedback: performance analysis and optimization. Wireless Personal Communications (Springer), pp. 1–20. Online First (2019)
15. Nguyen, T.-L., Do, D.T.: Exploiting impacts of intercell interference on SWIPT-assisted non-orthogonal multiple access. Wireless Commun. Mobile Comput. **2018**, 1–12 (2018)
16. Do, D.-T., Le, C.-B.: Application of NOMA in wireless system with wireless power transfer scheme: outage and ergodic capacity performance analysis. Sensors **18**, 3501 (2018)

17. Yue, X., Liu, Y., Kang, S., Nallanathan, A., Ding, Z.: Outage performance of full/half-duplex user relaying in NOMA systems. In: IEEE International Conference on Communications (ICC), Paris, France, pp. 1-6 (2017)
18. Ding, Z., Peng, M., Poor, H.V.: Cooperative non-orthogonal multiple access in 5G systems. IEEE Commun. Lett. **19**, 1462–1465 (2015)
19. Kim, J.-B., Lee, I.-H.: Non-orthogonal multiple access in coordinated direct and relay transmission. IEEE Commun. Lett. **19**, 2037–2040 (2015)
20. Liu, X., Wang, X., Liu, Y.: Power allocation and performance analysis of the collaborative NOMA assisted relaying systems in 5G. China Commun. **14**, 50–60 (2017)
21. Do, D.-T., Van Nguyen, M.-S., Hoang, T.-A., Voznak, M.: NOMA-assisted multiple access scheme for IoT deployment: relay selection model and secrecy performance improvement. Sensors **19**, 736 (2019)

Big Data Thinning: Knowledge Discovery from Relevant Data

Naji Shehab and Christos Anagnostopoulos

Abstract Using statistical learning theory and machine learning techniques surrounding the principles of Rival Penalised Competitive Learning (RPCL), this chapter proposes a novel approach aiming to aid *Big Data Thinning*, i.e., *analysing only the potential data sub-spaces and not the entire extensive data space. Data scientists, data analysts, IoT applications and Edge-centric services are in need for predictive modelling and analytics*. This is achieved by learning from past issued analytics queries and exploiting the analytics query access patterns over the large distributed data-sets revealing the most interested and important sub-spaces for further exploratory analysis. By analysing user queries and respectively *mapping* them into relatively small-scale predictive local regression models, we can yield higher predictive accuracy. This is done by thinning the data space and freeing it of irrelevant and non-popular data sub-spaces; thus, making use of less training data instances. Experimental results and statistical analysis support the research idea proposed in this work.

1 Introduction

Big data is a developing term that can be described as a large structured or unstructured body of data that has the ability to be mined for information, hidden patterns, correlations, and other advanced insights. With the use of machine learning techniques, such as predictive modelling and big data analytics, big data can be turned into big opportunities. However, analysing and creating large *global* predictive statistical learning models surrounding large volumes of data can be a difficult and resource-expensive task. This section shares the motivation and the rationale behind this chapter, and highlights the overall aims and hypotheses gathered in hopes of

N. Shehab (✉) · C. Anagnostopoulos
School of Computing Science, University of Glasgow, Glasgow G12 8QQ, UK
e-mail: naji.shehab@hotmail.com

C. Anagnostopoulos
e-mail: christos.anagnostopoulos@glasgow.ac.uk

© Springer Nature Switzerland AG 2020
G. Mastorakis et al. (eds.), *Convergence of Artificial Intelligence and the Internet of Things*, Internet of Things,
https://doi.org/10.1007/978-3-030-44907-0_11

introducing a new approach aimed at *thinning big data* through *knowledge discovery from relevant data*.

1.1 Motivation

Over the past few years, data has grown in enormous amounts; increasing both in size and growth rate. By estimating the world's technological capacity to store, communicate, and compute information, it was found that the world's data storage capacity increased at a compound annual growth rate of 25% per year between 1986 and 2007 [10]. Although not a new concept, the recent explosion in the amount of data generated to date is chiefly due to the rise of computers and technology capable of capturing it. We create as much data in two days as we did from the beginning of time until the year 2000. However, with the increase in data, came the increase in the demand for data analytics. The concept behind data analytics has been around for years; organisations and businesses dating back to the 1950s were using basic data analytics to discover explicit trends, informative conclusions, and supportive decisions. The more data you know, the more insights and more accurate prediction models can be generated. Hence, the term "Big data" arises; a term essentially referring to the collection of all data and the resultant ability to make use of it in several areas.

With "Big data" however, comes big responsibility to decode and analyse it. Although straightforward, the huge amounts of data produced over time are abstract and largely disconnected from their context. But relying solely on big data can result in inaccurate and misguided results that impact on business decisions and the production of innovation. Thus, navigating in this vast sea of data in the most efficient manner becomes extremely important in terms of cost, resources and overall efficiency of data collection and analysis [23].

Upon analysing the total work capacity of the world's installed computers servers, it was estimated that in 2008:

> The world's servers processed 9.57 Zettabytes of information, almost 10 to the 22nd power, or ten million million gigabytes [6].

Furthermore, learning from the past and taking cues from practical problem solving ideologies, it can be said that there is a clear need for a palette of methods that make use of big data in a refined manner through an interpretive lens. As more research delves into this topic, it has been repetitively found that certain sub-regions in big data could be irrelevant, valueless and too time-consuming to be practical. Studying the data ecosystem and thinning it to its most useful components will provide a more efficient framework for understanding user queries and returning results more quickly and more efficiently therefore requiring less resources that could otherwise be allocated to other more important data analysis.

The findings should make an important contribution to the field of Big Data, and Big Data Analyses [22].

1.2   Aims and Hypotheses

This proposal's main aim is to investigate the idea behind building *local* predictive models that root from user queries; where each predictive model respectively represents a query or a set of queries issued by IoT applications and/or data analysts in a data management ecosystem [17–21]. Each predictive model should be able to *predict the answer* of a specific data subspace, which is only pre-analysed due to its importance thus not explicitly accessing the data, or accessing *only* the *required* amount of *relevant* data [24–28]. By setting our focus around popular user queries, we can create higher predictive models that make use of less resources. To achieve this, the focus lies in two main research hypotheses.

- **Hypothesis 1**: Upon quantising a large-scale dataset, each respective cluster can be associated with an individual predictive local model. Given this, the local models should predict more accurately than its counterpart global model over the entire data space.
- **Hypothesis 2**: Upon observing user queries analytics, we can *learn* and make use of the principles of the RPCL and Learning Automata methodologies in order to extract the distributions and associations of the queries over the large-scale data space. Given this and the outcome of **Hypothesis 1**; we should be able to create respective local models over the past issued query sub-spaces providing high quality analytics.

1.3   Performance Assessment

Given **Hypotheses 1 and 2**, we are investigating, with the use of real data, the impact of the number of learned predictive models on the *scalability* of a data management system with the intention of requiring relatively smaller and more *relevant* data instead of a large volume of big and potentially irrelevant data.

1.4   Outline

The remaining sections in this chapter are structured as follows:

- **Section 2—Background and Relevant Research**: covers certain methodologies of machine learning that will be used throughout this chapter; providing the reader with background information to understand the process taken in the remaining chapters.
- **Section 3—Hypothesis 1**: begins by discussing the introductory approach taken to conduct, implement, and assess the *first* hypothesis. More specifically, it investigates the statistical properties of the dataset to be used throughout this chapter, and

explains a set of data analyses to be performed. It then goes into details explaining each step taken to implement the first hypothesis and presents the findings of the research conducted.

- **Section 4—Hypothesis 2**: begins by discussing the introductory approach taken to conduct, implement, and assess the *second* hypothesis. More specifically, it first informs the reader of a very high level design; formalising the architecture to be implemented. It then goes into details explaining each step taken to implement the design constructed and presents the findings of the research conducted.
- **Section 5—Conclusions and Future Work**: concludes the chapter by highlight-ing the findings of the research conducted, and discusses ideas for future work, where applicable.

2 Background and Relevant Research

This section is a summarisation of several research papers, books, and related machine learning algorithms. It provides background information and terminology used within the chapter. Section 2.1 concentrates on relevant competitive learning topics used, but focuses mainly on the **Rival Penalised Competitive Learning** algo-rithm. Section 2.2 provides an insight into **Learning Automata**. Finally, Sect. 2.3 introduces a smart cluster initialisation approach to be used throughout the quantising phases conducted.

2.1 Competitive Learning

Competitive learning (CL) is an adaptive form of unsupervised learning for data clustering analysis, in which nodes contest for the right to respond to a subset of the input data [14]. It can be widely applied in several vector quantisation processes such as classification, image and speech signal processing [11, 12], and data com-pression. A formally famous clustering algorithm, **k-means** [15], is a method of vector quantisation in which the input data set is partitioned into k clusters. Each cluster is represented by an initially seeded cluster head or mean centre, that adap-tively changes throughout the learning process. The algorithm computes the squared distance between each of the input vectors and the seeded centres, chooses the centre having the minimum distance (the winning centre), and moves the winning centre towards the respective input vector. Its high effectiveness, however, comes at a major cost of failure. It is well known that the **k-means** clustering algorithm has a crucial problem of selecting an appropriate k, the number of seeded cluster centres needed. This is a major drawback as the decision of k strongly affects the performance of the algorithm. Unfortunately, in many practical problems the pre-selection of k is a hard problem. Furthermore, the clustering algorithm heavily depends on the initial values of its seeded clusters. Inappropriate values could cause the "dead units" problem; that is, if a centre is initialised far away from the input data set in comparison with other centres, it may never win the competition. This results in the centre being redundant and never representing a cluster; hence, being classified as a dead unit. Focusing on the crucial problems described above, in hope of eliminating them, there have been several other advanced algorithms/techniques proposed such as: the **Frequency Sensitive Competitive Learning (FSCL)** algorithm [2], and the **Rival Penalised Competitive Learning (RPCL)** algorithm [1].

The **FSCL**, which was originally an extension of the **k-means** algorithm, showed notable improvement by introducing a new parameter, called the conscience (the relative winner frequency). Motivated by the earlier work of Grossberg [16], by introducing a conscience [9], each cluster centre holds a directly proportional relationship with its respective winning frequency. With this strategy implemented, the **FSCL** algorithm reduces the learning rate of frequent winners (reducing the chance to win the competition against other cluster centres); and, as a result, helps to avoid the dead units problem. It is evident that the conscience strategy and **FSCL** solves the dead units problem well, however, this is the case only when the number of clusters k in the input data set is known in advance by some prior knowledge [2]. This is the same problem that is addressed in the famous **k-means** clustering method: The performance of the clustering algorithm may be affected by the chosen value of k; if selected inappropriately, we may obtain very poor clustering results.

To tackle the selection of k, a new CL algorithm is proposed—**RPCL** [1]. This is done by utilising and adding a new mechanism into FSCL. For each input vector, the **RPCL** algorithm rewards the winning centre and penalises its rival (i.e., the second winner) with a de-learning rate that is smaller than that used by the winning centre. Interestingly so, experimental results show that **RPCL** surpasses and provides better convergence than the **k-means** and **FSCL** algorithm. This algorithm eliminates the "dead units" problem and preforms appropriate clustering without any prior knowledge of k.

2.1.1 The K-Means Algorithm

The classical iterative **CL** algorithm—**k-means**—calculates the distances between the randomly initialised centre vector set $w_{(1)}, \ldots w_{(k)}$ and a random input vector $\mathbf{x} \in \mathbb{R}_{(n)}$—where k represents the number clusters, and n represents the number of input vectors. After calculating the distances using a distance metric, such as the Euclidean distance, the centre $\mathbf{w_c}$ with the minimum distance is declared the winner:

$$\mathbf{w_c} = \arg\min_i \|\mathbf{x} - \mathbf{w_i}\|, i = 1, \ldots, k \tag{1}$$

The winning centre $\mathbf{w_c}$ is then updated by moving it closer towards the respective input vector $\mathbf{x}$; where λ represents a dynamic or constant learning rate:

$$\mathbf{w}_{c*} = \mathbf{w}_c + \lambda(\mathbf{x} - \mathbf{w}_c), \lambda \in (0, 1) \tag{2}$$

As a result, only the winning centre $\mathbf{w}_c$ is updated; where Eqs. (1) and (2) are often called the Winner-Take-All rule. This process is repeated until the algorithm converges (the learning rate λ tends to zero) or until some other stopping condition is met. While k-means is one of the simplest **CL** algorithms, there are other variations of **CL**. However, they are similar in concept and still have the "dead units" problem. To eliminate this we introduce a new mechanism, called the conscience.

2.1.2 The Frequency Sensitive Competitive Learning Algorithm (FSCL)

By focusing on eliminating the "dead units" problem from the classical **CL** algorithm, a new strategy is introduced by reducing the winning rate of frequent winners. Each centre is given a conscience that tracks the number of times it has won the competition against other centres and is penalised in proportion to its frequency of winning. Thus, all units participate in the quantisation (partitioning) of the data space. This can be achieved by modifying Eq. (1):

$$\mathbf{w}_c = \arg\min \lambda_i \|\mathbf{x} - \mathbf{w_i}\|, i = 1, \ldots, k \tag{3}$$

where λ_i represents the *conscience factor*, which is usually a small real number where $0 \leq \lambda_i \leq 1$, and is defined as:

$$\lambda_i = \frac{n_i}{\sum_{i=1}^{k} n_i} \tag{4}$$

where n_i is the cumulative sum of the number of occurrences the centre $\mathbf{w}_c$ has won the competition. After selecting the winning centre, the **FSCL** algorithm updates the centre by making use of Eq. (2) in similar fashion to **k-means**. This process is also repeated until the algorithm converges. In the **FSCL** algorithm, the competitive computing units are penalised in proportion to some function of the frequency of their winning (conscience). Because of their heavier conscience, high winning exemplars are discouraged from attacking new inputs, allowing all units to participate in the quantisation of the data space. As a result, this solves the dead units problem, however, this is the case only when the number of clusters k in the input data set is known in advance.

2.1.3 The Rival Penalised Competitive Learning Algorithm (RPCL)

By altering the Winner-Take-All rule, the **RPCL** algorithm adapts the approach of **FSCL** and performs appropriate clustering without any prior knowledge. This is done by *penalising* the winning centre's direct rival $\mathbf{w_r}$ (i.e., the second winner). The

RPCL algorithm finds the winning centre $\mathbf{w_c}$, similar to Eq. (3), in addition to a rival centre $\mathbf{w_r}$, and is defined as:

$$\mathbf{w_c} = \arg\min_i \lambda_i \|\mathbf{x} - \mathbf{w_i}\| : \mathbf{w_i} \in W, \tag{5}$$

$$\mathbf{w_r} = \arg\min_j \lambda_j \|\mathbf{x} - \mathbf{w_j}\| : \mathbf{w_j} \in W\backslash\{\mathbf{w_j}\}, \tag{6}$$

where both vectored centres are updated as such:

$$\mathbf{w_{c*}} = \mathbf{w_c} + \alpha_w(\mathbf{x} - \mathbf{w_c}) \tag{7}$$

$$\mathbf{w_{r*}} = \mathbf{w_r} + \alpha_r(\mathbf{x} - \mathbf{w_r}) \tag{8}$$

where α_w and α_r represent the learning rates in 0, 1 for the winner and rival units respectively in such a way where $r \neq w$. The set $W = \{\mathbf{w_i}\}$ refers to the set of all the representatives. In general practice it is found that $\alpha_w >> \alpha_r$ in most scenarios. This process can be visualised in the Fig. 1.

As shown in Fig. 1a, only one weighted centre is moved towards the cluster, while its direct rival is pushed back, away from the centre, by a certain kind of force. This

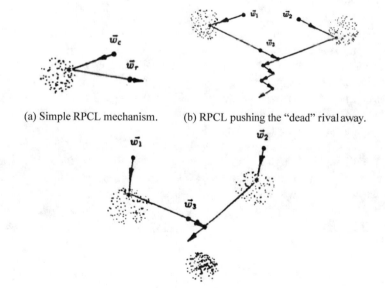

(a) Simple RPCL mechanism. (b) RPCL pushing the "dead" rival away.

(c) RPCL assisting the rival centre.

Fig. 1 Rival penalised competitive learning adapted from [1]. (a) The winning centre $\mathbf{w_c}$ is pushed towards the cluster, while its rival centre $\mathbf{w_r}$ is pushed away. (b) The rival centre $\mathbf{w_3}$ of both $\mathbf{w_1}$ and $\mathbf{w_2}$ is driven out along a zig-zag path. (c) $\mathbf{w_3}$, the rival of both $\mathbf{w_1}$ and $\mathbf{w_2}$, has an increased learning rate, pushing the rival centre closer towards its correct respective cluster

force is a counter balance to that generated by the conscience strategy of **FSCL**, which can be seen in Fig. 1b, c. In both cases, $\mathbf{w}_3$ is driven away along a zig-zag path as the random input samples are attracting the winning centres $\mathbf{w}_1$ and $\mathbf{w}_2$.

It is interesting to note that the **RPCL** mechanism, as shown above, has the ability to appropriately select the number of cluster representatives k by gradually pushing away extra units from the distribution of an input data set. Furthermore, as shown in Fig. 1c, the **RPCL** mechanism instinctively allows for a faster learning process: the de-learning of $\mathbf{w}_3$, caused by the learning of the winning centres $\mathbf{w}_1$ and $\mathbf{w}_2$, pushes $\mathbf{w}_3$ towards its correct respective cluster.

Overview

In Sect. 2.1 three competitive learning algorithms were discussed: **K-Means**, **Frequency Sensitive Competitive Learning (FSCL)**, and **Rival Penalised Competitive Learning (RPCL)**. Each algorithm is an adaptation of the next where **k-means** is known to be the simplest of the three. It aims to partition input data points into k clusters in which each input point belongs to the cluster with the nearest mean. Although effective, the **k-means** algorithm has some complexity in finding the optimal solution. Its clustering performance is heavily weighted on the pre-selection of k, and it does not tackle the "dead units" problem; that is, if a centre is initialised far away from the input data set in comparison with other centres, it may never win the competition, and as a result become a redundant centre.

The **FSCL** algorithm, that is an extension of **k-means**, is slightly more advanced. Adapting most of k-mean's implementation, it also makes use of a "conscience" parameter, also known as the relative winning frequency, whereby the learning rate of frequent winning centres are reduced; and as a result, the dead units problem is somewhat discoursed. This new strategy, however, was still heavily affected, and performed poor clustering, if the number of cluster k was selected inappropriately.

The **RPCL** algorithm, uses the technique of rewarding winning centres and penalising their rivals. By doing so it is able to automatically allocate an appropriate number of k cluster representatives over an input data set by gradually pushing away any extra units. Furthermore, since it is an adaptation of the **FSCL** algorithm, it also eliminates the "dead units" problem.

2.2 Learning Automata with Two Actions

By definition, a learning automaton is a branch of machine learning that dates back to the 1970s. In general, **Learning Automata** [13] has the ability of studying past experiences from an environment and selects an appropriate action based on that experienced learned. It possess a nature of adaptive decision-making and learns the optimal action through repeated interactions with its environment. All actions made are chosen according to a specific probability distribution $p_i \in \mathbf{P}$ (that provides the probabilities of occurrence of different possible outcomes) and is updated in the form of a reward or penalty.

Two different **Learning Automata** schemes are used in this research. The **Linear Reward-Penalty** scheme (L_{R-P}) and the **Linear Reward-Inaction** Scheme (L_{R-I}); both of which are defined in two actions. The latter only defers from the former whereby the probability distribution is only changed when there is a favourable result from the environment. Adapted from [13], the two actions that formalise the **Linear Reward-Penalty** (L_{R-P}) scheme are defined as such:

$$p_{i*} = p_i + \lambda(1 - p_i) : p_i \in \mathbf{P}, \tag{9}$$

$$p_{i*} = p_j(1 - \lambda) : p_j \in \mathbf{P} \backslash \{p_i\}, \tag{10}$$

where λ is a constant or dynamic learning factor that directly affects the significance of *reward* in the range ($0 < \lambda < 1$). Subsequently, the **Linear Reward-Inaction** Scheme (L_{R-I}) is another linear scheme formalised by two action and is defined by modifying (L_{R-I}):

$$p_{i*} = p_i(1 - \lambda') : p_i \in \mathbf{P} \tag{11}$$

$$p_{j*} = \frac{\lambda'}{k - 1} + p_j(1 - \lambda') : p_j \in \mathbf{P} \backslash \{p_i\}, \tag{12}$$

where λ' is another constant or dynamic learning factor that directly affects the significance of a *penalty* in the range ($0 < \lambda' < 1$).

Learning Automata will be used throughout this research as a form of a stochastic machine learning algorithm that selects the most optimal mapping of user query access patterns based on past interactions of issued queries with the environment of a query space. Upon the introduction of a change to the environment, a performance evaluation is performed that ultimately exhibits behaviour that will prove beneficial in selecting data subsets that are relevant to user queries. Hence, probabilistically and stochastically thinning a large amount of data.

2.3 The Improved Initialisation Algorithm: K-Means++

In most clustering algorithms, such as **k-means**, it is standard practice to choose the initial position of the selected number of k clusters uniformly at random from the distributed input data set. Although this methodology offers no reliable accuracy, its simplicity is very appealing in practice. This initialisation process of clusters can be classified as a NP-Hard problem, however [3] proposed a way of avoiding the sometimes poor clustering of the **k-means** algorithm by augmenting it with a simple randomised seeding technique, leading to a combined algorithm called **k-means++**. The intuition behind the **k-means++** algorithm is simple and its process can also be defined in the steps below:

1. Choose an initial centre c_1 uniformly at random from the distributed data set χ.
2. Where $D(x)$ is the shortest distance from a data point x to the closest centre already chosen. Choose the next centre c_i, with the probability of $\frac{D(x')^2}{\sum_{x \in \chi} D(x)^2}$ such that $c_i = x' \in \chi$.
3. Repeat step 2 until a total of k centres have been chosen.

With **k-means++** in use, we will be able to improve the quality of the local optimum, thus providing a smarter and more reliable form of initialisation.

Overview

This section familiarised the reader with the a number of research topics and ideas related to the chapter at hand. Throughout this chapter the **RPCL** algorithm will be used as the primary method for quantisation. **RPCL** was found to be an efficient and simple to implement competitive learning algorithm. It clusters data accordingly without the use of any prior knowledge and tackles the dead units problem. **Learning Automata** methodologies will be used to further investigate and build a stochastic system surrounding the study of query access patterns. Finally, **k-means++** was investigated and will be used as an efficient way to initially seed all quantisation processes.

3   Hypothesis 1

Upon quantising a large-scale dataset, each respective cluster can be associated with an individual predictive local model. Given this, the local models should predict more accurately than its counterpart global model over the entire data space.

To assert this conjecture, with the use of real data, this section investigates and compares the predictive performance of several *local* models, against the singular disparate *global* model.

3.1 Design

From a high-level point of view, the design flow for building Hypothesis 1 is fairly straightforward and intuitive. After performing a set of introductory analyses, a large-scale dataset is aimed to be quantised according to the **RPCL** algorithm, creating k number of clusters (where the number of k clusters is determined by the **RPCL** algorithm itself). By doing so, each cluster will represent a relatively smaller and more relevant set of data subspaces; holding a smaller percentage of the total amount of data. Subsequently, each data subspace is then split into training and testing subsets, converting them into predictive local models (where the number of *local* models is equal to k, the number of clusters formed). Each data subspace is split into two subsets of data in order to asses and evaluate each model's performance and generalisation capabilities on unseen data. This process is illustrated in Fig. 2 below as a set of layers.

To finalise Hypothesis 1, the predictive performance of each local model is measured against the base global model built, which in turn uses the entire data space. This evaluation is done using a continuous-valued set of estimation metrics such as **R-Squared (R^2)**, and **Root Mean Squared Error (RSME)**.

Fig. 2 The process of building local models through the illustration of multiple layers (Layer 1 shows a random input data pattern. Layer 2 shows the result of clustering an input pattern. Layer 3 shows that each cluster created is converted to a single local model; forming a set of local models)

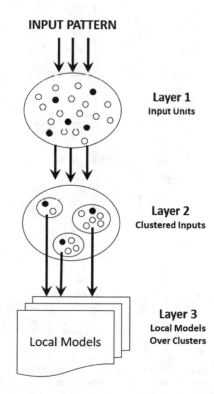

3.2   Implementation

3.2.1   Data Selection and Analysis

Throughout this chapter, a *large* combined cycle power plant dataset, that was collected over six years, was used. The dataset's features consist of hourly average
ambient variables: Atmospheric Temperature (AT), Ambient Pressure (AP), Relative
Humidity (RH), and Exhaust Vacuum (V) to predict the net hourly Electrical Energy
output (EP) of a power plant. To analyse the collected data further, **Exploratory
Data Analysis (EDA)** is performed. **EDA** is the process of understanding what the
data show and how it can be used to find patterns, relationships, or anomalies to form
subsequent analyses. The results of the **EDA** performed can be seen in Fig. 3 below.

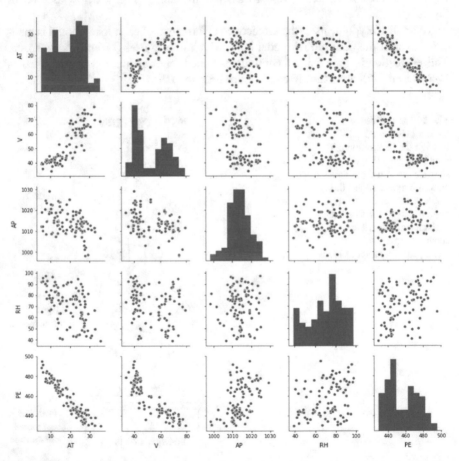

Fig. 3 A set of pairwise relationships in the dataset. Each variable in the dataset is shared in the
y-axis across a single row and in the *x*-axis across a single column. Axes are treated differently,
drawing a plot to show the univariate distribution of the data for the variable in that column

The plots shown display the distribution of all the single variables and their relationships with other variables, identifying trends and significant features with each
variable individually. The bar charts on each row represent the data distribution of
each respective ambient variable, while the remaining scatter plots represent relationships and patterns against every other ambient variable (where each point represents
an ambient variable pair). It can be seen, for example, that **AT** and **V** have a close-to
linear relationship (1st row, 2nd column). Furthermore, both of those variables also
hold a linear relationship towards the net hourly **EP** output (1st and 2nd row, 5th
column). As a result, by focusing on a simple 3-dimensional space, those three variables can be used in our extended investigations; where the two input variables **AT**
and **V** are used to predict values of **EP** in a linear fashion.

In addition to the exploratory analysis performed above, a set of scatter plots and
box plots over all variables were constructed to identify outliers (unusual observations that are far removed from the mass of data) in the dataset. The outliers were
investigated carefully and were removed, where it was found appropriate, to eliminate
the prospect of poor clustering.

To progress through onto the next stage, the *large* dataset was sampled to reduce the marginal cost of data processing. For future reference this dataset will
be represented as X. After establishing the selection of the input data, the dataset
X was ready to be quantised. This was done by implementing and making use of
the **k-means++**'s initialisation, and **RPCL** algorithm respectively. Thereafter, each
respective cluster was fitted into a linear regression model. The normalised sampled
dataset to be quantised can be seen in Fig. 4; where each point represents an input
pair.

3.2.2 Initial Seeding

Initialisation of cluster centres is critical to the quality of the local optima found.
As a result, to reduce the significance of sensitivity of initial seeding, we made use
of **k-means++**'s smart initialisation algorithm aiming to yield the most possible
effective improvement throughout the clustering phase. Although this is computationally costly relative to random initialisation, this process often results in more
rapid convergence of clusters.

Initial seeding was primarily achieved by uniformly choosing an input point at
random, from our dataset χ, to be the first cluster head representative (cluster centre)
c_1. Then, for each input data point x, the distance $D(x)$ to the nearest cluster head
was computed. Finally, the selection of the next cluster head was chosen with the
probability of x being proportional to $D(x)^2$ amongst the entire subspace of χ. By
doing so, every newly seeded cluster head representative was chosen in a stochastically biased manner by a set of previously selected cluster heads. This was then
repeated until k clusters centres were initialised over the data subspace, forming a
set of clusters C.

Throughout the initialisation scheme, a decision was to be made on the value of k,
the total number of cluster head representatives. Since no prior information existed

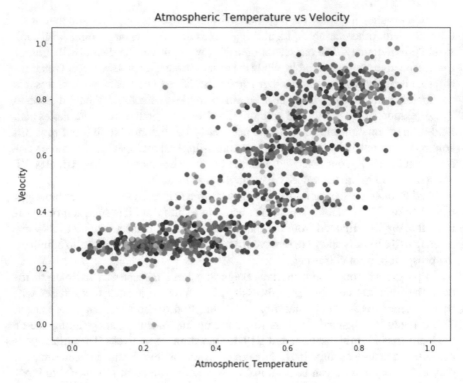

Fig. 4 A simple plot representation of the input points (AT, V) of the dataset that is to be quantised. The equivalent of layer 1 in Fig. 2

to predict an appropriate k at that stage, a large set of initial seeded clusters were constructed over a range of k values. Although this is sequentially expensive, the final resulting number of clusters would be decided during the actual quantising phase.

3.2.3 Data Quantisation Using RPCL

To quantise the dataset, shown in Fig. 4 above, we make use of the **RPCL** algorithm. Due to its nature, each resulting cluster would be in competition with one another, transitioning through a *learning* and *de-learning* phase.

For each introduction of a input data point x, only the first and second winning centres from the set of centres C were identified. Those winning centres were chosen according to Eqs. (5) and (6) stated above in Sect. 2.1.3; representing the closest and second closest cluster heads with respect to a specific input data point x. As a result of winning the competition, the first winning centre $\mathbf{w_c}$ transitions through a *learning* phase, and is attracted towards x; while the second winning centre $\mathbf{w_r}$ (i.e. the rival), transitions through a *de-learning* phase, and is instead repelled away from x. By setting the learning rates α_w and α_r to a a constant value of 0.1 and 0.05 respectively, the new positions of both centres $\mathbf{w_c}$ and $\mathbf{w_r}$ can be calculated. To do so, the level of attraction and repulsion of $\mathbf{w_c}$ and $\mathbf{w_r}$, respectively, were to be determined.

The level of attraction of $\mathbf{w_c}$ was determined by multiplying the learning factor α_w by the original distance calculated between $\mathbf{w_c}$ and x. While the level of repulsion of $\mathbf{w_r}$ is determined by multiplying the de-learning factor α_r by the original distance calculated between $\mathbf{w_r}$ and x. Subsequently, both winning centres $\mathbf{w_c}$ and $\mathbf{w_r}$ are updated according to Eqs. (7) and (8) respectively. This process was inherently repeated across all data inputs x. However, to progress through the quantising phase, an appropriate selection of k had to be investigated.

A.   Selection of K

RPCL has a unique characteristic where it holds the ability to select and estimate an appropriate number of clusters k. Due to its nature, this is achieved by pushing any extra unnecessary dead units away from the data subspace, resulting in some clusters attracting less data inputs. Subsequently, upon clustering over a set of k values, some cluster centres hold and represent a significantly small or negligible amount of input data points. To resolve this issue, a decision was made whereby any cluster holding a composition threshold of less than 5% of the total set of inputs is considered too small to be further examined. Taking this into consideration, in addition to RPCL's nature, the clustering algorithm is run on a set of k values until convergence and the highest valid k is selected (i.e. highest number of clusters possible, where each cluster represents more than 5% of the total set of inputs). The results of this convergence scheme over the dataset in question can be seen in Table 1.

It is evident, in both cases where $k = 10$ and $k = 11$, clusters exist below the accepted composition threshold . When $k = 11$ two clusters should be neglected, clusters 3 and 8, which hold a data composition of 2.6% and 4.6% respectively. When $k = 10$ only one cluster should be neglected, cluster 3, which holds a data composition of 4.7%. When $k = 9$, all cluster data compositions are above the selected threshold resulting in the highest number of valid cluster division. It is interesting to note that the RPCL algorithm, guides us towards an appropriate selection of k, where two clusters is neglected when $k = 11$, one cluster is neglected when $k = 10$, and none is neglected when $k = 9$ (i.e. RPCL is directing us towards a selection of k, where $k = (11 - 2) = (10 - 1) = (9 - 0) = 9$). As a result, we can confidently quantise our dataset using RPCL where $k = 9$.

B.   Results of RPCL Clustering

Having acquired the appropriate value of k over the selected dataset X, the results of the initial seeding process can be seen in Fig. 5, while the results of the quantisation process can be seen in Fig. 6. In both figures, the symbol represents each cluster head representative and is labelled accordingly from 1 through 9. From this, each cluster head representative can be seen to be initially positioned proportionally around the data map. Due to RPCL's nature, by introducing one data input at a time, each cluster head representative is put in competition against one another. The results of this competition is represented in Fig. 6 represents, showing a split of the entire data map into smaller subspaces. Where

Table 1 The resulting cluster data compositions, using the RPCL algorithm, when converged over a set of k values

Cluster Comp %	1	2	3	4	5	6	7	8	9	10	11
k = 9	20.1	9.3	5.1	14.5	11.8	7.8	9.0	7.7	14.7	–	–
k = 10	15.3	9.0	**4.7**	11.0	10.9	8.1	8.6	6.7	8.2	17.5	–
k = 11	14.7	7.1	**2.6**	11.1	10.5	8.0	7.2	**4.6**	8.3	17.8	8.1

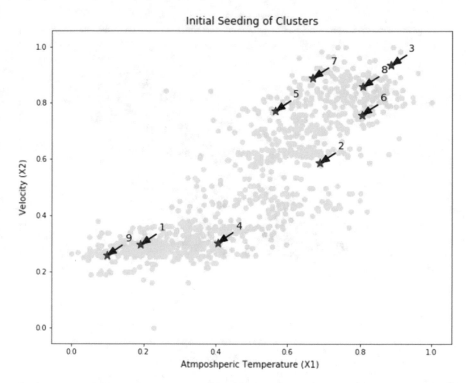

Fig. 5 A plot representation of the selected input data points with a set of 9 initial seeded cluster heads. Each seeded cluster head is represented by the symbol and is labelled accordingly from 1 through 9

each subspace is represented by its distinctive colour and unique cluster head representative.

3.3 Evaluation

In this section, the predictive performance of each model (local and global models) is examined in detail using a set of estimation metrics. This is followed by a formalised discussion. In general, when a model has a small and unbiased difference between the observed values (the validating or testing sets) and the model's predicted values, the model is said to fit the data well.

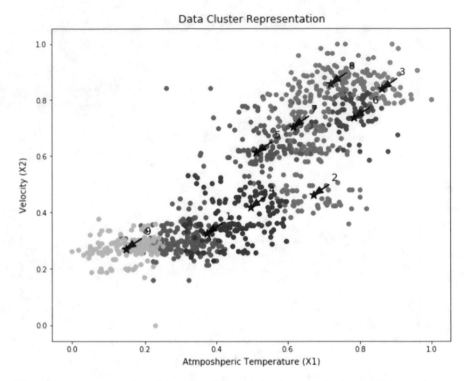

Fig. 6 A plot representation of the selected input data points quantised and split into 9 clusters. Each seeded cluster head is represented by the symbol and is labelled accordingly from 1 through 9; where each cluster head's data composition over the data map is identified by a unique colour

3.3.1 Building Predictive Global and Local Models

After quantising the large-scale dataset, each respective cluster can now be transformed into individual predictive local models. For each data subspace, that is respectively extracted from each cluster, a model is created with hopes of predicting well to new data. Forming a total of nine local models.

To build a predictive model, the extracted data is first split into training and testing data subsets, forming a total number of nine distinctive training and testings sets of data; where each of the nine training sets of data are then fitted into their own linear regression models. This approach, however, was found to have a caveat of overfitting or underfitting the predictive ability of each model. As a result, **K-Folds Cross Validation** is used instead, by splitting each respective subspace further into k different subsets (or folds). By doing so, each model will produce k predictive scores where the final score for each model will be presented by the average value against each of the folds.

Finally, in addition to creating nine local models, a base counterpart global model, built over the entire data space, is created in the same manner. The entire data space is first split into training and testing sets of data, and by using **K-Folds Cross Validation**, the best evaluated global model was found and used. To assert its predictive performance against the set of local models created, it is then evaluated against the *testing* data subsets of each corresponding local model.

3.3.2 Evaluation of Predictive Models

It is known that linear regression calculates an equation that minimises the distance between all data points and the fitted line created. To measure the success and performance of each model, a continuous-valued set of estimation metrics are used such as: **R-Squared** (R^2), and **Root Mean Squared Error (RSME)**.

Background Information

R-Squared is a statistical measure of how close the dependent variables (the real data) are to the fitted linear regression line. It represents an estimated strength of the relationship between each model and its respective response variable. Generally, the better the model fits the data, the higher the R-Squared value. This is formulated by:

$$R^2 = 1 - \frac{\sum (y - y')^2}{\sum (y - \bar{y})^2} \tag{13}$$

where y is the actual observed value, y' is the predicted value of y, and $\bar{y}$ is the mean of all the y values.

RMSE is a quadratics scoring rule that measures the average magnitude of errors in a set of predictions. **RMSE** minimises the conditional mean, producing negatively-oriented scores, which means the lower the values the better the model. It is formulated as the square root of the average of square differences between every prediction y and its respective set of observed value $\bar{y}$:

$$RMSE = \sqrt{\frac{1}{n} \sum_{n}^{i=1} (y_i - \bar{y}_i)^2} \tag{14}$$

Correspondingly, Table 2 shows the extracted performance of each predictive model in question, using the metrics specified above.

By analysing the extracted metrics measured above it can be seen that every local model, with the exception of local models 6 and 9, produced relatively lower **RMSE** scores when compared against its counterpart global model. With further inspection, every local model can be interpreted to produce, on average, **10%** less of the total error produced by its counterpart; thus, proving an increase in predictive power. This, however, is not the case when using the **R-Squared** statistical measurement.

Table 2 A table showing the results (the predictive performance) of each local model created according to their mean valued cross-validation R-Squared and RMSE measurements. These measurements are compared to the baseline global model

Local models	1	2	3	4	5	6	7	8	9	Global model
R^2%	33.89	27.96	48.10	36.72	48.44	43.10	26.99	26.63	32.34	91.33
RMSE %	6.12	6.84	5.34	6.65	5.27	7.58	5.98	5.51	7.40	7.00

It can be seen that the global model produced a very high and impressive score of **91.33%** while every local model produced a significantly smaller result, where **48.44%** was the highest score produced. This contradicting significance in results was not originally expected, however, the reasons to the findings above are further discussed in Sect. 3.3.3!

3.3.3   Discussion

The **R-Squared** metric, also known as the coefficient of determination, is only used when there exists some type of linear correlation between the variables it is being tested on (i.e. it only applies to linearly correlated variables). Upon further inspection of the **R-Squared** metric, it was found to hold a bias judgement in the evaluation constructed above. Given that there exists a linear correlation in the global model, the **R-Squared** metric was suitable, however, this was found not be the case when measuring the best line fit of each of the local models.

By quantising the entire dataset, we are *spraying* the data and as a result breaking the linearity that the dataset originally had. Subsequently, these *sprayed* data subsets were used to build our local models, and as a result, each respective local model cannot simply be determined by the best linear line fit. Hence making the **R-Squared** metric a very bias estimator by taking into consideration that the local models formed should produce a better linear fit (i.e. a better **R-Squared** score) over the subsets of data when compared against its counterpart global model (that evidently holds a much stronger linear relationship).

In addition to this, the global model exploits the fact that it holds a much larger set of data (approximately **90%** more than each local cluster created). Taking this into consideration, we are confidently able to perform a best line fit over such a *dense* and structured dataset. Since the metric in question is heavily weighted by the number of points, the global model achieved a very high **R-Squared** score. However, by quantising the dataset into smaller subset of data, we are reducing the density and mass of the input data used to build each local model. Thus strongly affecting the **R-Squared** measurement of all local models constructed.

To resolve the bias nature of the **R-Squared** metric, the introduction of a kernel regression could be exploited. It makes use of a non-parametric technique in statistics to estimate the conditional expectation of a random variable in a non-linear manner. This, however, is out of the scope of this chapter, and as a result, will be further investigated in future work!

3.3.4   Evaluation of Hypothesis

By ignoring the outputs of the **R-Squared** metric due its bias nature in this evaluation, and solely focusing on the **RMSE** results produced; it can be said that the set of local models achieved, on average, a **10% increase in predictive performance** when compared against the base global model. Furthermore, by examining Table 1 we are able to extract the data composition of each local model constructed. Since nine local models were created, it can be seen from the first row, where $k = 9$, that each local was constructed and composed of less than **15%** of the entire large-scale dataset. Thus supporting Hypothesis 1's conjecture: "Local models should predict more accurately than its counterpart global model over the entire data space."

Overview

This section investigated and asserted the conjecture of Hypothesis 1 by comparing the predictive performance of several quantised local models against a singular global model. It presented a set of introductory steps alongside decision-making analysis. Furthermore, implementation details regarding the two heavily used machine learn- ing methodologies (**RPCL, and k-means++**) were shared and discussed alongside any respective findings throughout the assertion. The next section concentrates on asserting the second hypothesis that exploits user behaviour to build more significant local models.

4 Hypothesis 2

Upon observing user queries analytics, we can learn and make use of the principles of the RPCL and Learning Automata methodologies in order to extract the distributions and associations of the queries over the large-scale data space. Given this and Hypothesis 1; we should be able to create respective local models over the past issued query spaces providing high quality analytics.

To assert this conjecture, with the use of real data and generated queries, this chapter exploits query analytics access patterns to reveal thereafter the most interested and important subspaces. Subsequently, upon investigating each query's corresponding association with a specific data region, queries are then mapped into in a single *local* model. The predictive performance of these *local* models over each query is then compared against the singular disparate *global* model.

4.1 Design

By analysing or simulating a number of user queries over a query space, each query can be classified under a specific query subspace. From this, queries of high importance and significance can be extracted. Subsequently, each query can then be mapped over a specific region of data (a subspace of data), and as a result, only necessary and relevant data are extracted. Overtime, with every introduction of a new query, the system built will go through a "learning phase" that increases the probability

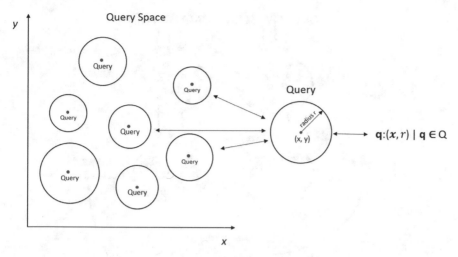

Fig. 7 A representation of a query space filled with a number of queries; where each query is represented as a 2-dimensional disc with a radius r

of a specific query (that is classified by a query subspace and its respective query head representative) to extract a specific region of data (that is classified by a data subspace or subspaces and their respective data head representatives). By doing so, only high probable data regions will be extracted/used, thus *thinning the data*.

4.1.1 Defining a Query

In a set of queries Q, a singular query $q \in Q$ can be represented as a real-valued vector that holds two components, a centre point x and a radius r. In terms of dimensionalities, if a query's corresponding centre point x is of dimension d, then any query q can be considered to be of dimension $(d + 1)$; where 1 represents the radius r (i.e. the centre point x is of one dimension less than that of the ambient space). This can be visualised as a hypersphere with a centre x and a radius r. However, to reduce the complexity of such a figure, in the case where a query q is 3-dimensional, it can instead be represented as a 2-dimensional disc with a radius r. This is illustrated in Fig. 7, that represents a set of queries Q in a query space.

4.1.2 Query-Data Space Projection

Upon defining a query space, Q, each query unit needs to be represented by a *representative query point*. This is done in order to group up more *similar* queries. To do so, a query space will be quantised, creating a number of query clusters; where each query cluster will be represented by the respective cluster head. The same procedure is performed over the data space, thus, acquiring two different sets of

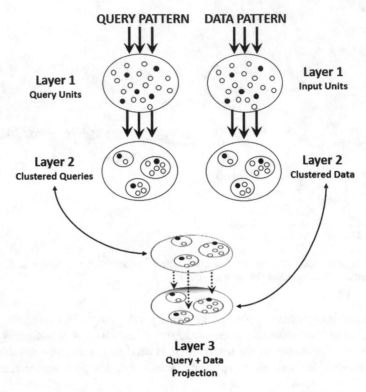

Fig. 8 The process of projecting a query space over a data space through the illustration of multiple layers. Layer 1 shows a random set of input query patterns (left) and input data patterns (right). Layer 2 shows the result of quantising both input patterns forming a clustered set of queries (left) and a clustered set of data (right). Layer 3 shows the intended projection, where specific quantised query subspaces can be mapped to specific quantised data subspaces

quantised/clustered spaces which allow projection to occur. Projection can exploit that there exists specific queries that can be correspondingly associated with specific subsets of data. This process is illustrated in Fig. 8.

5   Implementation

5.1   Generating Queries/Building the Query Space

Queries are usually issued by users, however, since they are not readily available for use in this research, we shall generate a set of queries instead, the so-called *query workload*. This is done in order to later **map a query sub-space to a set of data sub-spaces**.

Since the input data space being used is 2-dimensional, every input query must be of $(2 + 1)$ dimensions. In doing so every query $\mathbf{q}$ will, as a result, hold a 2-dimensional centre $\mathbf{x}$ and a radius r. Furthermore, by keeping the radius r constant throughout all queries, each query $\mathbf{q}$ can then be uniquely identified by its random positional component over the query space. Thus, to generate a set of queries, the 2-dimensional centre point $\mathbf{x}$, that can be represented as a positional (x, y) pair over the query space, is focused. This is achieved by building a number of 2-dimensional normal Gaussian distributions with varying mean value pairs; where each mean value pair represents the random positional (x, y) pair of a query $\mathbf{q}$. By selecting nine mean value pairs, and keeping the variance of each normal distribution constant, a set of nine query distributions were created.

In order to define a query space, only half of each of the nine generated query distributions are mapped. The remaining half will act as *"new"* queries issued by users. Those will be used to learn and exploit user behaviour in order to establish the most interested and important subspaces. The resulting query space generated can be visualised in Fig. 9.

Fig. 9 A simple plot representation of a generated query input pattern over the query space that is ready to be quantised. Each query input point is represented by its positional (x, y) value on the query space (cf. Layer 1 query units)

5.2 Query Quantisation Using RPCL

To group up more similar queries, the query space defined above will be quantised according to the **k-means++** and **RPCL** algorithm respectively, creating a number of query clusters and groups of query subspaces. As a result, the quantisation procedure that was used in Hypothesis 1 is to be adapted and applied in a similar fashion to the query space.

To further reduce the significance of sensitivity of initial seeding, the **k-means++**'s smart initialisation algorithm was used throughout the query clustering phase. However, unlike the data clustering phase, some prior information of the number of appropriate clusters k_q was known. To be exact, nine positional coordinates were used in the Gaussian distribution when generating the set of queries over the query space. Accordingly, nine corresponding query clusters were to be expected. The results of the initial seeding process can be seen in Fig. 10; which illustrates the formation of the nine query head representatives and their position proportionally around the generated query map.

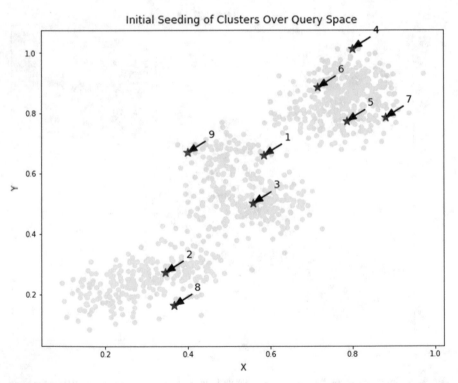

Fig. 10 A plot representation of the selected query points with a set of nine initial seeded cluster heads. Each seeded cluster head is represented by the symbol and is labelled accordingly from 1 through 9

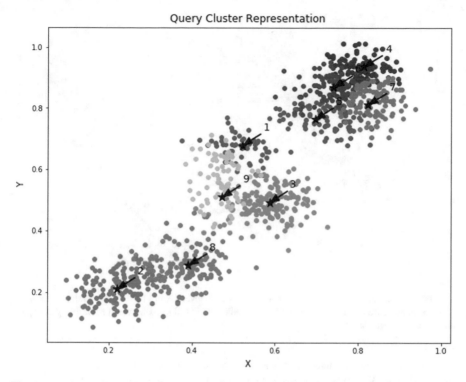

Fig. 11 A plot representation of input query points quantised and split into nine clusters. Each cluster head is represented by the symbol and is labelled accordingly from 1 through 9; where each cluster head's composition over the query map is identified by a unique colour (cf. Layer 2 clustered queries Fig. 9)

By re-implementing **RPCL** over the generated query map above, each query cluster head representative is similarly put in competition against one another. The results of this competition can be seen in Fig. 11.

5.2.1 Extraction of Relevant Data

In Sect. 3.2.3 and 4.2.2 above, a quantised query and data space of nine data subspaces were created respectively, thus establishing two different sets of quantised spaces. By acquiring both quantised spaces, a projection of one space over the other is possible and can be visualised in Fig. 12. This is done in order to exploit the fact there exists specific queries that can be correspondingly associated to specific subsets of data.

For example in Fig. 12 below, we could except that data representative/cluster 9 (identified in yellow on the left figure) would be extracted to build any sort of predictive analysis over any query that is associated with query representative/cluster number 2 (identified in orange on the right figure). Hence, providing a sense of mapping between each query and its associated subset of data.

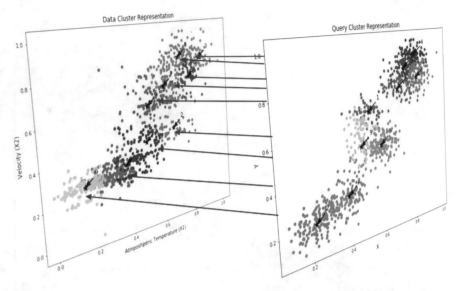

Fig. 12 The projection of the quantised query space retrieved from 11 over the quantised data space from 6 (lines represent how and what type of relations could exist between each query subspace and each data subspace)

However, our aim is to extract *only* the most significant and most interested subsets of data. To achieve this a series of steps have been constructed and defined:

- **Step 1**: Convert each query representative to a probability vector, using a probability distribution function, to represent the probable association of a certain data subspace to the specific query in question.
- **Step 2**: Make use of **Learning Automata** methodologies to study and analyse the behaviour of user issued queries by:

  - Classifying each query issued to the closest and second closest query representatives.
  - Identifying the data subspace associated with both query representatives.
  - *Rewarding* the closest query representative, for winning the competition, by increasing the probability of said query to be associated to said data subspace.
  - *Penalising* the second closest query representative, for losing the competition, by decreasing the probability of said query to be associated to said data subspace.

Step 1

In order to associate *only* high probable data subspaces to a specific query, each query representative in the quantised query space defined above will be converted to a probability vector **p** using a probability distribution function. Each probability vector **p** created will consist of n number of components/indices; where each component p_i (where $i = 1, \ldots, n$) will represent a data representative and initially hold a probable

Fig. 13 A bar chart representation of the initial state of the probability distribution of a query extracting data from a specific data subspace (bars represent the probable value of a query q_i to be associated to its respective data subspace/cluster representative)

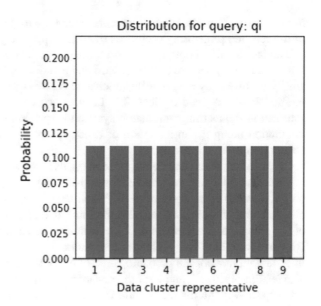

value of $\frac{1}{n}$. This is defined such that each individual component p_i **p** cumulatively sums up to 1.

Since there exists nine query and data representatives respectively over our quantised query and data spaces, a total of nine probability vectors are created; where each probability vector holds nine components, which, in turn, each hold an initial probable value of $\frac{1}{9}$. As a result, each query starts with an equal probable state of extracting specific data subsets over the data space. The initial state of each query probability vector can be visualised in Fig. 13.

Step 2

In Sect. 4.2.1, a set of queries Q were generated; half of which will now be used as a set of queries issued by users. In a similar competitive fashion to that of the **RPCL** algorithm, each query representative will be in competition for each query input vector introduced. When a query is issued, the closest and second closest query representatives on the query space are identified. This is done by measuring the Euclidean distance between each query representative and the input query's positional component. By doing so, we are able to identify the "winning" query representative q_w and its respective "rival" q_r. However, instead of pushing a query input vector towards or away from the identified query representatives, the probability vectors associated with the specific identified query representatives are respectively rewarded and penalised. More specifically, a single component in each of the selected probability vectors is rewarded/penalised.

To identify the specific single component (i.e. the data subspace associated with both query representatives), the winning query representative q_w is first projected onto the data space. Subsequently by using the same distance metric, the closest data representative d_i is identified; thus representing the respective component and index of the selected probability vectors as d_i and i respectively.

Having identified both query representatives q_w and q_r and their associated data representative d_i, we are able to reward and penalise q_w's and q_r's respective probability vectors $\mathbf{P}$ by adapting the stochastic behaviour of the **Learning Automata** methodologies defined in Sect. 2.2. In other words, by setting the random environment in the learning automaton system to represent the task at hand with every interaction being the introduction of a new query, we are able to select the most optimal action as a response.

To increase the favourable response of q_w, its respective probability distribution is rewarded by adapting the **Linear Reward-Penalty** (L_{R-P}) scheme; *increasing* q_w's probable value of component p_i in $\mathbf{P}$, and *decreasing* the probable value of every other component p_j in $\mathbf{P} : (i \neq j)$. Furthermore, to decrease the unfavourable response of q_r, its respective probability distribution is penalised by adapting the **Linear Reward-Inaction** (L_{R-I}) scheme; *decreasing* q_r's probable value of component p_i in $\mathbf{P}$, and *increasing* the probable value of every other component p_j in $\mathbf{P} : (i \neq j)$. The learning factors λ and λ' of the reward and penalty schemes were set at 0.5, and 0.1 respectively; giving the favoured response q_w a higher margin of reward against its rival q_r.

This process can be illustrated in Fig. 14; where a new query q^* is introduced and as a result two query representatives and their respective probability vectors are rewarded/penalised.

Correspondingly after this has been repeated over a wide set of user issued queries, each query representative will hold a certain probability density function over the set of data subspaces. It will reveal the most relevant subsets of data in the data space. The final results of this process (after rewarding and penalising multiple query probability vectors) can be visualised in Fig. 15. Subsequently, each query will then be converted into a single predictive local model.

5.3 Evaluation

In this section, the predictive performance of both sets of local query-based models and their respective counterpart set of global models are examined in detail using the same set of estimation metrics to that used in Sect. 3.3. This is followed by a discussion section.

5.3.1 Building Predictive Global and Local Models

In Hypothesis 1, nine local models were created to represent the nine quantised subspaces over the defined data space; where each model only used the corresponding data held strictly within its respective cluster. In a similar yet altered fashion, due to the fact that nine quantised *query* subspaces exist over the defined *query* space, a

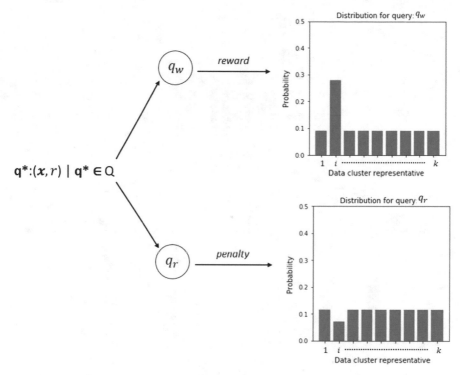

Fig. 14 A visualisation of the process and outcome of respectively rewarding the closest query representative q_w and penalising the 2nd closest query representative q_r

set of another nine local models will now be created. More specifically, nine query-based local models will be created. However, instead of building a predictive model strictly over a single data cluster, each local model is now built over the set of data clusters that the respective query in question is most associated with. This is achieved by investigating the probability distributions of each query and extracting the data clusters each query is associated with.

Upon further investigating each query's respective probability distributions, (as seen in Fig. 15), some distributions show more than just a single association of data regions. For example, in the resulting probability distribution in query representative 9, it can be said that two data regions/clusters, 4 and 5, are both relevantly needed to build the corresponding query-based local model. More specifically, roughly **56%** and **20%** of the query's relevantly associated data originates from data cluster representative 4, and 5 respectively. From this, two possible approaches of model building were formulated where any data cluster representative yielding anything higher than **10%** association will be considered *significantly sufficient* to be used and merged:

- **Option A**: Only use a subset of the merged sets, where each subset is extracted with direct relationship to its probable value of association. In other words, if a cluster c is associated to a query q with a probable value of p, only extract and

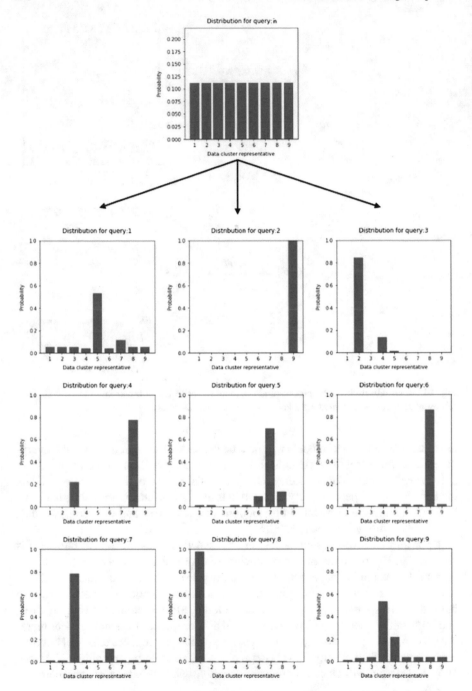

Fig. 15 The final representation of states of each query's probability distribution over specific data clusters

use its respective probable percentage (i.e. only $[p * 100\%$ of the data in cluster c) as input. This is done in order to further thin the amount of input data used to train our local models, however, for low probable values, this might introduce the concept of outliers and as a result affect the predictive performance of our models.

- **Option B**: Use the entire merged sets of data that are associated with a specific query, increasing the amount of input data needed to train our local models (when compared against option A), however, reassuring that all the necessary data is extracted to build a scoring predictive model.

Since option A might introduce the concept of outliers in our input dataset it was avoided; thus query-based local models were built according to option B. Subsequently, in similar fashion to Sect. 3.3, each subset of extracted data is split into **K-Folds Cross Validation** sets and the final scores for each model were presented by the average value against each of the folds.

Finally, in addition to creating nine query-based local models, the same base counterpart global model, that was trained and tested over the entire data space in Sect. 3.3 is used to assert its predictive performance against the set of local models created.

5.3.2   Evaluation of Predictive Models

Correspondingly, Table 3 shows the extracted performance of each predictive model in question according to the **RMSE** metric.

By analysing the extracted metrics measured above it can be seen that every local model produced, on average, **8%** less of the total error produced by its counterpart; thus, proving a slight increase in predictive power. Although this increase is not of large significance, the fact that these scores have been achieved with a much smaller subset of training input instances needs to be taken into consideration.

The extent as to how much less data is used can be examined in Table 4. It can be evaluated that each local model created made us of, on average, **83%** less data when compared against the global model built over the entire dataset.

Overview

This chapter investigated and asserted the conjecture of Hypothesis 2 by comparing the predictive performance of several quantised query-based local models against a set of respectively built global models. It presented a structured design of approach, and tackled a number of decision-making analyses. Moreover, it exploited the process of learning and analysing user behaviour, using **Learning Automata** methodologies to seek a more in-depth association between each query subspace and its corresponding data subspace/s. Research findings were investigated and shared throughout the assertion.

Table 3 Results (the predictive performance) of each local model created according to their mean valued cross-validation RMSE measurements. These measurements are compared to the baseline global model

Query-based models	1	2	3	4	5	6	7	8	9
Local RMSE %	5.29	7.72	7.22	6.55	5.72	3.19	6.93	6.37	6.38
Global RMSE %	5.26	8.04	7.34	7.32	5.52	5.18	8.21	6.34	6.72

Table 4 Data composition of each respective local model created in respect to the size of the entire global dataset

Query-based models	1	2	3	4	5	6	7	8	9
Data comp. %	20.8	14.7	23.8	12.8	16.7	7.7	12.9	20.1	26.3

6 Conclusions and Future Work

This chapter held a significant research aspect towards big data and big data analysis. It was formulated around two hypotheses, and made use of advanced machine learning techniques and statistical analyses to exploit a new approach to solving the big data issue. This section presents a final discussion summarising our goals and achievements, and further discusses ideas of future work where applicable.

6.1 *Generalisation of Findings*

This chapter presented a new approach to aid *Big Data Thinning*, by freeing the data space of any irrelevant and non-popular data subspaces. This has become evident through investigation of two main hypotheses.

In the first hypothesis (labelled above Hypothesis 1), a set of statistical evaluations were performed to exploit the fact that quantising a dataset and creating a set of small predictive local models performed better than any global model counterpart. By adapting a number of popular machine learning techniques, such as **RPCL** and **k-means++**, it was found that on average, the constructed local models produced a **10%** increase in predictive performance while making use of **90%** less data. This exerted the fact that smaller certain sub-regions in big data can be focused, reducing the total cost of building predictive models over an entire large-scale dataset, while still providing a high predictive score.

Influenced by the results seen in the research conducted around Hypothesis 1, the second hypothesis (above labelled Hypothesis 2) concentrated on adapting **Learning Automata** methodologies and schemes, through which we were able to build a stochastic learning environment of past user issued queries to exploit the most *popular* regions of data. To assert this conjecture, a set of statistical evaluations were performed over a number of (local) query-based predictive models. Each local predictive model was built over a specific data subspace, which was only pre- analysed due to its importance thus not accessing the entire data space; only the *required* amount of *relevant* data. In conclusion, it was found that on average each query-based local model produced an **8%** increase in predictive performance while making use of **83%** less data. This is not a significant increase in performance but it is a significant reduction in data use. By focusing on popular user queries, we were able to create higher predictive models that made use of less data resources; thus *thinning the big data*.

6.2 Future Work

Throughout the research conducted there existed some missed opportunities and further investigations that could otherwise be conducted in the near future.

6.2.1 Utilisation of a Dynamic RPCL Algorithm

RPCL was heavily used in this chapter, and was found to be a very effective algorithm for data clustering. Throughout the researching phase a more dynamic and efficient performing version of the original **RPCL** was found. Arthur and Vassilvitskii [7] proposes a "**Dynamically Penalised Rival Competitive Learning Algorithm (DPRCA)**" that moves undesired centres much faster than the original **RPCL** (showing results of better and faster convergence). This increase in algorithmic performance would come in very handy when dealing with big data!

6.2.2 Utilisation of a Scalable K-Means++

K-means++'s smart initialisation process was used throughout this chapter to improve the initial seeding process in all quantisation. Its major downside is its inherent sequential nature, which limits its applicability on larger sets of data. Bahmani et al. [8] proposes a "**Scalable K-Means**" approach, that drastically reduces the needed number of passes through the data, hence reducing the extra cost of proper initialisation.

6.2.3 Real World Application

This work dealt with generated queries which do not necessarily reflect the real world applications of these algorithms. Future work could investigate *real* sets of user-issued queries to understand the implications of the research done, and perhaps reduce the bias approach from generating a set of unreal queries.

Acknowledgements This research is funded by the EU-H2020 GNFUV Project (#Grant 645220) and the EU-H2020 MSCA INNOVATE Project (#Grant 745829).

References

1. Ahalt, S.C., Krishnamurthy, A.K., Chen, P., Melton, D.E.: Competitive learning algorithms for vector quantization. Neural Netw. **3**(3), 277–290 (1990). ISSN 0893-6080. https://doi.org/10.1016/0893-6080(90)90071-R
2. Anagnostopoulos, C., Kolomvatsos, K.: Predictive intelligence to the edge through approximate collaborative context reasoning. Appl. Intell. **48**(4), 966–991 (2018)
3. Anagnostopoulos, C., Triantafillou, P.: Efficient scalable accurate regression queries in In-DBMS analytics. In: IEEE International Conference on Data Engineering (ICDE), San Diego, CA, USA, 19–22 (2017)
4. Anagnostopoulos, C., Triantafillou, P.: Large-scale predictive modeling and analytics through regression queries in data management systems. International Journal of Data Science and Analytics (2018)
5. Anagnostopoulos, C., Triantafillou, P.: Query-driven learning for predictive analytics of data subspace cardinality. ACM Trans Knowl Discov. Data **11**(4), 47 (2017)
6. Anagnostopoulos, C., Savva, F., Triantafillou, P.: Scalable aggregation predictive analytics: a query-driven machine learning approach. Appl. Intell. **48**(9), 2546–2567 (2018)
7. Arthur, D., Vassilvitskii, S.: K-means++: the advantages of careful seeding. In: Proceedings of the Eighteenth Annual ACM-SIAM Symposium on Discrete Algorithms, SODA '07. Society for Industrial and Applied Mathematics, pp. 1027–1035. Philadelphia, PA, USA (2007). ISBN 978-0-898716-24-5. http://dl.acm.org/citation.cfm?id=1283383.1283494
8. Bahmani, B., Moseley, B., Vattani, A., Kumar, R., Vassilvitskii, S.: Scalable k-means++. Proc. VLDB Endow. **5**(7), 622–633 (2012). ISSN 2150-8097. https://doi.org/10.14778/2180912.2180915
9. Bohn, R., Short, J.E.: How much information? 2009 report on American consumers, vol. 01 (2009). https://www.researchgate.net/publication/242562463_How_Much_ Information_2009_Report_on_American_Consumers
10. Bohn, R., Short, J.E.: How much information? 2010 report on enterprise server information, p. 7 (2010). https://www.clds.info/uploads/1/2/0/5/120516768/hmi_ 2010_enterprisereport_jan_2011.pdf
11. Botoca, C., Budura, G., Miclau, N.: Competitive learning algorithms for data clustering. Facta Univ. Ser. Electron. Energetics **19**, 01 (2005). https://doi.org/10.2298/FUEE0602261B
12. Constandinos, X.M., George, M., Jordi, M.B.: Internet of Things (IoT) in 5G Mobile Technologies. Springer International Publishing AG (2016). ISSN 2196-7326. https://doi.org/10.1007/978-3-319-30913-2
13. Constandinos X.M. et al.: Socially-oriented edge computing for energy-awareness in IoT architectures. IEEE Commun. (2019)
14. Contandriopoulos, D., Brousselle, A.: Evaluation models and evaluation use. Evaluation **18**(1), 61–77 (2012). https://doi.org/10.1177/1356389011430371
15. Desieno, D.: Adding a conscience to competitive learning. In: IEEE 1988 International Conference on Neural Networks, vol. 1, pp. 117–124 (1988). https://doi.org/10.1109/icnn.1988.23839
16. Georgios, S. et al.: Elasticity debt analytics exploitation for green mobile cloud computing: an equilibrium model. IEEE Trans. Green Commun. Netw. (2019)
17. Grossberg, S.: Adaptive pattern classification and universal recoding: 1. Parallel development and coding of neural feature detectors. Biol. Cybern. **23**, 121–134 (1976)
18. Hilbert, M., López, P.: The world's technological capacity to store, communicate, and compute information. Science **332**(6025), 60–65 (2011). ISSN 0036-8075. https://doi.org/10.1126/science.1200970
19. Jun, L. et al.: D2D communication mode selection and resource optimization algorithm with optimal throughput in 5G network. IEEE Access, pp. 25263–25273 (2019)
20. Kolomvatsos, K., Anagnostopoulos, C.: Reinforcement machine learning for predictive analytics in smart cities. Informatics **4**(3), 16 (2017)

21. Lloyd, S.P.: Least squares quantization in PCM. Information Theory, IEEE Trans. **28**(2), 129–137 (1982)
22. Makhoul, L., Rpucos, S., Gish, H.: Vector quantization in speech coding. IEEE Trans. Neural Netw. **73**(11), 1551–1558 (1985). https://labrosa.ee.columbia.edu/~dpwe/papers/MakhRG85-vq.pdf
23. Narendra, K.S., Thathachar, M.A.L.: Learning Automata: An Introduction. Prentice-Hall Inc, Upper Saddle River, NJ, USA (1989). ISBN 0-13-485558-2
24. Nasrabadi, N.M., King, R.A.: Image coding using vector quantization: a review. IEEE Trans. Commun. **36**, 957–971 (1988). ISSN 0090-6778. https://doi.org/10.1109/26.3776
25. Rumelhart, D., McClelland, J.: University of California. Parallel Distributed Processing: Foundations. A Bradford book. MIT Press (1986). ISBN 9780262680530
26. Stelios, P., Evangelos, S., George, M., Constandinos, X.M.: A hyper-box approach using relational databases for large scale machine learning. International conference on telecommunications and multimedia TEMU 2014. IEEE Communications Society proceedings, pp. 69–73, 28–30 July, Crete, Greece
27. Xu, L., Krzyzak, A., Oja, E.: Rival penalized competitive learning for clustering analysis, RBF net, and curve detection. IEEE Trans. Neural Netw. **4**(4), 636–649 (1993). ISSN 1045-9227. https://doi.org/10.1109/72.238318
28. Yannis, N. et al.: Vulnerability assessment as a Service for Fog-Centric Healthcare ICT ecosystems. J. Peer-to-Peer Netw. Appl. Springer (2019)

Optimizing Blockchain Networks with Artificial Intelligence: Towards Efficient and Reliable IoT Applications

Furqan Jameel, Uzair Javaid, Biplab Sikdar, Imran Khan, George Mastorakis, and Constandinos X. Mavromoustakis

Abstract Blockchain is one of the new tools which is still experiencing a serious lack of understanding and awareness in general public. Besides, to say that blockchain is a versatile technology with many applications would be an understatement. The several features of blockchain technology have led to an immense amount of research interest in the blockchain systems, especially from the perspective of enabling the Internet-of-things (IoT). It is expected that its adoption will be gradual, starting with researchers and leading up to startups and companies who will discover its potential for change. Then the general public who will demand changes would use it as an everyday technology. Finally, organizations that had until then resisted change would adopt it at a large scale, thus, paving the way for globalization of blockchain. In this regard, it is important to understand that the blockchain is not a single object, a single trend or a particular feature, but a composition and collection of several autonomous as well as manually operated entities. Because of these modular features, the blockchain offers an infinite range of application choices. Another paradigm which is emerging in parallel is the so-called artificial

F. Jameel (✉)
Department of Communications and Networking, Aalto University,
FI-02150 Espoo, Finland
e-mail: furqanjameel01@gmail.com

U. Javaid · B. Sikdar
Department of Electrical and Computer Engineering,
National University of Singapore, 21 Lower Kent Ridge Road,
Singapore 119077, Singapore

I. Khan
Department of Electrical Engineering, University of Engineering & Technology,
814, Peshawar, Pakistan

G. Mastorakis
Department of Management Science and Technology,
Hellenic Mediterranean University, 72100 Agios Nikolaos, Crete, Greece

C. X. Mavromoustakis
Department of Computer Science, University of Nicosia and
University of Nicosia Research Foundation (UNRF), 46 Makedonitissas Avenue,
P.O. Box 24005, 1700 Nicosia, Cyprus

© Springer Nature Switzerland AG 2020
G. Mastorakis et al. (eds.), *Convergence of Artificial Intelligence and the Internet of Things*, Internet of Things,
https://doi.org/10.1007/978-3-030-44907-0_12

intelligence. The evolution of artificial intelligence is an incredible tale which started with neural networks in the last century and continued its way to the development of complex deep learning techniques. From the perspective of IoT networks, it has the potential to provide services like ultra-reliable and low-latency communications (uRLLC), long-distance and high-mobility communications (LDHMC), and ultra-massive machine-type communications (umMTC) which can be the key to realizing tactile Internet for next generation of wireless networks. Moreover, artificial intelligence techniques do not require a mathematically tractable model to optimize the performance of the devices. This means that they can be used for a number of environments that includes both indoor and outdoor communication scenarios. Although the applications of blockchain in IoT networks have much potential, there are several open issues and research challenges that need to be addressed first. In this context, one of the main issues in blockchain networks is branching, i.e., forking event. The forking not only introduces excessive overhead in the network but can also lead to potential security attacks. Thus, to overcome this interesting problem, we use deep neural networks in a distributed IoT setup. More specifically, the communication delay between the IoT miner and communication point is directly related to the rate of the link. We aim to maximize the rate of the communication link from IoT miner to the communication point with the help of neural networks. The numerical results show a significant improvement in the rate which can lead to a reduction in transmission delays. We anticipate that this work on artificial intelligence-enabled blockchain system would pave the way for future studies and research.

1 Blockchain

Blockchain is an online distributed ledger of cryptographically secured blocks [1]. It can be thought of as a special type of data structure in which data, usually termed as transactions, is linked together using cryptographic techniques. The techniques usually differ from application to application but a primitive example that is one of the foundational enablers of blockchain is Hashing. It can be defined as: "any function that can be used to map data of arbitrary size to fixed-size values." This means that for any hashing function, the size of the input is variable but that for the output, it is always constant.

Blockchain operates in a peer-to-peer (p2p) fashion, i.e., it depends on its users for its resources and operation. There is no central authority in a blockchain but instead, a decentralized mechanism is employed over its users; the nodes in the blockchain network. This is so that these nodes can work together and keep the blockchain secure from adversaries who want to double-spend their funds. A double-spending attack in a blockchain refers to an adversary who tries to manipulate the blockchain in such a way that funds used for one transaction can be reused for another one. Although double-spending attacks have been a problem for a long time in both centralized and decentralized payment protocols, the blockchain makes it infeasible unless and until an adversary or a group of adversaries gain control of 51% computational power of the blockchain network [2, 3].

The most widespread use of blockchain is in cryptocurrencies such as Bitcoin [2] and Ethereum [4]. However, many applications have adopted blockchain paradigm to provide decentralized and trust-free solutions such as blockchain-based data provenance for the Internet-of-things (IoT) [5] and trust management in VANETs [6], and mitigating IoT device-based DDoS attacks using blockchain [7] etc. Another use of blockchain, such as Ethereum [4], is with computer programs called 'Smart Contracts'. One of the advantages of using blockchain-based solutions is that they allow users to transact freely without relying on third parties which usually act as intermediaries between users and the system. Furthermore, two of the major concerns in blockchains are scalability and standardization which remain key developmental concerns [8] to date.

A blockchain typically consists of the following parts.

1.1 Blocks

A block is an instance which collates tuples of transactions together with respect to time using a timestamp. The first block in any blockchain is called the 'Genesis' block. All the succeeding blocks are added to it in a chronological order using a hashing function, i.e., the genesis block b_i is hashed and stored in the second block b_{i+1}. The hash of block b_{i+1} is stored into block h_{i+2} and so on. This way, each block b_j has a hash of the previous block b_{j-1} that can be verified and traced all the way back to the genesis block b_i. This forms the basis of a blockchain and can simply be formulated as:

$$\mathcal{B} = (b_0, b_1, b_2, \ldots) \tag{1}$$

$$b_i = \sum_{j=0}^{m} (tx, payload)_{ij} \tag{2}$$

where $\mathcal{B}$ represents a blockchain and b the chronologically appended blocks in it. Payload here represents data of concern with application-specific information. For reference, the index 0 is taken as the first (genesis) block. It is worth noting here that a blockchain can be verified recursively from its latest block back to its genesis block with the help of the hashing function because hash outputs are unique in nature and any change in the input will produce a different hash output. This way, a blockchain preserves data integrity since any change in any block instance will create a new hash and invalidate that block from the blockchain.

From Eq. (2), a block can contain m instances of transactions, data, and other information depending upon the nature of the application it is designed for. For adding new blocks in a blockchain, they must be mined first. Mining is the process of adding new blocks to a blockchain, thereby increasing its size and length. Miners, unlike users, are volunteers who provide their computational resources required by a blockchain to operate. Moreover, miners are rewarded in return for their contribution. For each block to be mined, the miners need to follow a consensus algorithm, for example, proof-of-work (PoW) in case of Bitcoin [2]. After successfully mining a block, it must then be broadcast in the network so that every node can accept and add it to their blockchain snapshot. A blockchain snapshot here refers to a copy of blockchain data which in turn usually contains transactions.

1.2   Transaction

A transaction tx is a set of instructions which changes the ownership of digital coins and/or digital assets from one user to another. A coin is a virtually valued currency whereas a digital asset can be a certificate, intellectual property, etc. The ownership of these coins/assets is changed by digitally signing a transaction using public key infrastructure (PKI).

A signed tx can contain coins/assets and other information with a digital signature created with a private key. After signing a tx, it is then broadcast in the network as on open transaction. Open transactions are those which have not yet been mined. The network can then verify this transaction signature using the corresponding public key and then mine it. Mathematically, a transaction can be formulated as:

$$tx = t_{in}Num \parallel t_{in} \parallel t_{out}Num \parallel t_{out} \parallel nonce \parallel data \parallel t \tag{3}$$

where t_{in} and t_{out} are the input and output vectors of a transaction which represent both the sender and recipient. These consist of tuples of elements and are used when the ownership of coins is digitally transferred from one user to another one. T_{IN} and T_{OUT} are the number of transactions relative to a timestamp t. The *nonce* field contains the challenge for miners to mine the transaction and validate it whereas the *data* field contains application-specific information.

All transactions relative to a timestamp are collated together which form tuples of transactions that need to be mined. When a transaction occurs and is broadcast in the network successfully, the ownership of coins change and the new owner is announced by the sender. The network participants collate it with other open transactions and verify its signature before mining.

1.3   Signing a Transaction

A blockchain platform typically uses PKI for data encryption and user authentication. PKI consists of a pair of keys: private and public which are used for producing digital signatures. The private key is meant for the user with which a transaction or any information of concern can be signed while the public key is meant for other users in a network to verify this signature. Before a transaction tx can be stored, it has to be validated first so as to prove its authenticity. Authenticity in a tx is achieved by digitally signing it with a private key for encrypting and then validating it with the corresponding public key. For a user i, it can mathematically be formulated as:

$$sign_i = f\left(private_{key_i}, payload\right) \tag{4}$$

$$verify_i = f\left(public_{key_i}, payload\right) \tag{5}$$

where $sign_i$ represents the signature of user i and is a function of $private_{key_i}$ and $payload$ which are the private key of and data to be signed by user i, respectively. Similarly, $verify_i$ is a function of $public_{key_i}$ and the same payload. The only way a user can verify the signature $sign_i$ is with the help of the corresponding public key of user i.

Moreover, by using digital signatures, the authenticity of transactions is guaranteed without relying on third parties. In the case of Bitcoin [2] and Ethereum [4], Elliptic Curve Cryptography (ECC) is used with the Elliptic Curve Digital Signature Algorithm (ECDSA). ECDSA is a cryptographic algorithm to ensure that funds can only be spent by their rightful owners while ECC is an approach to public-key cryptography based on the algebraic structure of elliptic curves over finite fields. A common comparison between ECC and Rivest Shamir Adelman (RSA), Diffie-Hellman (DH), and Digital Signature Algorithm (DSA) is given in Table 1 [9].

From Table 1, it can be seen that for a system security strength of 80 bits, ECC needs 160 bits while all of the rest need 1024 bits. Similarly, for a security strength of 256 bits, ECC needs 521 bits while the rest need 15,360 bits. This proves that ECC needs shorter key lengths when compared with the others to achieve the same security strength level. These specifications show that ECC can achieve higher levels of security using smaller key lengths. Note that smaller key lengths are translated into the lower computational overhead.

Table 1 Security strength comparison

Security strength in # of bits	ECC (bits)	RSA/DH/DSA (bits)
80	160	1024
112	224	2018
128	256	3072
256	521	15,360

1.4 Mining

Until now, the flow of data in a blockchain can be thought of as:

$$transaction_i \xrightarrow{append} transactions_{open} \xrightarrow{collate} block \xrightarrow{mine} blockchain$$

In a typical blockchain system, all transactions awaiting approval by the system are appended together as elements of open transactions. These open transactions are then collated together in a block by a miner. A block here can have fixed size and/or a specific time interval, i.e., mine a block if it reaches a certain size or after every time t. Block formation depends on the system design and the application as well that it is designed for.

Before mining a block, a miner first checks the validity of all the open transaction signatures using ECDSA and then follows a consensus algorithm to conform to the mining principles for adding a block to the blockchain. Therefore, mining can be defined as: "the process of adding a block to a blockchain while conforming to the desiderata of a consensus algorithm." For example, Bitcoin [2] uses proof-of-work (PoW) whereas Ethereum [4] uses both PoW and proof-of-stake (PoS) consensus algorithms, respectively. In this chapter, the case for PoW is considered.

PoW is a distributed consensus mechanism whose main function is to present a puzzle (an exceptionally difficult mathematical problem) to miners each time a block needs to be mined. The puzzle has a target value v and any block to be mined is hashed first using a hashing function to meet the required target v. The target value is a n bits number with a specific m number of zeros as significant digits. The choice of m and n defines the difficulty of the puzzle. To solve it, at most 2^n rounds of attempts have to be run.

The PoW is an exponential function which means a lower target value will require more attempts of hashing. The solution finding process is a probabilistic one since to find the target value, a miner has to hash a block iteratively. This is because each input produces a unique hash and different hashing attempts are required to solve the challenge. Blockchain makes double-spending attacks and spamming infeasible because of this very attribute. Mathematically, a simple PoW function can be formulated as:

$$pr(h \leq v) = v/2^n \tag{6}$$

where the probability of finding a proof h (mining) for a target v is $pr(h \leq v)$. When a miner finds the target value, it broadcasts its proof in the network. The other miners then re-run this proof and validate the block. Once the block is validated and accepted by the majority of the network, it is then finally added to the blockchain. It is worth noting here that although the process of finding a proof is time-intensive, the verification can be done easily in one run using the proof that produces the required target value.

2 Applications of Blockchain Technology

The previous section introduced the basic principles and components of a typical blockchain system along with some examples. In this section, although there are many applications and use-cases of blockchains, only some of the practical and commonly used ones that are transforming the society are discussed.

2.1 Local Energy Trading

The demand for power is an ever-increasing phenomenon, especially with the move towards urbanization, it has continued to escalate. The significant increase in energy consumption has brought big challenges to energy security and the environment [10]. To build a sustainable energy system and address the associated demand concerns, distributed generation models with integrated microgrids are gaining worldwide attention, i.e., smart grids.

In traditional power grid models, a majority of financial and energy trading infrastructures are centralized that rely on intermediaries and PKI for establishing trust in their systems. Although PKI can be used to provide a certain level of security in centralized architectures, ensuring the authenticity of public keys without relying on an intermediary is an issue. The intermediaries may handle accounts, digital assets, and transactions, process and analyze data, and provide security. However, such infrastructures are prone to the following security concerns:

1. *Single point-of-failure*: If a central server in a centralized architecture fails, it can disrupt the entire system and render its security services dysfunctional.
2. *Lack of user privacy*: Centralized structures rely on middlemen which if compromised, may reveal and predict the energy routines of users and their private information. Any leakage of such sensitive information may have consequential results [11].
3. *Restricted trading*: Conventional energy trading frameworks restrict their users to trade energy with the electric utility company only and not with other users, thereby restricting them.

These problems can be addressed using a decentralized design choice, blockchains. A trustless or semi-trustless system can help build an energy trading framework which can enable local energy trading among users, preserve their privacy, and provide them with transaction anonymity. It will also help in confirming the authenticity of user transactions without the need of an intermediary, opening a gateway to trust-free cryptographic transactions using cryptocurrencies [12].

2.2 Internet of Things

The IoT is one of the most important emerging technologies of the present era aimed at the integration of physical devices in a wide range of applications. The devices range from vehicles to bicycles, smart homes to smart cities, CCTV to smart cameras, sensors to radio-frequency identification (RFID) tags, etc. These devices can communicate and share information independent of human intervention. Figure 1 presents an overview of a traditional IoT network where different kinds of sensors and devices interact with a central server through a communication channel. A majority of traditional IoT infrastructures are centralized in nature and rely on third parties for establishing trust in their systems. The third parties may handle accounts and transactions, process and analyze data, and provide security. Such centralized architectures are prone to the following security concerns:

1. *Limited computational ability*: IoT devices are small in size. Their processing abilities are resource-constrained which limits them from performing computing-intensive tasks.
2. *IP protection*: IoT devices are user accessible devices which are vulnerable to intellectual property (IP) theft. Therefore, an IoT device needs to be protected against IP theft, tampering, cloning, and reverse engineering.
3. *Patching*: Contemporary IoT environments are unable to deliver timely updates and/or patches to devices which is a key concern in IoT architectures. This makes the devices prone and exposed to attacks for a longer time. Therefore, a platform where timely updates can be delivered to IoT devices is much needed.

Fig. 1 A conventional IoT architecture proposed by [5]

The aforementioned problems can be addressed using blockchains. A blockchain-based solution can register IoT devices and use its broadcast attribute for decentralization to deliver timely patches to them. It can also provide resource management and provisioning to maintain equilibrium between the operation of IoT devices. Furthermore, it can preserve the data integrity to provide IP protection and defense against tampering attacks.

2.3 Next Generation Payment Solutions

The operation of blockchain is decentralized in nature, i.e, it operates in a trust-free manner and does not rely on central authorities which usually act as third parties in traditional banking and payment solutions. This inherent attribute of blockchains can help enable future payment solutions where users and banks can transact directly with other users and banks, thereby eliminating third parties and the associated time, labor costs. The potential applications of such include but are not limited to cross-border and inter-bank payment protocols.

Furthermore, the following are some of the but not limited to, domains that can also find applications in blockchain-based solutions:

- Digital asset management
- Claims processing in insurance
- Supply chain management
- Healthcare
- Data persistence.

3 Fundamentals of Artificial Intelligence

Hardly any other area of computer science has continually grown as often as the area called artificial intelligence. It generally refers to the simulation of human brain processes through training and learning [13]. Although the recent hype has stirred significant research interest in artificial intelligence, the initial efforts on the development of a learning machine go back to the beginning of the 19th century [14]. The story begins with the British mathematician Alan Turing and one of the pioneers of theoretical computer science. As early as the 1930s, he wrote basic papers on the capabilities and limitations of computers. All this at a time when computers were still filling huge halls and their usage was very limited. Through his work, he laid the foundations for modern computer science. Basically, he asked the following questions:

1. Could a machine emulate the capabilities of the human mind?
2. Could a machine think and act like humans?

Sadly, he was largely left to theorize because he did not have the technology to take his ideas forward. The big development that allowed us to think of artificial intelligence as something of practical value, is the development of the digital computer. The modern silicon computer contains billions of components that are incredibly powerful and can perform several million operations in a second. Back in the 1950s, computer scientists, working with machines far less powerful than today's laptop, wondered whether a set of instructions could be found to program a computer that would cause it to exhibit intelligence.

With this intent, the scientists in 1964 created a computer program called "Eliza" at the artificial intelligence laboratory at MIT. Fundamentally, Eliza was a chatbot for a little conversation. It could use different tricks to ask the names and, later on, it would call those names. For too many people, at least superficially and for a short time, it appeared that it was exhibiting intelligence. Of course, if you interacted with it for more than a few minutes, you would realize that it was not much intelligent. Still, it seemed to *mimic* the intelligence [13]. Besides shifting the paradigm of the nature of computation, there's something else which has been altered for successful implementation of artificial intelligence. For this, let us consider the different writing patterns. Of course, there is tremendous variability in human handwriting and due to this reason, it is so difficult to program a computer directly to do something like recognizing human handwriting. In fact, there are several possibilities concerning handwriting and that's really what defeated old-fashioned artificial intelligence back in the 1950s and 1960s. Thus, there was a need to alter the paradigm of modern software to handle uncertain situations.

To understand this idea of uncertainty and to demonstrate a modern view of machine learning, we need to take a look at the concept of information [14]. The concept of information was provided by Claude Shannon back in the 1920s. In their book, Shannon and Warren Weaver defined information as A measure of ones freedom of choice when one selects a message [15]. This provided a mathematical basis for the concept of information and gave birth to the field of information theory. Information is an essential part of communication and it is broadly defined as the degree of uncertainty in the data. Today, we're seeing a shift from the software which is based on logic, that is everything is 0 or 1, to the software which deals with uncertainty [13]. The software is capable to quantify the amount of uncertainty and deal with it using previously available information.

This field of artificial intelligence was tremendously popular back in the 1950s and 1960s which was founded on the idea that computer scientists would be able to program computers to be intelligent and much of work was devoted to scientific research and natural language processing, as shown in Fig. 2. Technically speaking, the scientists were interested in writing a computer program or a set of instructions that would exhibit intelligence, only if, a computer follows these instructions at a very fast speed. At that time, there was a tremendous amount of hype about artificial intelligence and its potential. However, that excitement did not last for very long.

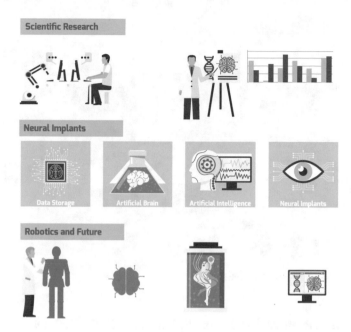

Fig. 2 An illustration of various areas of research interest in artificial intelligence

3.1 Artificial Intelligence in Modern Era

Despite the setbacks for research in the 1980s, the foundations for the machine learning approaches have been laid today. Departing from the conventional event-driven processes, as shown in Fig. 3, there are different objectives for the machine learning processes. The machine learning techniques are divided into three main categories: supervised machine learning, unsupervised machine learning, and reinforcement machine learning [16]. In the first system, behavior learning is initially offline, that is, separate from the application scenario. Only then will the learning be applied and not changed.

3.1.1 Supervised Learning

In supervised machine learning, a computer program gets known sample data and is trained on the desired interpretation and the associated output. The goal is to find general rules that connect the known input data to the desired output data and then use those rules to create new output with new input data [15]. The advantage of the classification is that it is always judged based on the interplay of several values. As a result, a person with low savings but high income in the class would

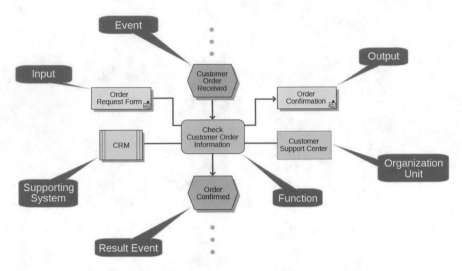

Fig. 3 Illustration of event-driven processes

be creditworthy. Regression allows predictions about continuous values, such as the income development of a person, while classification differentiates between classes, such as creditworthiness [17].

3.1.2   Unsupervised Learning

Unsupervised machine learning works without previously known association and labeling of input data. The possible results are completely open. Therefore, the computer program can not be trained, but rather has to recognize structures in the data and transform them into interpretable information.

3.1.3   Reinforcement Learning

In reinforcement machine learning, the third category of machine learning, a computer program learns directly from the experiences. It interacts with its environment and receives a reward for correct results. The program can be compared to a trained animal, for example by being rewarded in a game situation when it wins the game [18]. The goal now is for the program to remember the consequences of its action and use that knowledge to maximize its reward.

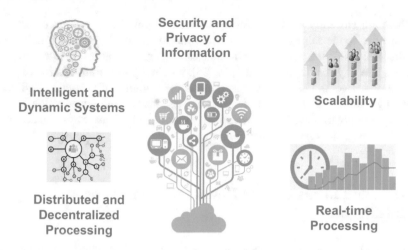

Intelligent and
Dynamic Systems

Security and
Privacy of
Information

Scalability

Distributed and
Decentralized
Processing

Real-time
Processing

Fig. 4 Applications of artificial intelligence in various domains related to IoT networks

3.2 Impact of Artificial Intelligence on Industries

Artificial intelligence has provided an immense development boost in business practice in recent years. While Industry 4.0 focuses on optimizing and automating production and logistics processes, artificial intelligence is also increasingly addressing administrative, planning and planning processes in marketing, sales, and management on the way to the holistic algorithmic enterprise [19]. For now, we would only take a look at some of the motivating factors for using artificial intelligence in the industry [20]. The usage of artificial intelligence technology in marketing, product automation, sales, and customer services are some of the key applications of artificial intelligence, as shown in Fig. 4.

Artificial intelligence-first, as a possible mantra of the massive disruption of business models and the development of fundamentally new markets, is becoming more and more prevalent. There are already many use cases across industries that demonstrate the innovation and design potential of 21st-century core technology like low latency vehicle to vehicle communications [21, 22]. Decision-makers from all industrial nations and sectors agree on increasing involvement of artificial intelligence technology in the production of goods. But there is no holistic evaluation and process model [23].

There is also an immense potential for change and design for our society. In Europe too, the first data science programs are already in place to ensure the training of junior staff. Nevertheless, the *War for Talents* is currently raging, as the staff pool is very limited, but the demand remains high in the long term. In addition, digital data and algorithms also enable completely new business processes and models [17]. The methods used a range of techniques from simple hands-on analysis with small data to advanced analytics with Big Data.

There are currently a lot of informatics-related remarks by experts on artificial intelligence. To the same extent, there are a variety of popular science publications and discussions of the general public that can incorporate artificial intelligence in domains like physical layer security [24, 25], wireless power transmission [26, 27], and sensor networks [28]. What is missing is the bridging of artificial intelligence technology and methodology to clear business scenarios and added value. IBM is currently moving massively with Watson in the companies but always remains at the teaser level of a clear business application.

4 Data and Infrastructure Desideratum of Artificial Intelligence

Data has and continues to contribute in many ways to the development and success of artificial intelligence in the business world [29, 30]. However, prior successful implementation of data platforms and artificial intelligence, some key aspects need to be explored further. In this regard, a variety of methods are used for the evaluation and analysis of data, especially in the area of industrial IoT applications [31]. The following section explains the synergy effects of data and artificial intelligence. These synergies can be divided into three areas, which are briefly explained below.

4.1 IT Infrastructures

With the advent of big data, many companies have been forced to adapt their IT infrastructures to changing circumstances. It was done by either investing directly in the hardware or through cloud services to handle the increasing amount of data. Only the large investments in IT infrastructures can enable companies to implement and apply new methods and complex systems [32]. The success of the IBM marketed artificial intelligence *Watson* would not be possible without the Hadoop cluster of 90 computers with 19 TB of memory. Such successes help to consistently work on the hardware-based improvement of the artificial intelligence [19].

4.2 Algorithms and Methods

In order to generate information and knowledge from data, different techniques are being used that were originally used in the artificial intelligence domain. Especially, a large amount of semi- and unstructured data could not be automated without machine learning, natural language processing, and computer vision. As long as data remains unused, it brings no added value. The need for processing and analyzing data-driven actions has been identified by companies, which has led to investments in the development of algorithms and methods beyond IT infrastructure for distributed mobility and exploitation of existing resource (e.g., interference) [33]. Of late, data scientists are being increasingly sought who, in addition to data search and processing, are involved in the further development of existing machine learning algorithms. Particularly, high investments in these areas come from companies who are sitting on the *fuel* of the future: the data. For example, the tech giants Facebook, Google, Microsoft, and IBM together invested more than ten billion US dollars in artificial intelligence research and development in 2015.

4.3 Training Data

It is no wonder that the major US corporations are investing heavily in the development of artificial intelligence. The biggest synergetic effect of big data and artificial intelligence lies in the fact that large amounts of data are available to model and train artificial intelligence [23]. The most acclaimed achievements of artificial intelligence in recent years have been achieved through deep learning techniques. For a long time, results in this area were achieved only with great effort and expert knowledge. Therefore, this approach led to a niche existence.

However, when very large data sets are used to train artificial neural networks, the results improve substantially. As a rule of thumb, a supervised deep learning algorithm with around 5000 labeled examples per category and a training dataset of at least 10 million labeled examples performs at least like a human in recognizing and classifying images. The advent of huge amounts of training led to significant improvements in machine learning, sparking enthusiasm for artificial intelligence.

5 Blockchain Network Architecture

We consider an artificial intelligence-enabled blockchain network consisting of M and N IoT miners and communication points, respectively, as illustrated in Fig. 5. The communication points can be seen as a static infrastructure such as base stations. These points are associated with IoT miners that can be computationally capable mobile devices that can be industrial robots gathering transaction data from other devices [34]. When mining computation is executed by IoT miners, an IoT miner can refer an independent computing node, and, also, an IoT miner can refer to the head node of a local computing cluster.

In Fig. 5, IoT miners in remote locations are able to directly communicate with each other. Therefore, each IoT miner is associated with a different communication point that is connected to a backhaul network. In the considered network, the ledger is located at the communication point while the IoT miners are used for computing. When transaction records are stored as blocks at the communication points, those

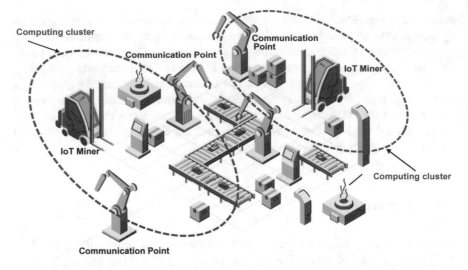

Fig. 5 Blockchain architecture and network setup for Industrial IoT scenario

blocks must be validated by proof-of-work schemes so as to guarantee that the transaction in the block is original [35, 36]. Then, the communication points can delegate the proof-of-work computation needed for this validation to the wireless IoT miners. Once each IoT miner completes its proof-of-work computation, an acknowledgment message is sent from the IoT miner to the associated communication point, called the source point. The source point propagates the reception of the acknowledgment message to other IoT devices through the backhaul links among the communication points. We assume high-bandwidth, fiber backhaul links between IoT devices and, hence, the message propagation latency in the backhaul network will be negligible.

In a blockchain system, the reception order of the acknowledgment messages from multiple IoT miners to the backhaul network should be identical to the order of completion of the proof-of-work computation. If the acknowledgment message sent earliest arrives at the communication points interconnected by the backhaul network later than other acknowledgment messages, it will lead to a forking event. In a blockchain system, the miner that completes the proof-of-work with the shortest delay will receive a unit of reward. However, when a forking event occurs, the miners can no longer discern which miner completed the current proof-of-work computation with the shortest delay. Therefore, if a forking event occurs, the miners must recover from it by repeating the computation of the proof-of-concept of the forked block to decide which miner will be rewarded. This recovery from a forking event will clearly increase the total latency required to complete the proof-of-work computation of a block.

6   System Model

We consider that all the IoT miners start their proof-of-work at the same time, having the same computing powers. These IoT miners keep executing the proof-of-work computation until one of the IoT miners completes the task by finding the hash value target. When an IoT miner executes the proof-of-work of the current block, the time for execution of the task will be an exponential distribution given as

$$f_C(c) = \lambda e^{-\lambda c}, \tag{7}$$

where the computing speed with power P_c and scaling factor λ_0 is referred as

$$\lambda = \lambda_0 P_c. \tag{8}$$

Once the IoT miner completes the proof-of-work, the acknowledgment of the message needs to be delivered to the communication point [37]. This is important so that other miners could stop their proof-of-work and save computation resources. We derive the transmission latency of the IoT miner when an IoT miner sends an acknowledgment message to associated communication point under the impact of interference from another IoT miner. The received signal at the communication point, which is sent by an IoT miner, can be written as

$$y_F = \sqrt{\alpha_N P} s_N h_{U_N} + \sqrt{\alpha_F P} s_F h_{U_F} + n_{U_F}, \tag{9}$$

where s_N is the interfering signal and n_{U_F} represent the additive white Gaussian noise (AWGN) with zero mean and N_0 variance. The received signal to interference and noise ratio (SINR) can be given as

$$\gamma_F = \frac{|h_{U_F}|^2 \alpha_F P}{|h_{U_N}|^2 \alpha_N P + N_0}. \tag{10}$$

Thus, the data rate from the IoT miner to the communication point can be written as

$$C = B \log_2 (1 + \gamma_F), \tag{11}$$

where B denotes the bandwidth of the system. The wireless transmission latency of IoT miner can now be written as

$$T_t = \frac{K}{C}, \tag{12}$$

where K represents the size of the acknowledge message and C is the data rate. It is evident that the total transmission latency can be reduced by increasing the data rate. In the next section, we provide a deep neural network-based technique for increasing the data rate, thereby, reducing the latency and forking event.

7   Optimization Using Deep Neural Networks

This section would present a deep learning-based resource optimization scheme for improving the achievable data rate of blockchain systems. More specifically, we use an artificial neural network to learn the dependencies between output (optimal power allocation factors) and inputs (channel gains). To attain this objective, we carefully train, and then, test our multi-layer neural network. As indicated above, this network consists of multiple neurons along with several hidden nodes, as illustrated in Fig. 6. Due to the low computational complexity of the trained neural networks, it is well suited for blockchain architecture to avoid forking events. This means that once the model has been trained on a set of channel inputs, the realtime evaluation phase involves vector multiplications and linear transformation without performance loss.

7.1   Problem Formulation

Let us now formulate the problem for maximizing the data rates. It can be seen that the optimal values of the power allocation factor would result in providing the

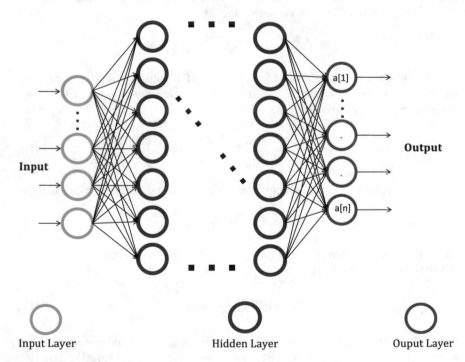

Fig. 6 Artificial neural network having input and output neurons with several hidden layers

maximum data rates. Thus, the optimization problem of the achievable rate becomes equivalent to

$$\max_{\alpha_F} \log_2(1 + \gamma_F). \tag{13}$$

7.2 Deep Learning Network Setup

The artificial neural network in our study has a single output and input layers while having multiple hidden layers. The major cause of using multiple hidden layers is to avoid the under-fitting of the data and maintain a relatively low level of complexity. Besides this, it is necessary to have multiple hidden layers in the neural network to understand the interplay of the inputs and outputs. As stated earlier, our inputs (channel realizations) and outputs (power allocation factors) need to cater to a variety of environments between communication points and IoT miners. The values of input parameters are fed into the network during the training phase of the neural network.

8 Numerical Results

This section provides details of the simulation setup and the relevant discussion on the obtained results. We performed the link-level simulation in Python. In this regard, it is worth highlighting that we use rectified linear unit (ReLU) activation function at each hidden layer. The ReLU activation function can be written as

$$x = \max(w, 0), \tag{14}$$

where x denotes the output of the function while w denotes the input. As the cost function, we used mean square error and apply the mini-batch algorithm for gradient calculation.

In Fig. 7, we demonstrate the utility of our proposed deep learning technique to maximize the rate of communication between the communication point and IoT miner. As a benchmark technique, we have compared our method with the random power allocation technique. For random power allocation, the power allocation factors are chosen randomly based on a uniform distribution. It can be seen from Fig. 7a that the proposed artificial neural network approach performs better as compared to the random selection of the power allocation factors. However, at larger values of transmit power, the performance gains are reduced. This is because of the increased interferences at higher values of the power which reduces the data rates for both schemes. Similar trends can be observed in Fig. 7b where performance decreases at larger values of transmit power. However, the proposed artificial neural network-based approach still outperforms the random power allocation. Furthermore, the increased number of hidden layers provides somewhat improved performance.

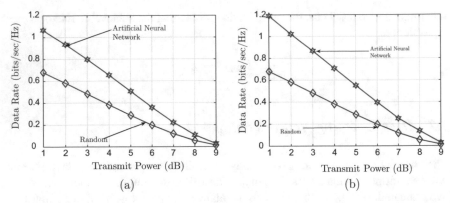

Fig. 7 Data rate performance comparison versus transmit power **a** 02 hidden layers **b** 04 hidden layers

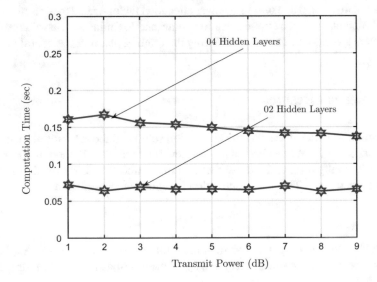

Fig. 8 Computation time against transmit power for 02 hidden layers and 04 hidden layers

In order to highlight the impact of the computational complexity of the proposed scheme, we provide a comparison of complexity by increasing the number of hidden layers. In particular, Fig. 8 shows the computational complexity of the scheme when the number of hidden layers is 02. This means that layer 1 has 200 neurons while layer 2 consists of 100 neurons. It can be seen that the computational complexity remains virtually unchanged due to the increased transmit power levels. Similar observations can be made for the second curve in Fig. 8 whereby the complexity has been given for 04 layers of the artificial neural network. It can be seen that the computational complexity of the 04 hidden layers is slightly more than the 02 hidden layers. This confirms the fact that there is a delicate balance between the obtained performance gains and the complexity level of the network.

9 Conclusion

The blockchain is a new database technology that leverages and takes full advantage of the Internet, free protocol, computing power, and cryptography. This distributed transactional database is comparable to a large accounting-based system in which each new transaction is written following the others, without the possibility of modifying or deleting the previous ones. Due to growing number of applications of blockchain in IoT networks, it is becoming increasingly important to analyze the impact of this technology in large-scale networks. With the objective of improving the performance of these networks, this chapter has provided an in-depth analysis of how artificial intelligence techniques (e.g., deep neural networks) can be used for optimizing the performance of blockchain systems. Furthermore, the results have shown that the deep neural networks can indeed improve the performance of IoT miners in blockchain networks.

References

1. Wood, G.: Ethereum: a secure decentralised generalised transaction ledger. Ethereum Proj. Yellow Pap. **151**, 1–32 (2014)
2. Nakamoto, S.: Bitcoin: A Peer-to-Peer Electronic Cash System
3. Karame, G.O., Androulaki, E., Capkun, S.: Double-spending fast payments in bitcoin. In: Proceedings of the 2012 ACM Conference on Computer and Communications Security, CCS'12, ACM, New York, NY, USA, pp. 906 917 (2012). https://doi.org/10.1145/2382196.2382292
4. Buterin, V.: Ethereum: a next-generation smart contract and decentralized application platform (2014). Accessed on 22 Aug 2016 . https://github.com/ethereum/wiki/wiki/White-Paper
5. Javaid, U., Aman, M.N., Sikdar, B., Blockpro: Blockchain based data provenance and integrity for secure IoT environments In: Proceedings of the 1st Workshop on Blockchain-enabled Networked Sensor Systems, BlockSys'18, ACM, New York, NY, USA, 2018, pp. 13–18. https://doi.org/10.1145/3282278.3282281
6. Javaid, U., Aman, M.N., Sikdar, B.: Drivman: driving trust management and data sharing in vanets with blockchain and smart contracts. In: 2019 IEEE 89th Vehicular Technology Conference (VTC2019-Spring), pp. 1–5 (2019). https://doi.org/10.1109/VTCSpring.2019.8746499
7. Javaid, U., Siang, A.K., Aman, M.N., Sikdar, B.: Mitigating Iot device based DDoS attacks using blockchain. In: Proceedings of the 1st Workshop on Cryptocurrencies and Blockchains for Distributed Systems, CryBlock'18, ACM, New York, NY, USA, pp. 71–76 (2018). https://doi.org/10.1145/3211933.3211946
8. Croman, K., Decker, C., Eyal, I., Gencer, A.E., Juels, A., Kosba, A., Miller, A., Saxena, P., Shi, E., Gün Sirer, E., Song, D., Wattenhofer, R.: On scaling decentralized blockchains 9604, pp. 106–125 (2016)
9. Aitzhan, N.Z., Svetinovic, D.: Security and privacy in decentralized energy trading through multi-signatures, blockchain and anonymous messaging streams. IEEE Trans. Dependable Secure Comput. **15**(5), 840–852 (2018). https://doi.org/10.1109/TDSC.2016.2616861
10. Sun, Q., Li, H., Ma, Z., Wang, C., Campillo, J., Zhang, Q., Wallin, F., Guo, J.: A comprehensive review of smart energy meters in intelligent energy networks. IEEE IoT J. **3**(4), 464–479 (2016). https://doi.org/10.1109/JIOT.2015.2512325
11. Anderson, R., Fuloria, S.: Who controls the off switch? In: 2010 First IEEE International Conference on Smart Grid Communication, pp. 96–101 (2010). https://doi.org/10.1109/SMARTGRID.2010.5622026

12. Beck, R., Stenum Czepluch, J., Lollike, N., Malone, S.: Blockchain—the gateway to trust-free cryptographic transactions. In: 24th European Conference on Information Systems (ECIS), Istanbul, Turkey, 2016, Springer Publishing Company, pp. 1–14 (2016) (ISBN to be added later)
13. Russell, S.J., Norvig, P.: Artificial Intelligence: A Modern Approach. Pearson Education Limited, Malaysia (2016)
14. Li, D., Du, Y.: Artificial Intelligence with Uncertainty. CRC Press (2017)
15. Dörner, S., Cammerer, S., Hoydis, J., ten Brink, S.: Deep learning based communication over the air. IEEE J. Sel. Top. Signal Process. 12(1), 132–143 (2017)
16. Jiang, C., Zhang, H., Ren, Y., Han, Z., Chen, K.-C., Hanzo, L.: Machine learning paradigms for next-generation wireless networks. IEEE Wirel. Commun. 24(2), 98–105 (2016)
17. Rondeau, T.W.: Application of artificial intelligence to wireless communications. Ph.D. Thesis, Virginia Tech (2007)
18. Silver, D., Hubert, T., Schrittwieser, J., Antonoglou, I., Lai, M., Guez, A., Lanctot, M., Sifre, L., Kumaran, D., Graepel, T., et al.: A general reinforcement learning algorithm that masters chess, shogi, and Go through self-play. Science 362(6419), 1140–1144 (2018)
19. Rondeau, T.W., Bostian, C.W.: Artificial intelligence in wireless communications. Artech House (2009)
20. Jameel, F., Chang, Z., Huang, J., Ristaniemi, T.: Internet of Autonomous Vehicles: Architecture, Features, and Socio-Technological Challenges. IEEE Wirel. Commun. 26(4), 21–29 (2019)
21. Jameel, F., Wyne, S., Nawaz, S.J., Chang, Z.: Propagation channels for mmwave vehicular communications: state-of-the-art and future research directions. IEEE Wirel. Commun. 26(1), 144–150 (2019)
22. Jameel, F., Javed, M.A., Ngo, D.T.: Performance analysis of cooperative V2V and V2I communications under correlated fading. IEEE Trans. Intell. Transp. Syst. 1–9 (2019)
23. Konar, A.: Artificial Intelligence and Soft Computing: Behavioral and Cognitive Modeling of the Human Brain. CRC Press (2018)
24. Jameel, F., Jayakody, D.N.K., Flanagan, M.F., Tellambura, C.: Secure communication for separated and integrated receiver architectures in SWIPT. In: IEEE Wireless Communications and Networking Conference (WCNC), pp. 1–6 (2018)
25. Jameel, F., Haider, M.A.A., Butt, A.A.: Physical layer security under Rayleigh/Weibull and Hoyt/Weibull fading. In: 2017 13th International Conference on Emerging Technologies (ICET), pp. 1–5 (2017)
26. Mukhlif, F., Noordin, K.A.B., Mansoor, A.M., Kasirun, Z.M.: Green transmission for c-ran based on swipt in 5g: a review. Wirel. Netw. 25(5), 2621–2649 (2019)
27. Jameel, F., Chang, Z., Ristaniemi, T.: Intercept probability analysis of wireless powered relay system in kappa-mu fading. In: 2018 IEEE 87th Vehicular Technology Conference (VTC Spring), pp. 1–6 (2018)
28. Jameel, F., Wyne, S., Krikidis, I.: Secrecy outage for wireless sensor networks. IEEE Commun. Lett. 21(7), 1565–1568 (2017)
29. Abduljabbar, R., Dia, H., Liyanage, S., Bagloee, S.: Applications of artificial intelligence in transport: an overview. Sustainability 11(1), 189 (2019)
30. Yao, M., Sohul, M., Marojevic, V., Reed, J.H.: Artificial intelligence defined 5g radio access networks. IEEE Commun. Mag. 57(3), 14–20 (2019)
31. Jameel, F., Javed, M.A., Jayakody, D.N., Hassan, S.A.: On secrecy performance of industrial internet of things. Internet Technol. Lett. 1(2), e32 (2018)
32. Jameel, F., Kumar, S., Chang, Z., Hamalainen, T., Ristaniemi, T.: Operator revenue analysis for device-to-device communications overlaying cellular network. In: IEEE Conference on Standards for Communications and Networking (CSCN). pp. 1–6 (2018)
33. Jameel, F., Wyne, S., Javed, M.A., Zeadally, S.: Interference-aided vehicular networks: future research opportunities and challenges. IEEE Commun. Mag. 56(10), 36–42 (2018)
34. Aman, M.N., Taneja, S., Sikdar, B., Chua, K.C., Alioto, M.: Token-based security for the internet of things with dynamic energy-quality tradeoff. IEEE IoT J. 1–1 (2018). https://doi.org/10.1109/JIOT.2018.2875472

35. Lei, A., Cao, Y., Bao, S., Li, D., Asuquo, P., Cruickshank, H., Sun, Z.: A blockchain based certificate revocation scheme for vehicular communication systems. Future Gener. Comput. Syst

36. Ren, Y., Leng, Y., Zhu, F., Wang, J., Kim, H.-J.: Data storage mechanism based on blockchain with privacy protection in wireless body area network. Sensors **19**(10), 2395 (2019)

37. Sun, Y., Zhang, L., Feng, G., Yang, B., Cao, B., Imran, M.A.: Blockchain-enabled wireless internet of things: performance analysis and optimal communication node deployment. IEEE Internet Things J

Industrial and Artificial Internet of Things with Augmented Reality

Pedro Gomes, Naercio Magaia, and Nuno Neves

Abstract Internet of Things (IoT) is a concept that proposes the inclusion of physical devices as a new form of communication, connecting them with various information systems. Nowadays, IoT cannot be reduced to smart homes as due to the recent technological advances, this concept has evolved from small to large-scale environments. There was also a need to adopt IoT in several business sectors, such as manufacturing, logistics or transportation in order to converge information technologies and technological operations. Due to this convergence, it was possible to arrive at a new IoT paradigm, called the Industrial Internet of Things (IIOT). However, in IIoT, it is also necessary to analyze and interact with a real system through a virtual production, which refers to the use of Augmented Reality (AR) as a method to achieve this interaction. AR is the overlay of digital content in the real world, and even play an important role in the life cycle of a product, from its design to its support, thus allowing greater flexibility. The adoption of interconnected systems and the use of IoT have motivated the use of Artificial Intelligence (AI) because much of the data coming from several sources is unstructured. Several AI algorithms have been used for decades aiming at "making sense" of unstructured data, and transforming it into relevant information. Therefore, converging IoT, AR, and AI makes systems become increasingly autonomous and problem solving in many scenarios.

1 Introduction

The industry is part of an economy that produces materials through mechanized and automated processes. In the past, there were several industrial revolutions that led us to the world we know now. The first one began with the introduction of mechanisms that produced energy from water and steam. The second one introduced electric power allowing the creation of first-line assemblies. And, the third one came

P. Gomes · N. Magaia (✉) · N. Neves
LASIGE, Departamento de Informática, Faculdade de Ciências, Universidade de Lisboa, Lisbon, Portugal
e-mail: ndmagaia@ciencias.ulisboa.pt

© Springer Nature Switzerland AG 2020
G. Mastorakis et al. (eds.), *Convergence of Artificial Intelligence and the Internet of Things*, Internet of Things,
https://doi.org/10.1007/978-3-030-44907-0_13

with the introduction of automation and the appearance of the first programmable logic controllers. Based on this, we can say that there has always been an interest in increasing productivity and efficiency in industrial processes and, as a result of the recent advances in information and communication technologies, the fourth Industrial Revolution, also known as *Industry 4.0* or *Industrial Internet*, commenced. This new revolution was initiated by the German government aiming at increasing its competitive and sustainable advantage. The concept of *Industrial Internet of Things* (IIoT) was an adaptation of the industrial revolution with the concept of the Internet of Things (IoT), hence being used interchangeably [1].

The latter concepts, i.e., IoT, proposes the inclusion of physical devices as a new form of communication and connecting them with various information systems. The most frequent use cases in IoT generally refer to devices with certain integrated functionality, whether simple or not, such as controlling an environment through actuators or obtaining sensor values from a machine [2]. Nowadays, we cannot reduce IoT to smart houses, and because of the recent advances in hardware and cloud computing, this concept has evolved from small to large-scale environments. There was also a need to adopt IoT in several business sectors such as manufacturing, logistics or transportation in order to converge information technologies and technological operations [1]. IIoT initially referred primarily to a structure in which a large number of devices or machines that were connected to each other were synchronized through software or existing Machine-to-Machine (M2M) technologies in industrial environments. However, at the moment, this concept is no longer only applicable to machines, but also to intelligent computers and people that enable intelligent industrial operations through advanced data analysis. In IIoT, IoT bases are used where data obtained from machines also play a fundamental role, as it was necessary to increase the level of automation, thus introducing intelligence into devices. Technologies such as ubiquitous computing, large storage systems and M2M connections have led to the creation of Cyber-Physical System (CPS) [3] in IIoT capable of deciding and triggering actions to be able to control these systems autonomously. To achieve this autonomy, it is necessary to interact with the physical and virtual worlds simultaneously. This interaction enabled a better communication between humans and machines, forming connected machine networks that follow the social paradigm, therefore creating new paradigms in product creation and, finally, services based on intelligent systems [1]. In addition, these technologies enabled the adoption of new concepts such as Artificial Intelligence (AI) and Augmented Reality (AR) in the industry.

1.1 Artificial Intelligence

IIoT has become increasingly dependent on big data and consequently AI, creating a new paradigm called *Artificial Internet of Things* (AIoT). When we speak of AI we are talking about computer-based techniques that use a lot of unstructured data to process, interpret and visualize aiming to obtain credible information. Both big data and AI enable predictive historical analysis and to obtain information about what is happening inside a machine. Furthermore, because of this analysis, there is the possibility of having machines and services that are more efficient, predicting failures and reducing unnecessary inefficiency or maintenance costs. Hence, an IIoT system that incorporates this type of intelligence can collect data from sensors and, if necessary react faster than a human. There are three AI techniques that have been proposed for IoT [4], namely supervised learning, semi-supervised learning, and unsupervised learning. The integration of AI-based algorithms into the different IoT architectures to process data in a way to take autonomous decisions caused the appearance of interconnected intelligent IoT systems. These systems allow combining innovative technologies such as AR.

In the context of IIoT, the cloud offers an accessible and elastic infrastructure in order to allocate resources necessary for the system to scale. This is an important point that differs from traditional databases because we do not have to worry about the resources available for storage. For example, if we need a system with the ability to increase or decrease the storage needs, the cloud is a good example, which made it quite an interesting technology to adopt in IIoT services. In addition, the cloud provides development environments to its users, enabling accelerated development of applications. For instance, Microsoft Azure provides support for .NET applications and big data tools. Nevertheless, there are not only advantages on using a cloud-based system. Latency, which is the time difference between the start of an event and the time when this event is received, is very much a problem when using the cloud. In systems where the goal is to store data for use in big data, that is not a problem; however, in industrial systems where real-time information is required, latency can greatly affect the system's performance. Cloud systems are usually located on the Internet, where traffic can be affected by network quality, and where IIoT data are subject to latency. So, what will be the best methodology to satisfy storage requirements and to solve the latency problem? The answer is to use fog or edge computing. Fog and edge computing are cloud infrastructures located closer to the edge of the network [1]. In fog computing, the "intelligence" is transferred to the local network, and a pre-processing of the data is performed by an IoT node or gateway.

1.2 Augmented Reality

Since the late 1960s, the use of computer-aided design (CAD) has provided major advantages to industrial sectors such as automotive and aeronautics. The integration of CPS in the industry made possible to create and model any system or equipment in 3D, giving rise to a new concept called Cyber-Physical Equivalence (CPE) [1]. The latter refers to the fact in which the virtual and physical can interact simultaneously in real time. This new concept is also related to the concept of Digital Twins, where there is a virtual simulation of the system with the ability to update and change according to the changes at the physical level. A digital twin is not just an exact virtual copy of a product, but a combination of the two worlds allowing to analyze data, to monitor systems, to test new configurations, and to control the life cycle of a product. Thus, how could it be possible to add value to this combination? It is now possible to take advantage of this virtualization and use technologies that make use of AR to visualize these digital systems.

According to [5], AR is defined by three fundamental requirements: (i) combine real and virtual content, (ii) interactive in real time and (iii) accurate alignment of real and virtual objects. These three characteristics also define the technical requirements that an AR system must have. An AR system can combine real objects with virtual ones allowing interactions with the user, and a tracking system is able to find the position of users and virtual object. The convergence of the concept of digital twins with technologies that make use of virtual or augmented reality can increase the value of this combination by enabling the latter systems to visualize data in real time where, for example, one can operate a piece of equipment without being close to it, or even have the ability to predict a problem.

Considering the enormous potential of AR and IoT, several challenges have arisen to integrate both technologies. AR has as one of its objectives to provide an intuitive interface for IoT, using a virtual overlay of information. Therefore, one of the main challenges for those developing IoT or AR applications is to understand how this connection can be made more intuitive, enabling consumers to interact in a natural way with the virtual object, thus causing state changes in the same object. Service providers using AR also have the challenge of using real-time IoT data to debug and repair faults, hence providing a higher quality of service (QoS) and shorter response times.

More and more companies are taking advantage of the "power" of AI, IoT and AR in intelligent factories to reinvent themselves and gain competitive advantage. For example, mining companies are adopting strategies based on IoT, AR, and AI to increase their effectiveness in mining operations. They can obtain the extracted quantity in real time, perceive the fatigue of their employees or make a daily analysis of the equipment available in addition to these functionalities [6]. Companies in the energy sector use real-time data acquisition and analysis to optimize energy consumption and create sustainable smart cities. Moreover, because of the need to interconnect all components of a factory or a city, companies are forced to improve and innovate through this convergence. Converging IoT, AR and AI make systems increasingly autonomous and problem-solving in many scenarios.

2 Industrial Internet of Things

The impact of IoT in our lives is increasing as most devices already connected in a network to communicate with each other. IoT-based technologies adopted in many sectors such as medical care, security, surveillance, and product management allow many new services to meet the needs of their users. An IoT system must take into account characteristics such as heterogeneity, scalability, interoperability, ubiquitous data exchange, security and privacy, among others due to the plethora of different devices and protocols [7].

As previously stated, IIoT brings together advances of two revolutions that have emerged over decades, that is, the industrial and internet revolutions. These revolutions together with the technological evolution have given rise to three key elements that define IIoT: (i) intelligent machines, (ii) advanced analytics and (iii) people at work. Intelligent machines allow machines to have software and sensors connected to the Internet. Advanced analytics make use of physical data and AI algorithms capable of analyzing data. The last element is to connect people in the office or in an industry so that they can interact with each other at any time or even provide support remotely [8]. The combination of these elements offers new business opportunities, enabling a new perspective and analysis of data that traditional methods did not have. IIoT makes use of IoT applied in the industry since it is based on the characteristics such as heterogeneity, scalability and interoperability [7], that a system must consider. In addition, it uses a large amount of aggregated data, where it is crucial to get better visibility and perception of systems that are often linked to the cloud. Because of the adoption of IoT in the industry, we have been able to transform business operations or processes through analytical mechanisms used in big data and AI. These processes are optimized through gains in operational efficiency, productivity and automation of the equipment, thus generating more profit. Although what was initially said may seem like an M2M system applied to industry, the difference with IIoT is in the operation scale as it is necessary to consider large data streams stored in the cloud in real time and massive data analysis. This analysis thus allows obtaining information and statistics from data aggregated through AI algorithms. In many industry sectors, the introduction of AI into an IoT system has enabled the systems to perform predictive maintenance, i.e., through certain data patterns, events or problems that could stop a piece of equipment have been predicted.

IIoT can be seen in layers. Sensors and devices of pieces of equipment can be found in the lower layer, and gateways or communication hubs, which are responsible for communications between the sensors and the cloud (that will be the third layer), can be found in the layer above. In the fourth layer, there is big data and AI software, which can be utilized to analyze and optimize the desired information. At the first tier, we can integrate customer relationship management (CRM) software, which cannot only help to plan and control processes that are happening at the factory more efficiently, but also inform a customer if changes on the product are required. In Fig. 1, we can see the elements that are currently present on IIoT. These elements are the result of a constant evolution that happened in both the information systems and hardware since this was the only way to create the concept of IIoT. Systems such as CPS, Industrial Automation and the cloud are already known and part of the IIoT concept, however, due to their development and integration with other systems over the years it has been possible to introduce technologies such as AR and AI in order to evolve the IIoT concept. Various business sectors' digital transformation also provided the need to converge technologies already present in IoT with others such as AI and AR. Therefore, the emergence of the IIoT concept is a consequence of the need to innovate and obtain new business models for the benefit of companies.

Fig. 1 Elements of the Industrial Internet of Things

2.1 Cyber-Physical Systems

As we can see in Fig. 1, one of the main components that constitute IIoT is a CPS. Initially, embedded systems were used in small scales and independent of each other, but with advances in technology and communication, it was possible to adopt networked devices. This adoption gave rise to a new concept, called CPS, which functions as the link between a system/equipment and the real world. The term CPS appeared in 2006 and there is no clear definition, however, it is important to note the difference between an embedded system and a CPS. In an embedded system, only information systems are incorporated into physical devices meanwhile CPS are automated systems that allow connecting processes that occur on the physical world with computing and communication infrastructures, whose purpose is to control these same processes and adapt the system to new conditions [9]. CPSs are the integration of computing, networking, and physical processes capable of monitoring and controlling them. Unlike embedded systems, CPSs are designed to be connected in a network with all devices integrated with physical actuators that by acting change the environment. CPS permits the digital world to interact through computers and software with real-world processes. This interaction makes it possible to manage and control processes running in real time. To simplify, one can think of a CPS as containing embedded systems, but giving importance to the domain of communications with the physical world.

With these systems applied to IIoT, industries do not only see what happens but also interpret data through their analysis. Consequently, a CPS has the ability to do predictive analysis, as it was possible to create different scenarios where the probability of an occurrence in a piece of equipment could be estimated. It is also important to consider that these systems can take adjustment measures at any time without human intervention. The introduction of these systems can also be very advantageous, for example, in hazardous environments such as mining that requires constant analysis of workers' biometric data, and intervene promptly if necessary. The ability to expand the interaction of the physical world by means of computer-generated interactive systems such as virtual and augmented reality is a key factor in establishing a two-way link between the physical and digital. To connect a CPS with other computational elements as intelligent mechanisms or 3D models of a system, makes it possible to increase efficiency, usability, autonomy and adaptability.

2.2 Industrial Augmented Reality

Since the 1960s, computer vision has been very important in the industry sector, especially when it was adapted to the control and projection of equipment and products. It is not imperative that an IoT system has integrated visual computing, however, IoT systems using CAD have integrated systems that allow the development of new solutions. It is hence possible to make the system more complete and integrated and to create new business models.

As stated before, a CPS "bridges" the physical and digital worlds by building greater flexibility in the virtual simulation of products or processes before and during an operation. This flexibility together with CAD allows simulating all processes that cover the Product Life cycle Management (PLM), from the design phase of the product where the customer has difficulties in realizing the necessary requirements until the maintenance part. The simulation of a process permits the reduction of production errors or even increase the efficiency of the product development since there is a constant simulation during PLM phases. This modeling/simulation was only possible with the appearance of the CPE [1]. Differently from CPS, CPE is an approach where the virtual world and production environments are synchronized, and not just create a production process in a digital image. The creation of CPE is directly linked to the concept of digital twin. A digital twin is a virtual model of a process, product, or service that overlaps the virtual world with the physical to allow data analysis and system monitoring to avoid problems, optimize maintenance costs or even the planning of future processes. A digital twin can update the system as changes occur on the physical level. According to [10], "Digital twins are becoming a business imperative, covering the entire life cycle of an asset or process and forming the foundation for connected products and services". In an early phase of PLM, this model aims to help companies improve the customer experience by better understanding the needs or requirements of a product, reducing future errors. In the final phase, workers can be assisted with monitoring and analysis capabilities. GE Renewable Energy has developed a digital wind farm with the aim of improving performance, reducing risk and costs [11]. This was made possible through the simulation of a digital environment so that each wind turbine could be configured before construction, thus gaining 20% efficiency by analyzing data from the real environment.

Due to the constant evolution and increase of computational power in technologies such as smartphones and smart glasses, there has been an interest in using both technologies in the industry. Because of this interest, there was a convergence between the digital twin and AR applications, making it possible to create new use cases in the industry. AR enables the visualization of a virtual object in the real world through a device. In 2019, IDC estimates the world's augmented and virtual reality expenditures of almost $20.4 billion [12]. Just think how virtual simulation models can accelerate the entire production chain, or even maintenance processes, from the visualization of the 3D model in AR applications. Furthermore, a virtual layer integrated into the IoT system to obtain and visualize information in all types of industry and environments through AR devices could be added.

The simulation and control product life cycle is also one of the main uses of AR in the industry and is defined in five parts: product design, manufacturing, commissioning, and inspection and maintenance [13]. The first phase, the drawing, is focused on realizing the idea and the requirements that a product should have. Initially, when requirements are specified a user may not have the perception of what they really want, and this phase is important because through AR he can visualize the product or project and change according to the imposed requirements. Consequently, it can significantly reduce error costs that may arise in the future. In the assembly stages, AR and IoT together can assist the assembly processes instead of using typical traditional systems (i.e., manuals). For example, the AR system can show hidden structures (i.e., wires) inside the walls. After the creation of a product or process, it is necessary to verify if there were modifications during the assembly process and AR there could help to verify these requirements. In the maintenance step, a product or system having the ability to use IoT and AR data can make the response shorter in the event of a problem or even provide remote maintenance. In addition to these examples, AR also has another important role in interactive machine maintenance training. AR can provide hands-on training where users are given visual instructions of actual objects. Previously, technicians needed to participate in training courses in order to be able to solve equipment problems. However, with AR it is possible to accelerate the training phase, since manuals are not used anymore, and they were replaced by 3D drawings where you can view the instructions or see where each part is located. Machines and production lines generate a large amount of data that must be read in seconds in order to obtain relevant information in real time. This requires not only new technologies that enable the exchange of large information flow but also new ways of visualizing information. In this context, virtual analysis has come together to "separate" technologies such as big data, IoT, the cloud, and AI algorithms in order to contradict problems previously difficult to solve and analyze or reveal hidden patterns that were not obvious at first sight.

For example, Boeing developed one of the first industrial applications of AR [14], which aided airplane assembly processes to increase worker efficiency and reduce costs. Another example of applications where it is possible to converge AR with IoT is in emergency services. Firefighters are already using devices and equipment built into helmets that receive data such as temperature smoke detectors, and presence status. This information allows a firefighter to see if there are people in a building during a fire.

AR does not only depend on IoT but also on AI to be effective, because only through this convergence it is possible to predict, for example, which interface is shown to the user depending on the surrounding environment. This is possible thanks to the implementation of AI for object recognition and tracking as well as gestures recognition. Recognition and tracking allow people to interact with objects in the virtual space or even move an object in the real space. They do not only recognize an object, but also the gestures or interactions of other objects. For instance, transport companies can use AI to predict weather events in order to divert routes and display information to the driver through AR [15]. The combination of AR with AI also provides users in retail or marketing areas, where users can get information on what they are buying, with a digital representation of the products before they buy or even to obtain recommendations based on their preferences. The purpose of this experience is to enable the customer to reduce uncertainties when they want to buy a product and to allow the company to collect data about the customers themselves.

In summary, all these examples of applications that support industrial processes have given rise to a new concept called Industrial Augmented Reality (IAR). This definition began in the early 1990s with the Boeing project but the convergence of the latest advances in electronics, sensors and paradigms, such as IoT and AI, have led to the development of more advanced applications for industrial systems. IAR has proven to be a widely used tool across several business areas, enabling smart factories where workers monitor and interact with real-time systems. Although there are already some examples of applications that use AR as technology to support industrial processes, IAR is still very early in its development. Computer graphics, AI and object recognition are examples of technologies that allow the use of AR in the industry.

In [41], a project was designed with the goal of training people in the process of assembling and maintaining equipment through a multimodal AR system. In this project, it was possible to conclude that an AR platform could reduce learning and training time compared to traditional methods. Moreover, people made fewer mistakes and achieved better performance times. Accordingly, specific training strategies based on AR benefited not only people that were learning new skills but also reduced costs associated with those processes.

2.3 IoT Communication Systems

Wireless networks play a key role in IIoT as they interconnect devices present in the industry. They are also responsible for making the traditional industry more "connected", along with technologies such as CPS, the cloud, AI and AR.

At the beginning of the third millennium, wireless networks generated a growing interest from an industrial perspective. However, they were considered slow and insecure and many security departments avoided their use. This initial wireless technology only allowed limited bandwidth and coverage, and although security was not yet a priority, it was an important issue to consider in the future. The wireless medium is "open" and because of that, anyone can listen to any frequency as long as it is transmitted through the air, being this the main reason for them being discouraged in the industry. Another important aspect of this technology was the need for an Access Point (AP) to have a network identifier (SSID) so that wireless devices could identify and connect to the network [16]. The first APs were open, i.e., without authentication credentials, and the data was either unencrypted or protected by a very weak security protocol called the Wired Equivalent Protocol (WEP). With the improvements in security, specifically Wi-Fi Protected Access II (WPA2) encryption and authentication [17], and speed over time, these types of networks became essential in many industry sectors.

In M2M communications, sensors capture events that are transmitted over the network to a system that converts these events into information. Due to enhancements on Wi-Fi technology, M2M communications are no longer just one-to-one being also possible to transmit data to multiple devices. However, what happens when there are no Wi-Fi networks available, that is the case, for instance, of remote areas? To answer the latter, please consider sensors whose key requirements are low power consumption and long-range data transmissions. Most IoT applications use short-range frequencies, about 10–100 meters, and in many use cases, this range is not enough, as sensors might be located in distant areas. Traditional communication technologies such as Wi-Fi, Bluetooth or ZigBee, can only easily handle applications with a reduced transmission range. Conceptually, radio wave indicates that the reach is inversely proportional to the frequency, that is, the lower the frequency, the higher the reach. Hence, if a lower frequency is used, the range increases and the signal penetrate obstacles easily. The latter motivated the creation of low-power wide area network (LPWAN) such as IEEE 802.11ah, SigFox, and LoRa, which scatter large scopes and consume very little energy, thus meeting the requirements imposed by IoT systems. The main benefits of LPWAN are of being able to transmit a signal through dense locations and having reduced power consumption, which permitted sensors to have a longer "life" (i.e., 5–10 years). These networks use low frequencies allowing optimizing energy. LPWAN besides having a good range, is also better than Wi-Fi as the signal is not interrupted if there are obstacles in front [18]. There are other wireless technologies that met IoT requirements such as IEEE 802.15.4, low-power Bluetooth, ZigBee-IP NAN, and cellular M2M.

On the other side of the spectrum, there is also a variety of mobile networks that have gained attention in IoT applications, as they are far-reaching and do not use much energy. That is the case of 2G, 3G, and 4G mobile networks. Although the latter were not optimized for this purpose as they were not prepared for Machine Type Communication (MTC) [19], there was an interest in adapting them to IoT systems, thus, creating the NarrowBand-IoT (NB-IoT). This standard was created by the Third Generation Partnership Project (3GPP) [20] to address the need for very low data rate devices that needed to be connected to mobile networks. Despite being similar to Sigfox and LoRa, NB-IoT uses a faster modulation rate consequently requiring a higher power consumption [21]. With the emergence of 5G this problem can be overcome, given that it provides faster data rates with low latency and improved coverage for MTC communications, hence enabling the adoption in IoT applications. 5G networks are still in their infancy being the successors of 4G LTE network [19], nonetheless, they will not be affected by Wi-Fi signals, buildings, etc. Additionally, 5G networks will significantly contribute to IoT by linking billions of smart devices where they can interact with an intelligent environment, without human intervention through faster communication links, increased frequency range, and guaranteed energy efficiency. This network will have speeds faster than the current ones (4G LTE), that is, speeds up to 10 Gbps and will allow to connect numerous devices at the same time.

It is not easy to attain a network frequency that can have characteristics such as speed, long distance, transmission power, reliability, and latency, because while some give priority to range and latency other cherish speed. In IIoT, there are still many challenges related to the type of communication necessary in a system, because one may have to consider technical aspects that can be deterministic for a system to be able to respond to all eventualities.

3 Artificial Internet of Things

Nowadays, AI is not only reduced to the ability to imitate human thought or to act as such but rather as a complete set of operations resulting from the computational capabilities available today. AI can take advantage of information technologies, including data collection and processing (i.e., machine learning, natural language, etc.). The concept of IoT is essentially to get data from connected devices and take advantage of them to transform data into information that can be used by people or devices. However, the great challenge is when we analyze data to obtain information after data analysis. The adoption of interconnected systems and the use of IoT have motivated the use of AI because much of the sensor data coming from several sources is unstructured. What would you do without the ability to automate information? Would it be possible to analyze large amounts of data without using AI, given that old information management approaches cannot manage unstructured data?

Several AI forms have been used for decades, with the aim of "making sense" of unstructured data, transforming it into relevant information. Therefore, the combination of IoT and AI is no longer surprising. Nonetheless, the big challenge is to analyze data and acquire information.

Big data refers to a large set of structured or unstructured data from various data sources that cannot be stored in structured databases [16]. This information coming from big data increases the efficiency of machine learning and AI algorithms, as the greater the amount of data, the greater the reliability of the information generated. In addition, it is thanks to big data and its data analysis enabled the efficiency increase of AI and its commercial value. IoT systems have sensors that interact with the environment through data analysis, hence deciding to send an event to an actuator that will act faster than a human. In many industry sectors, the time a machine is inactive causes a reduction in productivity as well as costing a lot of money. The fact that IoT and AI exist mainly in industrial environments makes it possible to create computer systems with the ability to perform predictive analysis and reduce maintenance costs. Predictive maintenance makes use of AI analytics with IoT data aiming to predict in advance equipment failure and scheduling a maintenance procedure. A suitable approach to converge AI with IoT is the development of robots or systems whose purpose is to supervise independently tasks or events without human intervention, or even to create concepts where intelligence can be part of a network of sensors. There is a need to give machines the ability to understand the information they collect in order to create patterns that allow them to act in case of problems. Future AI trends include solutions where there is integration with IoT technologies, security, and cloud computing. AI techniques and IoT systems along with cloud computing can be used to provide analytics and decision making in a more efficient and autonomous manner. The integration of AI-based algorithms to process data and make decisions in different architectures of an IoT system allows combining new technologies like bots and AR to create new solutions [22]. An IoT system that does not integrate these technologies in which it can learn, interact and increase its capabilities to respond to an event, is a system where there is only data transmission without any meaning.

3.1 Bringing Intelligence to the Data

As we attempt to make machines take autonomous and more efficient decisions than a human, AI inevitably becomes a necessity. In the fourth industrial revolution, industry systems are transformed into digital systems. IoT and big data assume an important role to enable companies to deal with the large volume and variety of data coming from the equipment. Here, IoT describes a network of interconnected devices constantly sending data, making it primarily responsible for the big data concept. In the year 2020, it is estimated that around 50 billion devices will be connected to the Internet [23]. The digital transformation in the industry from sensors to IoT systems fostered the creation of new business models. Emerging technologies, e.g. the cloud, enabled major analysis hence providing ubiquitous access to information.

In addition, it allows companies to adopt IoT systems, big data, and AI to increase their competitiveness through predictive analytics. One of the main challenges that today is imposed from the analysis of big data and AI algorithms is the visualization of results, since the latter is considered during decision-making.

Big data can be characterized by three concepts: volume, variance, and velocity [24]. The volume refers to large amounts of data that are generated from multiple sources. The variety refers to the use of various types of data from different sources to analyze a situation or event. The velocity corresponds to the time that a system takes to interpret the data and give a response. In IoT, a system can have multiple devices generating a flow of constant data that not only result in a large volume but also incorporates several events. As more IoT systems decision-making depends on data, its velocity is also a factor that influences decision-making. Some data, e.g., the one obtained from M2M sensors, require real- or near-real-time analysis, therefore placing additional pressure on storage systems and big data algorithms. An example of the importance of the latter is the stock and financial market data, where answers in milliseconds are required. This is a decisive factor in the Industrial Internet because devices tend to send values to a processing domain and wait for the response. This response can sometimes be influenced by several factors on the network such congestion which may influence response times.

AI has become imperative in IoT systems because they both depend on one another and cannot exist without big data and intelligent data analysis creating a new concept called *Artificial Internet of Things*. AI technologies rely not only on large amounts of data but on the quality of these data since a better quality is related with a better prevention or detection of patterns. The growth of AI is proportional to the growth of big data as the availability of data volume is what permits the implementation of AI in IoT systems. The greatest benefit in adopting the concept of AIoT is the possibility of creating a cognitive computing system capable of detecting patterns in the data and consequently predicting flaws or anomalies that a human would have difficulty detecting.

There are several types of AI algorithms that are used to label information from unstructured data. Supervised Learning is an AI algorithm that assumes that the data used to train an algorithm is labeled with its information and is typically used in small-scale systems. The data provided do not need to be labeled but require more accurate learning techniques [4]. Clustering or semantic analysis are examples of algorithms where data is not labeled and can be applied depending on the problem. These techniques allow unstructured data to create standards or to group automatically data following a degree of similarity, where this criterion of similarity is part of the definition of a problem. By creating this pattern of data from sensors, an anomaly occurrence can be prevented since data being received does not fit the pattern, or vice versa, or even the data may be within a pattern that will eventually create a problem.

Although IoT depends on big data and AI to add commercial value to both technologies, visual analysis through AR could help in the convergence of these technologies. The industry can be a very a demanding and challenging scenario for visual analysis because modern machines and production lines can generate many data simultaneously and the only way to make sense of this data is through their convergence.

3.2 Edge and Fog Computing

Given a large amount of data generated by IoT devices, the cloud architecture often cannot handle all sorts of information. While the cloud provides us with services such as computing, storage, and communication, these features when centralized can cause problems such as latency or poor performance. Both problems can be determinant when asserting a system's efficiency. Aiming to address both problems, two concepts are presented, namely edge and fog computing.

Edge computing is not a new concept but has generated a growing interest as it helps industrial organizations to turn large amounts of data, e.g., from sensors, into intelligence, but closer to the data source. Edge computing is a decentralized computing infrastructure in which computing resources can be used "at the edge" of the network [25]. The edge is a logical layer rather than a physical one, therefore, this concept can be implemented depending on business logic. The key benefits of edge computing are to improve performance, decrease data privacy and security concerns and reduce operational costs. Edge is not only a way to collect data before you send it to the cloud, but also a way to analyze, process and act on these data. Thus, it is important to optimize industry data. Only the aggregate data should be sent to the cloud, hence reducing bandwidth and improving the response time.

For instance, in a wind farm, if wind speed changes direction, the edge software can analyze this data in real time and adapt the turbines to optimize production. This data could also be sent to the cloud. However, the latter data could be less productive and costly. Another important point is privacy and security. Edge computing allows businesses to reduce Internet connections, thereby reducing security risks by using local computing. The last point emphasizes the reduction of operational costs. Bandwidth and latency are expensive features, and this is where edge computing comes into play by reducing these costs as well as by minimizing communication needs. If a company is located at a remote location and it needs to ensure that it has control over the production rate, the alternative is to build an edge computing infrastructure, where it is possible to process data in real time, maintaining the flexibility that this approach offers [25]. So, it is possible to increase operational efficiency and reduce maintenance costs. Edge computing creates an infrastructure to handle a large amount of data by allowing reducing events response times, and by avoiding being constantly dependent on the cloud. In short, it is possible to avoid large bandwidth requirements, which also allows to analyze data locally, and at the same time, to ensure data security.

Fog computing is a concept in which intelligence can also be introduced close to the edge, being sometimes confused with edge computing. However, both concepts are different. Fog computing is a standard that defines how edge computing should work or be implemented thus facilitating computing, storage and network operations between cloud computing devices and data centers [26]. From a different perspective and depending on the requirements, fog can even be a cloudlet, an edge device or a micro-datacenter.

In 2016, Bombadier, an aerospace company, chose to use sensors in its aircrafts [26]. This innovation enables real-time performance data and efficient problem solving without needing to land the aircraft. By putting edge processing closer to sensors, one can reduce the need to send constantly data to the cloud or server. It is, therefore, possible to identify urgent problems that may arise, minimizing response time. Although being possible to process data on the edge when there is a need to do a predictive analysis based on data previously collected to determine whether an equipment can or not fail. Though, sometimes this analysis should be done in the cloud. The reason behind is the analysis and algorithms that require a tremendous amount of data.

Both concepts have their disadvantages because by putting too much processing at the edge, it is easy to overload devices or storage. Moreover, the limited redundancy is another problem. Both edge and fog computing are providing IoT architectures with more options, removing the boundaries of centralized servers, in addition to making IoT more distributed and flexible.

Figure 2 shows a possible architecture of edge and fog computing to facilitate the perception of the difference between them. The cloud features are at the top layer. In the middle layer, we have the architecture complexity, as it includes nodes in the network attached to other nodes in the fog network. These nodes consist of various devices on the network such as gateways and switches. Finally, the lower

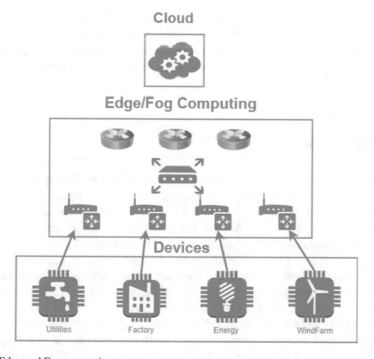

Fig. 2 Edge and Fog computing

layer represents sensor devices of an infrastructure. The lower layer sends data to the middle layer where it is processed by the network nodes. This processing is performed near the devices, hence reducing a transmission step that the network is subject to. After this processing is performed data is sent to the cloud [27].

In [28], an architecture called EXEGESIS is presented, which focuses on the concept of edge and fog computing. EXEGESIS is composed of three layers (mist, vFog, Cloud), where in the lower one are located the devices (i.e., sensors, actuators, mobile devices, etc.) that create connections between them forming a "neighborhood". This layer follows a hybrid peer-to-peer (P2P) approach consisting of regular nodes (RMNs) and supernodes (SMNs). SMNs create a logical fog topology along with the features provided by RMN and interact with the virtual fog (vFog) located in the above layer, and allow dynamic interconnections between various neighborhoods. Finally, in the upper layer, there is the traditional cloud that establishes connections with vFog, in order to provide computing resources and facilitate interconnections between various vFog elements. With this architecture, one can take advantage of existing cloud architectures just by changing the edge of the network to make the best it.

4 Applications

Although many IoT solutions are consumer-oriented, this trend is changing since the growth of IoTs is dramatically changing various industry sectors. This trend gives companies access to AI systems, communication mechanisms and new ways to interact with the digital world so that there is an improvement in the efficiency of their services and productivity.

4.1 *Manufacturing Sector*

The manufacturing industry is one of the sectors that benefited most with the introduction of IoT. Manufacturers in all areas, including automotive and electronics, invest in embedded devices to control and automate their equipment. Imagine, for example, that a customer could point a camera to a mobile device and see the digital model of an equipment and its associated values or problems.

Caterpillar (CAT) [29], a manufacturer of construction machinery equipment, has transformed its business by introducing IoT, AI and AR technologies and solutions. The company began equipping their machinery equipment with sensors as well as integrating software tools and analysis based on the cloud. This software helped its customers to reduce operational costs, decreasing maintenance time, and performing predictive analytics by making their machines autonomous. Bucklar, which is a CAT customer, saved about $600,000 in production costs through predictive maintenance. Another customer in the construction sector increased the use of their machinery equipment by 15% through analysis of data coming from the equipment [30]. Caterpillar Liveshare is an example where one can use AR as a collaboration tool allowing remote technicians and experts to connect through video and audio to solve a real-time problem. This AR video platform uses voice, 3D animations, annotations where users can draw, highlight and place cues on real-world objects and share screens [29]. Video streaming via Wi-Fi or mobile phone is also possible. If Wi-Fi or mobile coverage is low, technicians can send instant images instead of video streams, therefore reducing bandwidth. CAT has developed this platform to reduce the time to solve problems, increase service revenues, and increase customer satisfaction.

In 2014, *Bosch* Automotive Service Solutions introduced a platform called Common Platform Augmented Reality (CAP) that provided a multi-platform system capable of creating AR applications. The Bosch CAP allowed the integration of visual and digital contents in the creation process, through live images and a database in which content corresponding to the AR application was extracted [31]. This visual overlay is done through 3D models, photos, texts (information and explanations) and the tracking of objects. For maintenance, servicing or repairing, the CAP is used to view hidden cables or components and repair instructions that permitted to view the required work steps and the relevant tools. It was also possible to check the status of parts with real-time information. The CAP allowed to magnify and rotate objects and provided additional information such as training videos. Its main goal was to reduce time by taking faster decisions and to increase the assembly process, save costs and reduce error rates.

ThyssenKrupp is one of the leading elevator construction and maintenance companies in the world and makes use of AI, AR, and IoT in their equipment. The constant pressure of Asian markets has led the company to create new services such as predictive maintenance, automation of warehouses and assistance to the maintenance processes in order to be able to differentiate itself from its competitors. ThyssenKrupp's primary goal is to reduce significantly maintenance costs, to ensure that their elevators worked properly, and independently control their inventory. Working alongside Microsoft, the company was able to connect sensors in their elevators to the Azure cloud [32]. Through this system, they were able to capture data such as engine temperature, shaft alignment, speed and door operations. When a technician heads to a place, through the HoloLens mixed-reality headset, he is able to visualize all the equipment and the values of its components. Tools like AI enabled the perception of certain patterns in values that could cause problems in their equipment, hence accomplishing a predictive analysis.

Moreover, as part of the company's digital transformation, the need to control everything that goes in and out their warehouses arose. This transformation has eliminated physical inventories saving hours of work, and a virtualization of the warehouse where each piece could be easily located. This virtualization brought a new concept, called digital warehouse, which enabled a 30% improvement in terms of materials processing efficiency [33].

4.2 Aerospace Sector

The acceptance of IoT technologies in the aerospace industry has been slow because of the existence of complex systems whose costs are hard to estimate. However, the interest of airlines in these technologies increased due to their apparent benefits. An aircraft that has a system that could predict problems before they actually happen would save maintenance costs. The goal of aerospace IoT is to establish an integrated sensor infrastructure with computational intelligence, reliable information and component manufacturing in a more conscious manner. For this, the large volume of data produced by sensors is used to provide alerts or give information in real time through a descriptive (what happened) and predictive analysis. It is also necessary to take into account network latency and data security. Therefore, in IoT, edge analysis is being used, thus avoiding the need to send data continuously to the cloud.

Airbus is the largest aeronautical company in Europe and is responsible for assembling aircrafts and helicopters in various parts of the world. In order to create a new aircraft production process by means of digital tools, the Mixed Reality Application (MiRA) was created in 2009. MiRA is fully integrated with its information systems and combines real images with digital models on a tablet equipped with a camera. This application increases productivity on production lines using AR to visualize parts and detect errors. Equipment such as tablets and sensors, and AI software specially designed for this application were used to reduce the time required to check, recover and mount most probably damaged parts. With this application, it was possible to reduce from 300 h of fuselage (that is, the main carcass of the aircraft) support to 60 h. With the use of IoT, it is possible to increase efficiency, safety and control of flights, because of a better data interpretation.

4.3 Logistics Sector

Logistics is one of the areas that continually needs improvements to support the growing demand for products. Logistics companies depend on networks to manage large amounts of data volumes, allocate assets and keep track of deadlines. AI can organize and optimize these networks for levels of efficiency that a human cannot do so effectively. AI is able to adjust behaviors in the various sectors to reduce costs, time and increase productivity with cognitive automation [34]. Cognitive automation refers to the combination of AI and robotic processes for intelligent process automation, where AI can extract, perceive and learn unstructured data, and the robotic process executes rules-based workflows.

Logistics companies often rely on third parties such as sub-contractors, airlines, etc., leading accounting departments to process millions of invoices. AI technologies can make this work easier by extracting account information, dates, and unstructured data and inserting it directly into accounting software. Another important point related to logistics issues where AI can be useful is in customs. Customs clearance processes usually require many documents, tax returns, invoices, brokerage costs, etc. that need to be validated by employees. In other words, they are complex processes that require a lot of effort over many hours of work, because it is difficult to maintain the level of concentration at the end of the day, and, consequently, there are errors that can be costly. The solution to this is to use AI to facilitate these processes. Through training algorithms, one can use legislative materials, standards, and manuals to automate customs declarations [34].

As an example of the logistics sector, please consider DHL. *DHL* is an international logistics company that is trying to implement AR and AI in its industry, in areas such as storage operations, transportation optimization, delivery and assembly/disassembly services. According to [35], any paper approach is slow and prone to errors, and because of this, an AR-based software was created to allow real-time object recognition, barcode reading, internal navigation and information integration for a warehouse management system. With this system, each worker can visualize the best route to the location where the object is, and use bar code for automatic readings and recognition of places. It is also possible to visualize a storage process in order to understand how the warehouse is organized. The AR can optimize the transportation of cargo through its verification, therefore allowing knowing if there is available capacity in a truck to transport. An important point in the delivery service is to minimize transportation time, as it is many times affected by transit. Recently with the new cities infrastructures, it is possible to have digital maps with traffic patterns that allow improving the overall routing. As a result, the DHL group has implemented a feature to optimize transport using both AR and AI, allowing the driver to view an outdoor navigation system with real-time traffic information and if necessary to optimize this route. In the delivery service when using AR, it is possible to perceive which driver is appropriate to a given cargo and to calculate and visualize the free space in each transport. DHL is responsible for storing Audi components as well as for assembling their components. To optimize this assembly process and reduce error rates, an AR system powered by AI can provide an intuitive way to visualize the process to provide instructions for each step of work.

4.4 Water Sector

Thames Water is one of the largest British water-recycler companies that, through the implementation of sensors, remote communications, and big data algorithms can anticipate equipment failures and respond more efficiently to any critical situation that may arise due to adverse weather conditions. Additionally, it has installed smart meters for customers have access to water consumption through the telephone and gain control over their spending. With the water monitoring, it is possible for consumers to inform the water company quickly if there is a problem and most probably reduce consumption.

The company believes that it has found a way to adopt intelligent systems through IoT to make simpler and more efficient measurements. The goal of adopting IoT technology would be to make water consumption more sustainable and to have a closer relationship with the customer, and to provide more accurate data, hence, being able to improve the company's water supply system knowledge, for example, to where the water goes, or if leaks exist. The company estimates that by 2025 the number of readings per day through these meters rises to 35 billion.

The idea of combining computers, sensors, and networks to control systems has been around for many years. Usually, technologies such as IoT and AR are associated with applications for personal use such as home automation, and not so much with the industry. However, these technologies together with AI can create new business models or improve existing ones.

As was seen in the examples above, there are several industry sectors in which the adoption of systems where IoT, AR, and AI converged was notorious. In these same examples, it can be seen that there are several industry sectors where this convergence is more pronounced. From manufacturing to logistics, new solutions based on current market technologies were implemented fostering even more the digital transformation. The convergence of IoT with AI and AR allows solving problems that used to be hard, optimizing costs that were formerly invisible, and more importantly, to predict future problems. IoT leverages on AI to become efficient because the IoT concept alone, i.e., without a system that interprets data, seems weak. AR benefits from IoT and AI to display intuitively necessary information, also enabling a better interaction between physical and digital systems.

The latter examples have shown that there is a great deal of concern with the adoption of new technologies such as AR, IoT and AI by the industry. However, despite the increasing tendency in adopting them in various industry sectors, Fig. 3 indicates that this tendency will become even more prominent over the years [36]. Artificial

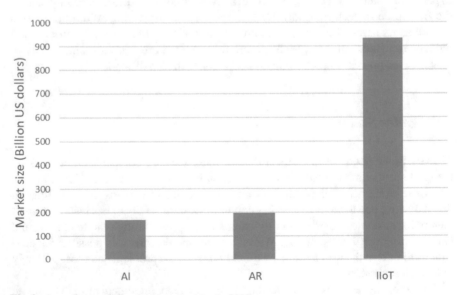

Fig. 3 An estimate of the global market size in 2025

intelligence has been one of the fast-growing technologies in recent years [37]. The development of cloud and big data infrastructures, and AI-based solution have led to a strong growth of the AI market, which has grown due to the significant impact that AR has had on industrial applications. Industry such as logistics, manufacturing, and aerospace are taking advantage of what this technology can provide to increase operational efficiency. It is estimated based on this growth that the market size will be close to 200 million in 2025 [38].

5 Future Directions

The digital transformation forced industry and companies to create their business models based on a competitive strategy and with this transformation, there was a need to create new or converge existing technologies.

Initially, both IoT, AI, and AR emerged as separate concepts to address various problems. However, there was a need to converge them to create new business models. In addition, this convergence leads to a greater connection between customers and companies, product innovation, new business models, process automation, labor replacement, and more efficient decision-making. Although there may be a need to value IoT data, in the future it is expected that AI algorithms can be further embedded in IoT system architectures and subsequently create more AIoT applications.

Future trends are based on creating autonomous decisions through AI along with IoT systems combining with new technologies such as bots and AR/VR. Thus, IoT applications can continuously learn, interact, ensure security and then carry that knowledge to edge processing [39]. Apart from the industrial sector, where there is already a convergence of IoT and AI, it is also in projects of intelligent cities that arise paradigms in which the goal is to optimize energy consumptions.

IoT systems are widely used in projects where there is a need to implement AI in edge and fog computing systems to address the enormous amount of data generated by devices. An example of this is in [40] in which an energy management architecture with edge computing based on a deep reinforcement learning (DRL) network is presented, in which both the cloud and edge servers implement machine learning algorithms. Network delay (jitter, latency) in data delivery, e.g., during data transmission, is a problem in disruptive systems and affect these systems' QoS [41, 42]. To address the latter, cloud services have been moved to the network edge. However, this may not fully solve the problems of IoT systems in environments such as factories, buildings, and smart cities where resources such as bandwidth are sufficient to be able to respond to large amounts of data. Many of the network technologies available today are not appropriate for the future, so the solution is to rely on the introduction of future mobile networks (e.g., 5G) [43]. In the coming years, IoT systems will be even more complex being that the reason why 5G will affect these systems by allowing data transfer rates faster than 4G. It will also support a large number of devices at the same time over a period of time.

Edge computing, big data, AI and AR along with IoT will certainly be among the major developments in the coming years. The ability to take advantage of large amounts of data, combine AI methods and to be able to view this information intuitively with AR, making convergence important for a new digital transformation.

6 Conclusion

The business world as we knew it a few years ago is not the same as today. There are industry sectors such as logistics, manufacturing, etc. that already started adopting IoT, AI, and AR. The combination of the latter technologies in the various sectors is no longer so surprising because of the digital transformation, and their introduction is nearly a requirement as it allows creating value, that is, new products and services. Some of the main concerns of companies will always include improving productivity, reducing costs and time, as only then they can become competitive in face of market demands. As these technologies are adopted, more data will be generated and consequently, it will be possible to improve many business processes. By analyzing these technologies separately, one may see how they depend on each other. Although IoT together with the power of the cloud has already made a global impact on the industry by allowing thousands of devices to be connected, the IoT concept alone seems weak. At a basic level, it is just a system with sensors that sends data to a storage system. Thus, there is a need to converge IoT with AI because data that is sent from several sources can increase the efficiency of machine learning algorithms. AR is a complement to extend information and interact with a digital system more efficiently. From the choice of articles in a warehouse to aiding a maintenance process, AR can take advantage of the information from IoT systems with AI. As mobile devices become smaller and more capable, it became possible to adopt AR in the industry as a means of visualizing a system. Some argue that this convergence may have negative consequences such as job cuts; however, that does not seem the case. Converging these three would only allow to optimize various factors in the industry, reduce manual work that no one wants to do, and above all increase human potential with their help.

According to the International Data Corporation (IDC), we are on the third evolution platform, which relies on pillars such as mobile, big data, analytics, cloud and social technologies [44]. However, some companies claim that they are already reaching the fourth platform with the convergence of these technologies. The fourth platform is the consequence of companies always trying to innovate in order to make their business models more productive and more differentiated from their competition. This is why the elements illustrated in Fig. 1 are already the consequence of this desire to innovate. Even though AI and AR belong to the "supposed" fourth platform, they are already being introduced in some sectors of the industry.

References

1. Posada, J., et al.: Visual computing as a key enabling technology for Industrie 4.0 and Industrial Internet. IEEE Comput. Graph. Appl. **35**(2), 26–40 (2015)
2. Rashid, Z., Melià-Seguí, J., Pous, R., Peig, E.: Using augmented reality and Internet of Things to improve accessibility of people with motor disabilities in the context of Smart Cities. Futur. Gener. Comput. Syst. **76**, 248–261 (2017)
3. Wang, L., Törngren, M., Onori, M.: Current status and advancement of cyber-physical systems in manufacturing. J. Manuf. Syst. **37**, 517–527 (2015)
4. Arsénio, A., Serra, H., Francisco, R., Nabais, F., Andrade, J., Serrano, E.: Internet of intelligent things: bringing artificial intelligence into things and communication networks. Stud. Comput. Intell. **495**, 1–37 (2014)
5. Azuma, R.T.: A survey of augmented reality. Presence: teleoperators and virtual environments. In: Proc. Work. Facial Bodily expressions Control Adapt. Games (ECAG 2008), vol. 1, pp. 355–385 (1997)
6. Accenture: The Connected Mine Solution. Available https://www.accenture.com/us-en/service-connected-mine-solution (2019). Accessed 27 Jan 2019
7. Miorandi, D, Sicari, S., De Pellegrini, F., Chlamtac, I.: Internet of things: vision, applications and research challenges. Ad Hoc Networks (2012)
8. Evans, P.C., Annunziata, M.: Industrial internet: pushing the boundaries of minds and machines. Gen. Electr. (2012)
9. Jazdi, N.: Cyber physical systems in the context of Industry 4.0. In: Proceedings of 2014 IEEE International Conference on Automation, Quality and Testing, Robotics, AQTR (2014)
10. Poddar, T.: Digital twin bridging intelligence among man, machine and environment. In: Offshore Technology Conference Asia 2018, OTCA (2018)
11. GE: Digital Wind Farm. Available: https://www.ge.com/content/dam/gepower-renewables/global/en_US/downloads/brochures/digital-wind-farm-solutions-gea31821b-r2.pdf. Accessed: 13 Mar 2019
12. IDC: Worldwide Spending on Augmented and Virtual Reality. Available: https://www.idc.com/getdoc.jsp?containerId=prUS43860118 (2018). 31 Mai 2018
13. Fite-Georgel, P.: Is there a reality in Industrial Augmented Reality?. In: 2011 10th IEEE International Symposium on Mixed and Augmented Reality, ISMAR 2011, pp. 201–210 (2011)
14. Caudell, T.P., Mizell, D.W.: Augmented Reality: An Application of Heads-Up Display Technology to Manual Manufacturing Processes, vol. 2, pp. 659–669 (2003)
15. Business, A.: Augmented Reality needs AI in Order to be effective. Available https://aibusiness.com/holographic-interfaces-augmented-reality/ (2018). Accessed 13 Mar 2019
16. Gilchrist, A.: Industry 4.0: The Industrial Internet of Things (2016)
17. Lashkari, A.H., Danesh, M.M.S., Samadi, B.: A survey on wireless security protocols (WEP, WPA and WPA2/802.11i). In: Proceedings—2009 2nd IEEE International Conference on Computer Science and Information Technology, ICCSIT 2009, pp. 48–52 (2009)
18. Frenzel, L.: What's the difference between IEEE 802.11ah and 802.11af in the IoT? Electron. Des. **65**(8), 30–31 (2017)
19. Li, S., Da Xu, L., Zhao, S.: 5G Internet of Things: a survey. J. Ind. Inform. Integr. **10**, 1–9 (2018)
20. The 3rd Generation Partnership Project: 3GPP: A Global Initiative. Available: https://www.3gpp.org/ (2019)
21. Mekki, K., Bajic, E., Chaxel, F., Meyer, F.: A comparative study of LPWAN technologies for large-scale IoT deployment. ICT Express **5**(1), 1–7 (2019)
22. Serrano, M., Dang, H.N., Nguyen, H.M.Q.: Recent advances on artificial intelligence and internet of things convergence for human-centric applications. In: ACM International Conference Proceeding Series (2018)
23. Badouch, A., Krit, S.D., Kabrane, M., Karimi, K.: Augmented reality services implemented within smart cities, based on an internet of things infrastructure, concepts and challenges: an overview. In: ACM International Conference Proceeding Series (2018)

24. O'Leary, D.E.: Artificial intelligence and big data. IEEE Intell. Syst. **28**(2), 96–99 (2013)
25. Satyanarayanan, M.: Edge computing. Computer **50**(10), 36–38 (2017)
26. Jablokow, A.: Edge Computing Vs Fog Computing. Available https://blog.ipswitch.com/edge-computing-vs-fog-computing (2018). Accessed 14 Mar 2019
27. Singh, S.P., Nayyar, A., Kumar, R., Sharma, A.: Fog computing: from architecture to edge computing and big data processing. J. Supercomput. **75**(4), 2070–2105 (2019)
28. Markakis, E.K., et al.: EXEGESIS: extreme edge resource harvesting for a virtualized fog environment. IEEE Commun. Mag. **55**(7), 173–179 (2017)
29. Caterpillar: Welcoming Augmented Reality to the Industrial World. Available https://www.cat.com/en_US/articles/customer-stories/built-for-it/augmented-reality.html (2018). Accessed 10 Dec 2018
30. Shields, N.: Caterpillar is embracing the IoT to improve productivity. Available https://www.businessinsider.fr/us/caterpillar-is-embracing-the-iot-to-improve-productivity-2017-11 (2017)
31. Jetter, J., Eimecke, J., Rese, A.: Augmented reality tools for industrial applications: what are potential key performance indicators and who benefits? Comput. Human Behav. **87**, 18–33 (2018)
32. Petrov, E.: ThingWorx, Augmented Reality, and IoT in the workspace. Available https://design-engine.com/thingworxaugmented-reality-and-iot-in-the-workplace/
33. Buntz, B.: Digital transformation strategies from ThyssenKrupp North America's CEO. Available: https://www.iotworldtoday.com/2017/12/05/digital-transformation-strategies-thyssenkrupp-north-america-s-ceo/. Accessed 20 Mar 2019
34. Gesing, B., Peterson, S.J., Michelsen, D.: Artificial Intelligence In Logistics (2018)
35. Glockner, H., Jannek, K., Mahn, J., Theis, B.: Augmented Reality In Logistics (2014)
36. Research, G.V.: Industrial IoT Market Size. Available https://www.grandviewresearch.com/press-release/global-industrial-internet-of-things-iiot-market (2019). Accessed 22 Mar 2019
37. Research, A.M.: Artificial Intelligence (AI) Market Overview. Available https://www.alliedmarketresearch.com/artificial-intelligence-market (2019). Accessed 22 Mar 2019
38. BIS Research: Augmented reality (AR) market size worldwide in 2017, 2018 and 2025 (in billion U.S. dollars). Statista. Available https://www.statista.com/statistics/897587/world-augmented-reality-market-value/ (2018)
39. Magaia, N., Pereira, P., Correia, M.: REPSYS: A Robust and Distributed Reputation System for Delay-Tolerant Networks. In: Proceedings of the 20th ACM International Conference on Modelling, Analysis and Simulation of Wireless and Mobile Systems—MSWiM '17, pp. 289–293 (2017)
40. Liu, Y., Yang, C., Jiang, L., Xie, S., Zhang, Y.: Intelligent edge computing for IoT-based energy management in smart cities. IEEE Netw. **33**(2), 111–117 (2019)
41. Mavromoustakis, C.X., Batalla, J.M., Mastorakis, G., Markakis, E., Pallis, E.: Socially oriented edge computing for energy awareness in IoT architectures. IEEE Commun. Mag. **56**(7), 139–145 (2018)
42. Magaia, N., Sheng, Z., Pereira, P.R., Correia, M.: REPSYS: A robust and distributed incentive scheme for in-network caching and dissemination in Vehicular Delay-Tolerant Networks. IEEE Wirel. Commun. Mag. pp. 1–16 (2018)
43. Magaia, N., Sheng, Z.: ReFIoV: a novel reputation framework for information-centric vehicular applications. IEEE Trans. Veh. Technol. **68**(2), 1810–1823 (2019)
44. Four pillars: Daraja Academy. Available http://daraja-academy.org/pillars-foundation-success-daraja/ (2017)

IoT Detection Techniques for Modeling Post-Fire Landscape Alteration Using Multitemporal Spectral Indices

Despina E. Athanasaki, George Mastorakis,
Constandinos X. Mavromoustakis, Evangelos K. Markakis, Evangelos Pallis,
and Spyros Panagiotakis

Abstract It is challenging to detect burn severity due to the long-time period needed to capture the ecosystem characteristics. Whether there is a warning from the IoT devices, multitemporal remote sensing data is being received via satellites to contribute to multitemporal observations before, during and after a bushfire, and enhance the difference on detection accuracy. In this study, we strive to design an infrastructure to fire detection, to perform a qualitative assessment of the condition as quickly as possible in order to avoid a major disaster. Studying the multitemporal spectral indicators such as Normalized difference vegetation index (NDVI), Enhanced vegetation index (EVI), Normalized burn ratio (NBR), Soil-adjusted vegetation index (SAVI), Normalized Difference Moisture (Water) Index (NDMI or NDWI) and Normalized wildfire ash index (NWAI), we can draw reliable conclusions about the seriousness of an area. The scope of this project is to examine the correlation between multitemporal spectral indices and field-observed conditions and provide a practical method for

D. E. Athanasaki (✉)
Department of Infrastructure Engineering, University of Melbourne, Melbourne, Australia
e-mail: dathanasaki@student.unimelb.edu.au

G. Mastorakis
Department of Management Science and Technology, Hellenic Mediterranean University, Agios Nikolaos, 72100 Crete, Greece
e-mail: gmastorakis@hmu.gr

C. X. Mavromoustakis
Department of Computer Science, University of Nicosia and University of Nicosia Research Foundation (UNRF), 46 Makedonitissas Avenue, P.O. Box 24005, 1700 Nicosia, Cyprus
e-mail: mavromoustakis.c@unic.ac.cy

E. K. Markakis · E. Pallis · S. Panagiotakis
Department of Electrical & Computer Engineering, Hellenic Mediterranean University, Estavromenos, 71004 Heraklion, Crete, Greece
e-mail: markakis@pasiphae.eu

E. Pallis
e-mail: pallis@pasiphae.eu

S. Panagiotakis
e-mail: spanag@hmu.gr

© Springer Nature Switzerland AG 2020
G. Mastorakis et al. (eds.), *Convergence of Artificial Intelligence and the Internet of Things*, Internet of Things,
https://doi.org/10.1007/978-3-030-44907-0_14

immediate fire detection to assess its burn severity in advance. Furthermore, a quantified mapping model is presented to illustrate the spatial distribution of fire severity across the burnt area. The study focuses on the recent bushfire that took place in Mati Athens in Greece on 24 July 2018 which had as a result not only material and environmental disaster but also was responsible for the loss of human lives.

1 Introduction

1.1 Remote Sensing: Delivering on the IoT

Several situations require constant observation due to the high likelihood of sudden changes, which could be costly if not detected on time. In these cases, remote sensors and IoT are complementary methods for early detection. For instance, satellite remote sensors can detect changes in temperature and humidity, while ultra-spectrum sensors can correlate those variations with the probability of bushfire in the area. By analyzing the data, the system can warn of impending predicaments [1].

Nowadays, there are not only cheaper and more efficient sensors but also high-speed data acquisition platforms. Connected devices collaborating to data flow platforms are the evolution in remote sensor technology and their use in IoT [2]. The IoT was built around the idea of data coupling and is supported by broadband networks and MEMs. The IoT sensors are usually electromagnetic sensors that record physical measurements. Depending on the type of the device, IoT sensors can transmit numerous information such as wind speed, water pressure, dissolved oxygen, heart rate, and so on [3].

1.2 Study Area

As is well known, fires are dangerous both for human life and for their properties. Their financial cost is one of the highest recorded over time. The severity of fires, especially in areas with dry climates, has led to enormous ecological disasters. The bushfires play an important role in the shaping of the structure of a place and the ecological balance of the area, as large amounts of pollutants affect the composition of the soil, water, and air [4].

The purpose of this study is to help the comprehension of:

(a) How the wildfire affects an AoI by changing their composition according to the spectral response.
(b) How the spectral response is related to the burnt area. Specifically, we are creating patterns for identifying the relation between Wildfire-derived Ash Index (NWAI) and differentiate Normalized Burn Ratio (dNBR)
(c) Visualizing that information by creating burn severity map according to dNBR.

The Initial Hypothesis Ho: Wildfire-derived ash Index (NWAI) is related to differentiate burn ratio (dNBR).

2 Materials and Methods

2.1 Area of Interest

Mati is a village on the east coast of the Attica region, 29 km (18 mi) east of Athens, Greece. The coordinates of Mati are 38°02′37″N 23°59′42″E with the highest elevation at 10 m (30 ft.) and lowest elevation at 0 m and total area of about 100 km². The dominant vegetation consists of pines and cypresses, and a small number of olive trees and fruit trees.

Mati was severely hit by a wildfire on the 23rd of July 2018 which left more than 100 people dead and over 200 injured. Thousands of vehicles and houses were destroyed, leaving Mati and nearby villages almost entirely burned [5]. According to the estimations of the European satellite service Copernicus, approximately 60,000 acres have ignited [6].

During the day, exceptional conditions were observed, with gusts of wind reaching record speeds of up to 120 km per hour. The actual temperature during the day was up to 35 °C, while the humidity level was lower than 47% [7–9].

In an indication of the size of the tragedy, more than half of the burnt area, 6,934 acres was a residential area. The fire also destroyed 2,018 acres of forest and almost 1,000 acres of low vegetation [10].

Fig. 1 Image of Mati, Athens in July 2018 before the bushfire and in August 2018 after the bushfire. *Source* Satellite images ©2018 DigitalGlobe, a Maxar company

Fig. 2 Satellite images prefire and post fire as they published by the Greek Government. Retrieved from: http://www.skai.gr/news/greece/article/379676/to-mati-prin-kai-meta-tin-fotia-mesa-apo-doruforikes-eikones/

Fig. 3 Attica, Athens Mati. Retrieved from: http://www.skai.gr/news/greece/article/379676/to-mati-prin-kai-meta-tin-fotia-mesa-apo-doruforikes-eikones/

2.2   Remote Sensors on Landsat 8 OLI/TIRS

The LANDSAT 8 satellite holds two main sensors, the Operational Land Imager (OLI) and the Thermal Infrared Sensor (TIRS).

OLI sensor gathers images from nine spectral bands through different wavelengths of visible light. It also collects near-infrared and short-wavelengths of light to observe a 185 km wide swath of Earth in 15–30 m resolution covering broad areas of the Earth's surface. It has adequate resolution to identify characteristics such as urban areas, fields and woodlands.

The Thermal Infrared Sensor (TIRS) measures earth surface temperature in two thermal zones using quantum physics technology to detect radiation.

TIRS uses Quantum Well Infrared Photo-detectors (QWIPs) to detect large wavelengths of light which their power depends on the surface temperature. Those thermal infrared wavelengths are beyond the range of visible light. QWIPs are a new low-cost technology developed by NASA's Goddard Space Flight Center in Greenbelt, Md [11].

Fig. 4 3D representation of the fire in eastern Attica by Copernicus. Publish on 31 July 2018 http://www.skai.gr/news/greece/article/379853/trisdiastati-apeikonisi-tis-fotias-stin-anatoliki-attiki-apo-to-copernicus/

2.3 Dataset

Bushfire severity was calculated across the study area from July to August 2018 by Landsat 8 imagery using Spectral indices.

Spectrum Images of Landsat 8 OLI/TIRS C1 Level-1 from Landsat Collection 1 Level-1 were acquired over the study area immediately before the bushfire on 18 July 2018 and right after the bushfire on 18 August 2018, using EarthExplorer (http://earthexplorer.usgs.gov), with land cloud cover and scene cloud cover less than 10%, accessed on 29 August 2018. The selection standard is that the compared datasets should be close to each other to eliminate the effect of season difference on vegetation and soil changes.

2.4 Classification

Three different sampling sites were classified as Burnt, Unburnt, and Near sea Burnt area according to the official publication from the Ministry of the Environment, the Ministry of Infrastructure and Transportation of Greece and form the Greek firefighting forces [12, 13]. More than 180 random points were selected preceding each site of the region.

The database was created using the coordinates and including 5 different indices values of every point.

Fig. 5 Satellite image of
Mati, July 2018 by ENVI

Fig. 6 Satellite image of
Mati, August 2018 by ENVI

2.5 Variation Detection Processing

2.5.1 Spectral Indices

Spectral indices are combinations of spectral representations of two or more wave-lengths that indicate the features of an area of interest (AoI). Spectral indices are the technique that characterizes the dynamics of the AoI using the capture of abrupt

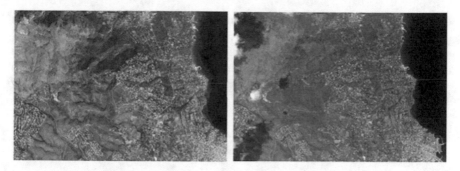

Fig. 7 Before and after fire effect in Mati Greece on 24 July 2018. Retrieved by, Press Conference of the Ministry of the Environment of Greece. http://www.skai.gr/news/greece/article/379676/to-mati-prin-kai-meta-tin-fotia-mesa-apo-doruforikes-eikones/

changes. Those can enhance the desired outcomes and reduce unwanted interferences such as atmospheric noise [14].

No spectral index is effective in all environments for the detection of the burn severity. Hence, a comparison is essential to adequately understand their limitations and effectiveness in identifying the fire effects. Spectrum index differences were calculated by subtracting the July 2018 (pre-fire) index values from the corresponding values in August 2018 (post-fire) index. We are comparing 5 spectral indices for Burn Severity detection [15].

The following Table 1 illustrates the Landsat 8 band designations [16].

Table 1 Landsat 8 band designations for the Operational Land Imager (OLI) and Thermal Infrared Sensor (TIRS)

	Bands	Wavelength (μm)	Resolution (m)
Landsat 8 Operational Land Imager (OLI) and Thermal Infrared Sensor (TIRS) Launched February 11, 2013	Band 1—Coastal aerosol	0.43–0.45	30
	Band 2—Blue	0.45–0.51	30
	Band 3—Green	0.53–0.59	30
	Band 4—Red	0.64–0.67	30
	Band 5—Near infrared (NIR)	0.85–0.88	30
	Band 6—SWIR 1	1.57–1.65	30
	Band 7—SWIR 2	2.11–2.29	30
	Band 8—Panchromatic	0.50–0.68	15
	Band 9—Cirrus	1.36–1.38	30
	Band 10—Thermal Infrared (TIRS) 1	10.60–11.19	100
	Band 11—Thermal Infrared (TIRS) 2	11.50–12.51	100

1. *Normalized difference vegetation index (NDVI) versus Enhanced vegetation index (EVI)*

   NDVI is a traditional vegetation index and is calculated using the reflection coefficient of the red and infrared bands. NDVI is significantly reduced after a fire, so it has often been used for burning area mapping and burning severity classification [9]

$$NDVI = \frac{(Band5-Band4)}{(Band5+Band4)}$$

$$dNDVI = NDVI \text{ prefire } - NDVI \text{ postfire}$$

EVI is estimated using the reflection of blue, red and infrared. EVI is a sensitive indicator for the shell structure parameters (e.g. leaf index) because it helps to minimize the effect of canopy background from the soil and degraded vegetation and reduces the atmospheric variation (Gao et al. 2000; Miura et al. 2001). Particularly, EVI is more effective for distinctive fragments of coniferous trees, such as pines and cypress at the sub-pixel level, than NDVI to Black Hills coniferous forest [17].

$$EVI = \frac{(Band5-Band4)}{(Band5+(6*Band4-7.5*Band2+1))} * 2.5$$

$$dEVI = EVI \text{ prefire } - EVI \text{ postfire}$$

2. *Normalized burn ratio (NBR)*

   This index highlights burnt sections in zones which are larger than 500 acres. The formula is related to NDVI besides that it relates near-infrared (NIR) and

Fig. 8 Enhanced EVI of subset of Mati, Athens, July 2018, before bushfire

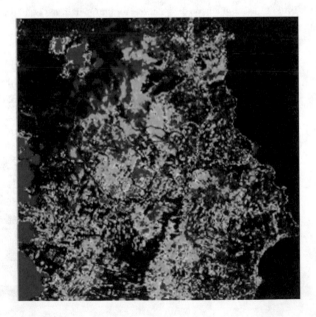

Fig. 9 Enhanced EVI of subset of Mati, Athens, August 2018, after bushfire

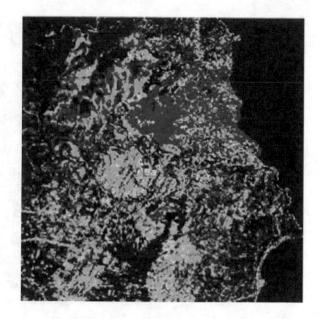

infrared (SWIR) wavelengths [18].

$$NBR = \frac{(Band5 - Band7)}{(Band5 + Band7)}$$

$$dNBR = NBR\ prefire\ -\ NBR\ postfire$$

Fig. 10 Enhanced NBR of subset of Mati, Athens, July 2018, before bushfire

Fig. 11 Enhanced NBR of
subset of Mati, Athens,
August 2018, after bushfire

3. *Soil-adjusted vegetation index (SAVI)*

Empirically originated NDVI outcomes have proven to be unstable, varying according to soil color, soil moisture and saturation effects from high-density vegetation.

Huete [4] tried to improve NDVI by developing a vegetation index that represented the difference of red and almost infrared extinction via the vegetation canopy, creating the Soil-Adjusted Vegetation Index. SAVI is a transformation technique that reduces the effects of soil brightness on spectral vegetation indices including a red and near-infrared (NIR) wavelength.

$$SAVI = \frac{(Band5 - Band4)}{(Band5 + Band4 + L2)} * L1$$

where L1 = 1.5 is a canopy background adjustment factor. An L2 = 0.5 in reflective space has been found to reduce variations in the luminance of the soil and eliminates the need for additional calibration for different soils [19]. The difference SAVI is

$$dSAVI = SAVI \text{ prefire } - SAVI \text{ postfire}$$

4. *Normalized Difference Moisture (Water) Index (NDMI or NDWI)*

The Normalized Difference Moisture (Water) Index (NDMI or NDWI) is a satellite-derived index from the Near-Infrared (NIR) and Short Wave Infrared (SWIR) channels. The SWIR reflectance reflects changes in both the vegetation water content and the spongy mesophyll structure in vegetation canopies

Fig. 12 Enhanced SAVI of subset of Mati, Athens, July 2018, before bushfire

Fig. 13 Enhanced SAVI of subset of Mati, Athens, August 2018, after bushfire

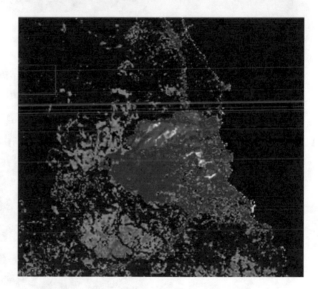

while the NIR reflectance is affected by leaf internal structure and leaf dry matter content but not by water content. The combination of the NIR and SWIR removes variations which are induced by leaf internal structure and leaf dry matter content, improving the accuracy in retrieving the vegetation water content. The amount of water available in the internal leaf structure largely controls the spectral reflectance in the SWIR interval of the electromagnetic spectrum. Therefore, SWIR reflectance is negatively related to leaf water content [20]. NDWI is computed using the near-infrared (NIR) and the short wave infrared (SWIR)

Fig. 14 Enhanced NDMI of subset of Mati, Athens, July 2018, before bushfire

reflectance:

$$NDMI = \frac{(Band5 - Band6)}{(Band5 + Band6)}$$

Fig. 15 Enhanced NDMI of subset of Mati, Athens, August 2018, after bushfire

and

$$dNDMI = NDMI \text{ prefire } - NDMI \text{ postfire}$$

5. *Normalized wildfire ash index (NWAI)*

NWAI is derived by the NDMI (Normalized Difference Infrared Moisture Index) standardized on the range values between 0 and 1, only from data within the boundary of the area burnt:

$$NWAI = \frac{(NDMIi - NDMImin)}{(NDMImax - NDMImin)}$$

where

NDIIi   is the value of each cell in the image,
NDMImin   is the minimum value and
NDMImax   is the maximum value within the area burnt

The NWAI is computed only within the burnt area satellite images before and after the bushfire.

The differenced NWAI is calculated by the formula.

$$dWAI = 0.05 + \frac{(NWAIprefire - NWAIpostfire)}{(NWAIprefire + NWAIpostfire)}$$

3   Results Verification

Fire or Burn severity was defined based on the degree of vegetation consumption and ground fuel. In this study, Burn severity is the magnitude of ecological change caused by bushfire [21]. These are short-term natural and chemical variations during the disturbance. Generally, Burn severity relates to the amount of time required to recover until reach the pre-fire conditions. The longer the recovery period is, the higher the burn severity is. Traditionally, burn area mapping is conducted immediately after fires [22].

Furthermore, the ash was sampled from sites that were classified as low, high and extreme fire severities. The ash covered the whole range of the burnt area.

There weren't any signs of post-fire redistribution of ash by water erosion at the study sites. Some redistribution of ash by wind and leaching is likely to have occurred during and after the fire.

3.1 Results

A t-test was applied to observations with 95% of confidence, to eliminate the outliers. The below graphs were generated using the mean values of the observations from both Burnt and Unburnt areas. All graphs and the qualified map were produced using R-Language.

The Normalized Burn ratio (dNBR) indicates the severity in both Burnt and Unburnt areas.

The samples from the identified burnt area give a mean value higher than 0.4, while the unburnt area has a mean value smaller than −0.004. Both results were expected since the burnt area is in moderate to high severity.

Enhanced Vegetation Index indicates that both areas were richer in vegetation before the fire. The noticeable difference between pre-fire and post-fire mean values lead to the conclusion of a significant vegetation alteration as displayed in Fig. 14.

Furthermore, the Soil-Adjusted Vegetation alteration graph illustrates the alteration before and after the fire. Those results are also expected since some variations on the soil—vegetation values were anticipated.

The Normalized Difference Moisture Index gives information about the humidity and the moisture of an area. Before the fire, the burnt area has a higher amount of moisture than the unburnt area. This reasonably occurred because of the dense vegetation in the area. The amount of moisture after the fire decreased dramatically in the burnt area while stayed stable in the unburnt area.

We identified the severity of our observations according to the table provided by Chafer 2008.

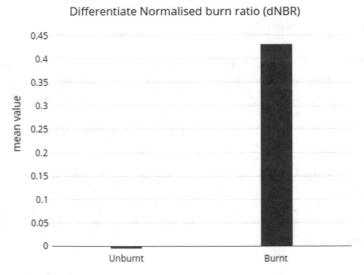

Fig. 16 Differentiate Normalized Burnt ratio (dNBR) of Unburnt and Burnt area

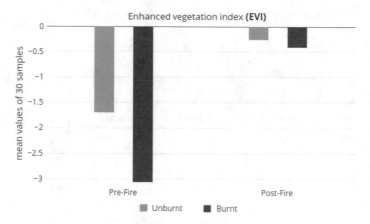

Fig. 17 Enhanced vegetation index of Unburnt and Burnt area

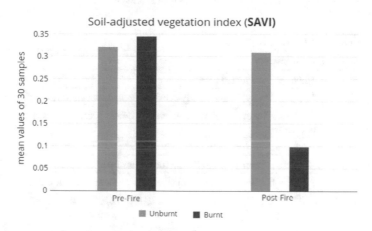

Fig. 18 Soil-adjusted vegetation Index of Unburnt and Burnt area

Following in the graph, 10% of the Burnt area lays in Low severity, about 35% in Moderate severity, more than 45% in High severity and nearly 5% in Extreme-high severity.

At the same time, the Ash-Index graph displays the mean value of the ash amount in both sites after the fire. The amount of the ash in burnt is more than twice higher than the in unburnt area.

The following image illustrates the relation-correlation graph for the dNBR against dWAI indices. By calculating the slope for the whole observation, it is obvious that in moderate severity, the relation is positive with a slope near to 0.45.

To support the results, the mathematical validity of this analysis was examined for normality while the dNBR values were tested, using a one-factor ANOVA with $a = 5\%$ critical value out of 180 randomly selected points of the burnt area.

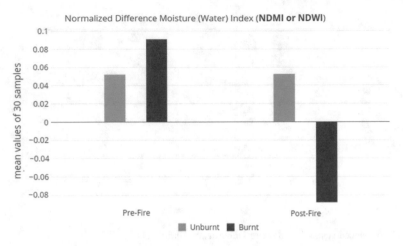

Fig. 19 Normalized difference moisture index of unburnt and burnt area

Table 2 Values of dNBR with Burn severity

Values of dNBR	Burn severity
<−0.1	*Unburnt*: unaffected area
−0.1 to +0.1	*Low fire severity*: low woody debris scorched, but canopy unaffected
0.1 to 0.27	*Moderate fire severity*: medium woody debris scorched, but canopy unaffected
0.27 to 0.44	*High fire severity*: canopy scorched
0.44 to 0.66	*Very high severity*: all available fuels consumed
>0.66	*Extreme fire severity:* all available fuels consumed including stems <1 cm thick

Conclusively, a Severity detection map follows, to quantify the spatial distribution of severity.

The red dots indicate the high severity areas while the yellow ones indicate the moderate severity and the blue is the unburnt area after the fire. Our observation tends to be similar to the defined burnt area according to the Greek Ministry of Environment and the National University of Athens [23].

3.2 Significance of Results

Using satellite images and remote sensing analysis methods, we can detect the difference in aspects of vegetation, moisture, soil quality and ash amount for pre-bushfire and post-bushfire landscape regions.

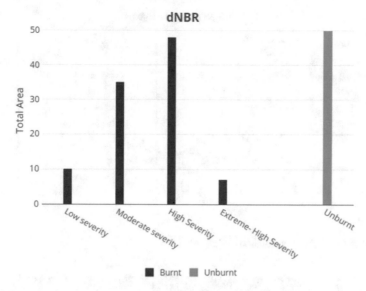

Fig. 20 Classification of dNBR in burnt and unburnt area

Fig. 21 Wildfire-derived ash Index of prefire and postfire area

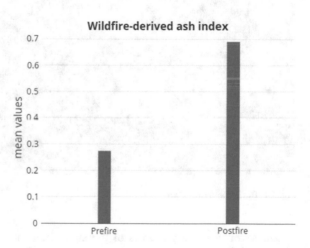

The approaches of Enhanced Vegetation index and Soil adjusted index could be helpful for the recovery work of agriculture and forestry.

Analysis of the Normalized Difference Moisture index and Ash index can be used not only in studying but also in many applications of environmental science, with possible solutions to the impact of the bushfire on human's health in the nearby region. From the calculation of the difference Normalized Burn Ratio dNBR, burnt severity can be detected. Burnt severity detection and analysis is a significant result with the use of remote sensing method.

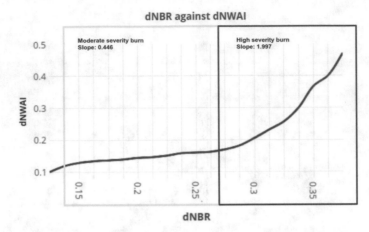

Fig. 22 dNBR—dNWAI correlation

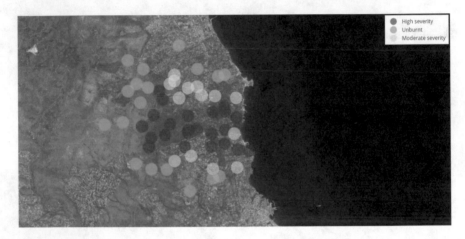

Fig. 23 Burnt Severity Map

The severity density map is of the significant meaning of all the environment recover work including vegetation coverage, soil quality recovery, mineral and biomass recovery, coastal environmental engineering management and air quality recover.

Fig. 24 Boundaries of the fire-affected area. Retrieved by https://edcm.edu.gr/images/docs/2018/Newsletter_Attica_Fires_2018_v11.pdf

4 Conclusion

Landscape alterations in Mati Athens around a bushfire are detected in July 2018 by comparing and analyzing satellite images, gathered by Landsat 8 remote sensors using spectral response.

The observable variations were detected in vegetation, moisture, soil quality and ash amount, in both burnt and unburnt areas. Our observations and calculations lead us to the conclusion that the dereference Ash-index (dWAI) is significantly correlated to the difference normalized burn Ratio (dNBR). The larger the severity is, the stronger the correlation there is. Eventually, the quantified map which illustrates the spatial distribution of the severity matches to the officially identified one from the Ministry of Environment of Greece.

Immediately after a disaster, mobile and portable communications are likely the only available connectivity option. Satellite networks are an integral part of the growing IoT ecosystem. By 2020 it is estimated that more than 20 million "things" will be connected worldwide. There is not yet a unique communication technology to handle this expected IoT/M2M connection by itself. For that reason, satellites are necessary for the needs of the IoT systems due to their resilience, Global footprint, and their Broadband, Nanoband and Broadcast capability. Additionally, satellite technology can offer a variety of frequencies, orbits, and speeds.

References

1. Chafer, C., Santin, C., Doerr, S.H.: Modelling and quantifying the spatial distribution of post-wildfire ash loads. Int. J. Wildland Fire (2016)
2. Koutantou, A.: 'Armageddon' fire in Greece kills at least 80, many missing. Retrieved from https://in.reuters.com/article/us-greece-wildfire/im-looking-for-my-mum-says-young-woman-as-number-of-dead-from-greek-fire-rises-idINKBN1KF1DR. Accessed 25 July 2018
3. Feilding, C., Glenn, S., Josh, H.: Greece wildfires: satellite imagery shows devastation in Mati—visual guide. Retrieved from https://www.theguardian.com/world/ng-interactive/2018/jul/25/how-the-fires-in-mati-greece-spread-a-visual-guide. Accessed 30 Aug 2018
4. Newsroom: According to scientist, record wind gust speeds helped rapid spread of Attica fires. Retrieved from http://thegreekobserver.com/greece/article/47008/according-to-scientist-record-wind-gust-speeds-helped-rapid-spread-of-attica-fires/ Accessed 30 Aug 2018
5. Accuweather: Retrieved from: https://www.accuweather.com/en/gr/athens/182536/july-weather/182536. Accessed 30 Aug 2018
6. Weather and science facts: Average Humidity for Greece in July. Retrieved from https://www.currentresults.com/Weather/Greece/humidity-july.php. Accessed 30 Aug 2018
7. Copernicus: Σχεδόν 13.000 στρέμματα γης έκαψε η φωτιά στην Αττική. Retrieved from https://www.protothema.gr/greece/article/808261/copernicus-shedon-13000-stremmata-gis-ekapse-i-fotia-stin-attiki/. Accessed 30 Aug 2018
8. Δελτίο Τύπου από το Υπουργειο Μεταφορων και Υποδομών τηε Ελλάδας, Retrieved from http://www.yme.gr/index.php?getwhat=7&tid=21&aid=6233&id=0. Accessed 15 Sept 2018
9. Greek firefighting forces, information for the three BushFires in Greece on 24/07/2018. Retrieved from https://www.fireservice.gr/el_GR/-/enemerose-gia-tis-dasikes-pyrkagies-tes-24-07-20-1. Accessed 15 Sept 2018
10. Lentile, L.B., Holden, Z.A., Smith, A.M.S., Falkowski, M.J., Hudak, A.T., Morgan, P., Lewis, S.A., Gessler, P.E., Benson, N.C.: Remote sensing techniques to assess active fire characteristics and post-fire effects. Int. J. Wildland Fire 15, 319–345 (2006)
11. Chen, X., Vogelmann, J.E., Rollins, M., Ohlen, D., Key, C.H., Yang, L., Huang, C., Shi, H.: Detecting post-fire burn severity and vegetation recovery using multitemporal remote sensing spectral indices and field-collected composite burn index data in a ponderosa pine forest. Int. J. Remote Sens. 1366-5901 (2011). ISSN: 0143-1161 (Print)
12. Huete, A.R.: A soil-adjusted vegetation index (SAVI). Remote Sens. Environ. 25(3), 259–309 (1988). https://doi.org/10.1016/0034-4257(88)90106-X
13. LSRS, Land Surface Remote Sensing. Retrieved from https://rdrr.io/cran/LSRS/. Accessed 28 Aug 2018
14. Chafer, Chrihzs J.: A comparison of fire severity measures: an Australian example and implications for predicting major areas of soil erosion. CATENA 74(3), 235–245 (2008)
15. Ganie, M.A., Nusrath, A.: Determining the vegetation indices (NDVI) from Landsat 8 Satellite Data. Int. J. Adv. Res
16. National University of Athens: Finds about the fire in Mati, Scientific report (version1.3). Newslett. Environ. Disaster Crisis Manage. Strat. (2018)
17. Iyer, A.: The Evolution of Remote Sensing: Delivering on the Promise of IoT, Datanami. A Tabor Communications Publication (2018)

18. ISO/IEC TR 19759:2005: Software Engineering—Guide to the Software Engineering Body of Knowledge (SWEBOK). Retrieved from: https://www.iso.org/standard/33897.html. Accessed 25 Aug 2019
19. IoTUK (2018),: Satellite Technologies for IoT Applications 11. Available on: https://iotuk.org.uk/wp-content/uploads/2017/04/Satellite-Applications.pdf
20. Internet of Things (IoT) and the Role of Satellites: Available on: https://www.esoa.net/cmsdata/positions/1695%20ESOA%20IOT%20%20Sat%20Brochure%20Proof%204.pdf (2018)
21. L3Harris GeoSpatial Solutions: Spectral Indices. Available on https://www.harrisgeospatial.com/docs/SpectralIndices.html (2018)
22. Thermal Infrared Sensor (TIRS). Available on https://landsat.gsfc.nasa.gov/thermal-infrared-sensor-tirs/
23. Landsat 8 band designations. Available on https://www.usgs.gov/media/images/landsat-8-band-designations

Internet of Things and Artificial Intelligence—A Wining Partnership?

J. Semião, M. B. Santos, I. C. Teixeira, and J. P. Teixeira

Abstract Hardware/Software (hw/sw) systems changed the human way of living. Internet of Things (IoT) and Artificial Intelligence (AI), now two dominant research themes, are intended and expected to change it more. Hopefully, for the good. In this book chapter, relevant challenges associated with the development of a "society" of intelligent smart objects are highlighted. Humans and smart objects are expected to interact. Humans with natural intelligence (*people*) and smart objects (*things*) with artificial intelligence. The Internet, the platform of globalization, has connected people around the world, and will be progressively the platform for connecting "things". Will humans be able to build up an IoT that benefit them, while keeping a sustainable environment on this planet? How will designers guarantee that the IoT world will not run out of control? What are the standards? How to implement them? These issues are addressed in this chapter from the engineering and educational points of view. In fact, when dealing with "decision making systems", not only design and test should guarantee the correct and safe operation, but also the soundness of the decisions such smart objects take, during their lifetime. The concept of Design for Accountability (DfA) is, thus, proposed and some initial guidelines are outlined.

J. Semião
University of Algarve, Faro, Portugal
e-mail: jsemiao@ualg.pt

J. Semião · M. B. Santos · I. C. Teixeira (✉) · J. P. Teixeira
INESC-ID, Lisbon, Portugal
e-mail: isabel.teixeira@tecnico.ulisboa.pt

M. B. Santos
e-mail: marcelino.santos@tecnico.ulisboa.pt

J. P. Teixeira
e-mail: paulo.teixeira@tecnico.ulisboa.pt

M. B. Santos · I. C. Teixeira · J. P. Teixeira
IST, University of Lisboa, Lisbon, Portugal

M. B. Santos
Silicongate, Lisbon, Portugal

G. Mastorakis et al. (eds.), *Convergence of Artificial Intelligence
and the Internet of Things*, Internet of Things,
https://doi.org/10.1007/978-3-030-44907-0_15

1   Introduction

The Internet has become the platform of human *globalization*. Internet allowed, unlike anything else before, a massive interconnection of people around the world. We can now say that there is an *Internet of People* (*IoP*), namely through social networks, wondering around and looking for education, business, pleasure and other more or less recommendable purposes. At present, humans are developing hardware/software (hw/sw) systems providing them a form of intelligence—the *Artificial Intelligence* (*AI*). Such hw/sw systems are now referred as smart objects, or "*things*". The Internet becomes progressively *the* platform for connecting "things". We can now say that there is an *Internet of Things* (*IoT*). In fact, a broader term for IoT has also been referred, the *Internet of Everything* (*IoE*), denoting physical devices and everyday objects connected to the Internet and outfitted with expanded digital features (like AI), extending the IoT emphasis on machine-to-machine (M2M) communications to describe more complex systems that also encompass people and processes.

Strangely, the Internet has relevantly modified *people's* values and way of thinking. For instance, *privacy*, once one of the most protected values, has been dropped out by social networks. In fact, in social networks, people expose themselves to the world in a way that would be unthinkable some decades ago. Hence, as some others aspects of modern life, that motivate people to share, not rationally explained, feelings and actions (e.g., the emotions associated to sports practices), the Internet is shaping the world. It gathers people in "one place", it modulates people's way of thinking.

History shows that modeling people's way of thinking according to some unique predefined pattern, can bring a dangerous and scaring outcome. The Babel tower is an interesting example of this, registered thousands of years ago in one of the most well preserved books in History. The evaluation of the situation was stated in the words: "now, nothing that they design to do will be impossible to them". Internet, like AI, does not differentiate good from bad. Its use for good or for bad greatly depends on the user. In fact, when a project falls in the hands of unscrupulous people, the outcome can be very negative. In reality, the Internet, being a valuable asset, can be used for excellent or for terrible ends. To avoid that this use depends entirely on the human user, some regulation is under way, by implementing Ethic values in the form of standards.

Now, that humans are developing smart objects, and connecting them through the Internet, the relevant question is: How to prevent the use of this global platform in IoT systems for bad purposes?

From the engineering point of view, the first goal is to guarantee the correct *functionality* of the hw/sw systems. However, in IoT systems driven by AI, it is important to make sure that these intelligent objects' *decisions* are also bounded by *Ethic values*. How can adequate Ethic values be implemented, regulated by Law, and monitored in the IoT world? One way to achieve this goal is by the establishment of *standards*. Design engineers already implement standards in their traditional system design. The establishment of standards for the IoT world is a must, particularly, with the advent of the G5 technology [1]. Accordingly, and wisely, IEEE is working on a "Global Initiative on Ethics of Autonomous and Intelligent Systems" [2]. Similar effort is underway at the Montreal University [3], and in the European Union [4]. All these efforts make clear that these issues need to be taken into account in engineering education.

The availability of highly complex hw/sw systems with increased functionality, and simultaneously with the decrease of physical systems dimensions, allowed advanced semiconductor technologies to be widely used in commodities, and in the improvement of the potential of communication systems. It turned possible to build up one of the most powerful communication system men built, the Internet, and thus remotely control systems mission functionality with minor costs for the consumer.

At the beginning, electronics-based hardware subsystems were used only as the physical platforms to run the software, which drove programmed algorithms and heuristics to perform pre-specified functions. Artificial Intelligence (AI) introduced two additional features, that turned hw/sw systems into *smart objects* [5]. First, the ability *to learn*, by modifying Data Bases (DB) through Machine Learning (ML) algorithms [6, 7]. Second, the ability to make *autonomous decisions*, making these objects smart actors, progressively choosing and modifying their own role. With the Internet, the global platform allowing massive communication among smart objects, the IoT discipline emerged. It is being viewed as a new gold rush, not only for wealth generation, but also as a powerful mean to influence the decision centers of this world.

IoT introduced a new paradigm in computing. Reportedly, by 2020, the number of IoT devices could reach 24 billion [8]. IoT is stimulating the fourth industrial revolution, bringing significant benefits by connecting people, processes, knowledge and data [9–11]. The possibility of interconnecting a huge amount of smart objects, with increasing *local* AI, is opening new avenues of research. Innovative IoT applications are being developed across various markets, from smart cities [12] down to health systems [13], automotive applications [14], aerospace, e-government [15] and so on [16]. Simultaneously, but not surprisingly, *cybersecurity* has become a critical challenge in IoT systems [17, 18].

Therefore, IoT and AI are now two dominant research themes in our scientific world. Their possible synergies are stimulating the imagination of many, and driving new research and development applications.

Nevertheless, are there *risks* on the convergence of IoT and AI? Does the IoT-AI convergence always lead to a winning partnership? We should keep in mind that AI and Autonomous Intelligent Systems (A/IS) are assumed "to behave in a way that is *beneficial to people* beyond reaching functional goals and addressing technical problems", as stated in [19].

The purpose of this chapter is to highlight relevant *challenges* associated with the development of a "society" of intelligent smart objects, trained to perform Machine Learning (ML) [6, 7] and decision-making. Moreover, challenges associated with two intelligent communities—*people* and *things*—and their interactions are also highlighted, leading to the conclusion that an intelligent *strategy* for AI-driven IoT systems must be developed. For this purpose, two key areas are considered here:

engineering education, and smart object design, test and safe operation. The concept of Design for Accountability (DfA) is also introduced.

The chapter is organized as follows. Technology background is reviewed in Sect. 2. The basic characteristics of the existing intelligent society built of *people*, and how they influence the implementation of objects intelligence, is considered in Sect. 3. Key challenges in the convergence of IoT and AI, together with the interaction of two intelligent communities—*people* and *things*—are identified in Sect. 4. In order to devise a strategy for AI-driven IoT systems, Sect. 5 considers two key areas of urgent research: Sect. 5.1—Engineering education, in order that system designers may include smart solutions for eventual shortcomings, and Sect. 5.2—Design for Accountability (DfA), in such a way that smart objects and IoT systems may monitor safe operation, and evaluate the soundness of the decisions they make, along product lifetime. Finally, Sect. 6 summarizes the main conclusions of the work.

2 Technology Background—Five Decades that Changed the Electrical Engineering Paradigm

Let us take a brief look into electronics-based technology evolution during the 20th century. Also, some technology limitations are highlighted, in order to justify measures that must be taken into account in order to guarantee correct and safe systems operation, particularly, when objects (controlled by electronic systems) are expected to take decisions.

The first breakthrough has been reached when the ability of electronics-based systems to perform a required functionality was discovered. This feature has been driving innovation since the use of vacuum-based technologies down to solid-state technologies. Vacuum tubes, like diodes and triodes, were used since the yearly days of the 20th century. The solid-state Bipolar Junction Transistor (BJT) was discovered in 1948, after World War II. Metal-Oxide-Semiconductor (MOS) technology, although discovered decades before, only emerged as a dominant technology after the BJT technology.

Since we live in an analogue world, with physical entities assuming continuous, real values, we started with *analogue* electronics-based devices, circuits and systems, devised and implemented in domain applications. For instance, in order to generate audio signals strong enough to be heard by a large audience, amplification systems have been built, composed by sensors (microphones), a signal amplifier, and actuators (loudspeakers).

However, analogue signals can easily be influenced by *noise* signals, either captured by sensors, or generated by signal distortion, introduced by the physical system performing the desired functionality. Consequently, a simple measure of analogue signal quality is the Signal-to-Noise (S/N) ratio. Often, S/N ratios in analogue physical systems are poor. However, what if we could transform continuous signal amplitude values into a set of discrete values? Using base 2, any integer value can be described by a sequence of two symbols: zero (0) and ones (1).

Hence, we have moved from analogue to *digital* data processing. The flexibility of the transistor, that can operate either in an analogue region, or as a switch, made everything easy. Using DC power supplies, ON transistors act like an almost short circuits (zero resistance), while OFF transistors operate like open circuits (infinite resistance). Circuit variables may easily assume '1' or '0' *logic* values, according to two discrete electric voltage levels, V_{DD} (the power supply voltage value) or 0 V (the Ground value). Digital electronics emerged, and digital data storage and processing can now be performed easily with virtually no noise. Digital electronics lead to the development of *computers*, in which a Central Process Unity (CPU) acts like the maestro in an orchestra, allowing software programs to be used to produce machine language, and to drive the underlying system hardware.

Now, having the ability to build digital systems, the next challenge was: how do we build complex systems, in a cost-effective way, so that hw/sw systems could reach the market at affordable costs?

Research on manufacturing technologies led to the development of simple Integrated Circuits (ICs). Progress in manufacturing IC technologies, namely in lithography, made possible to lay out and interconnect a large number of physical circuit elements. Very Large Scale Integration (VLSI) technologies emerged, as the move from micron-size to nano-size lithography made possible to build hardware silicon chips with millions of transistors. Gordon Moore, Intel co-founder, foresee in 1965 that computing would dramatically increase in power, and decrease in relative cost, at an exponential pace. The famous *Moore's Law* [20] hold for five decades, leading to extremely complex integrated systems, using semiconductor nanotechnologies. State-of-the-art IC technology goes down to 5 nm, i.e., the dimension of ten Si atoms! A positive feedback boost IC design, since increasingly sophisticated Electronic Design Automation (EDA) tools run in platforms that benefit from the progressive high performance achieved by recently designed chipsets, only possible to design with the support of EDA. Interconnection technologies, to interconnect complex devices, or modules, has also evolved from few modules, few interconnections to many, complex modules. Computer systems use *bus-based architectures*, in which control, and dialogue protocols define master-slave architectures. More complex interconnection architectures try to mimic the complexity of the interconnection among human brain's neurons, leading to the concept of *neural networks*.

A key concern in communication, as in digital operation, is power consumption. High speed processing (in the GHz range) means a significant power consumption. That is why, many portable smart objects require batteries with large autonomy. And *power management* becomes a key attribute of a good design.

Semiconductor Yield by existing manufacturing technologies is never 100%. Hence, there is a need to *test* individual components, which exhibit a spread in process parameters. Defining acceptable variable ranges, outliers must be discarded, while "good" chips are sold. Such go-no-go test needs to be performed in a cost-effective way. Hence, not only short sequences of test vectors need to be applied to each manufactured component, but also the identification of defective parts need to be very accurate, as field returns are costly and erode customer's confidence. A functional test becomes inefficient. For instance, to test a 64-bit multiplier module, all combinations of two 64-bit inputs would have to be applied. Therefore, a *structural test* is the more adequate solution [21]. As an exhaustive test is prohibitive in time and cost, circuit and systems topology needs to be activated and checked for correct operation with high-quality, short test sessions.

Hence, two disciplines emerged, in connection to IC design: *Design for Testability* (*DfT*) and *Built-In Self-Test* (*BIST*).

DfT techniques aim at making the test of a given design more cost-effective. For instance, if the activation or the monitoring of the system response is difficult to perform, by a given system architecture, a different architecture may be chosen or additional controllability and/or observability nodes may be required. Moreover, system partition (in test mode), e.g., breaking complex modules in simpler submodules, may be introduced.

As the hardware silicon real estate has become cheap, test functionality may be designed on-chip, so the mission functionality may be tested by the component itself. Such BIST techniques are useful, not only for production test, but also during product lifetime, when the system is in the field, in its real-world application. This additional feature is very attractive, as physical systems suffer from *aging effects*, especially if the *things* (the smart objects) are supposed to operate for long periods of time, e.g., in automotive applications. Aging effects may induce a soft degradation (i.e., lower performance), or a hard failure (due to a physical defect that compromises correct operation) after a long period of time. Self-test and aging monitoring are especially important in *safety-critical applications*. Using the same example, automotive electronics failure may cause harm, namely, people's death. As many systems have idle periods of time, such time periods can be used to perform self-test, and inform the system manager if everything is OK, or if system maintenance is required, or even if the system must be shut down, as safe operation is at risk. Safety-critical applications [22] may justify adding redundancy and voting or the deployment of built-in sensors in IoT applications [23, 24].

It is true that design and test methodologies developed for traditional hw/sw systems, where the software is developed in a *deductive* way, are necessarily different from design and test methodologies of AI-driven objects and IoT systems, where the software is *inferred* from data and some rules. Yet, there are key aspects that must be preserved, in both application domains, and probably emphasized in IoT, namely, the controllability and observability of the systems inputs and outputs and internal critical nodes.

3 Human Background

Why do we dedicate a Section to the Human background? Basically, to identify the basic characteristics of *people's* (the natural, intelligent society) behavior, in order to compare it with the AI-driven IoT society, under construction.

Humans are a magnificent form of intelligent life. In fact, neurosciences are beginning to unveil the wonder of the complex functionality of the human brain, associated with other parts of the human body. Individuals have something like 100 billion neuron cells, each one with the possibility to establish thousands of connections (synapses) with other neuron cells, creating a neural network, allowing data storage (memory), data processing, reasoning processes, deep learning, and decision-making. Decisions are made, usually, using a rational process, basic in logic. However, and not less important, decisions, in the Human domain, are also influenced by another mysterious human characteristic: *feelings*. Either what we like it or not, feelings may significant influence on decisions, regardless of what we know it should be done.

"All men are born *equal*" [25], as quoted by Thomas Jefferson in the US Declaration of Independence, after Montesquieu. However, are they really born equal? All men and women are really born equal, in the sense that they belong to the same species. However, they start to differ in their genetics, namely in their DNA. *Each individual is unique in the Universe*. Differences tend to increase, as human life, experience and learning tend to progress, during their lifetime. Such differences are heavily influenced by their historical context, culture and society. Life span is a few decades. One generation tends to pass to the following generation their values and culture. Data, information and knowledge each individual grasps, together with a reasoning process, and feelings, are key attributes in their decision process. Communication is a key attribute. We are influenced by others, as this conditions our data gathering.

Humans are not born as adults. A baby has a magnificent way of *learning*. For instance, in 1–2 years, one or more human languages are captured, regarding sounds, words, their meaning, sentences, grammar in such a way that the baby's brain learns to talk, building his own sentences, and starting to communicate with other humans. Usually, parents love and protect their children, monitoring the learning process. *Decision-making* is progressively allowed, as they observe that teen agers are now able to start making their own, autonomous decisions. Hence, the freedom to make decisions is progressively allowed, so they learn how to behave in society, to make a positive input to it, and hopefully, to be happy.

Humans also have another interesting feature: *conscience*. In fact, we all have inherent to us, a sense of what is right and wrong, and a sort of moral/ethical judge inside us, that either praise us when we do good deeds, or condemn us when we do bad things. Conscience is also influenced by the social environment. A lovely baby can become a vicious thief, causing pain to those who love him (or her). Nevertheless, the decision-making process of each individual is also biased by his internal conscience, unless conscience is overwritten by mental training (brain washing).

Artificial Intelligence (AI) tries to mimic the human mindset, through the use of Machine Learning (ML) techniques applied to some data set. This is the reason why the following reflections are included in this chapter.

As a starting point, it is interesting to remember that the use of data sets to base decision-making is not a breakthrough of AI and ML. In fact, human knowledge is built through *inference*. Observation and data gathering is used with reasoning to devise a *theory*, a possible explanation of a given reality. *Abstract thinking* allows humans to identify a plausible *cause* for a given *effect*. This reflects the ability to identify patterns, things which are common to given data subsets, and the ability to analyze, rearrange, store and use available data. Then, each theory needs to be validated, or discarded. *Experiments* need to be carried out, so that the obtained results can be compared with the ones predicted by the theory. This is the base of the *scientific method*, so successfully applied to move forward, and to reach new levels of knowledge and understanding. If a theory is validated, then various *applications* are considered to take advantage of the new knowledge, and *decisions* are made. The first part of this process is referred as *creativity*. The second part is called *innovation*. Usually, innovation generates wealth and welfare in human society.

Humans have a *critic spirit*: when we analyze a given reality, we are usually able to reach a conclusion, whether a given outcome is good or bad, whether it accomplishes a desired goal, or not, whether the process needs to be improved or not.

Humans are organized as *societies* (families, tribes, nations). They build Cultures. Institutions. Law and Ethics. *Ethics* usually are identified as *fundamental principles* that rule life in society. One of such fundamental principles is the right to make individual or collective decisions. We call it *freedom*. Of course, individual freedom may collide with other individual's freedom. Human freedom is limited by Law. Humans may make decisions; however, not all decisions are legitimate. From time to time, or suddenly (e.g., traffic police stop operation) human's actions are under scrutiny, to see if people obey the Law. Laws usually are associated with penalties, or punishment, or restrictions, if the Law is violated. Humans also have inherent to them the concept of *discipline*, for continuous improvement.

Humans, individual or collectively, evaluate *the paths* they are pursuing, and either continue in that direction, or abandon the path and choose another path. By *trial and error*, incremental or significant improvements are made. Such improvements are reflected in the Laws humans formulate, to guide their paths in life. Interestingly, while the Physical world is deterministic and reliable (governed by immutable Physical and Chemical Laws), humans are non-deterministic, due to decision freedom. The outcome of their decisions is sometimes unpredictable (differences emerge, like between, e.g., election polls and results).

Nevertheless, *Humans are accountable*, no matter in what sphere they act: within the family, on the job, in their community, to the State, and so on. Sooner or later, they have to be made *responsible* for their acts, or for the lack of adequate action. Only small children are not accountable (their parents are for them), due to the fact that their autonomous decision-making, for obvious reasons, is very limited.

Looking at human History, *communication* evolved slowly, in small communities. Communication was basically *direct* among people, face to face. Knowledge was

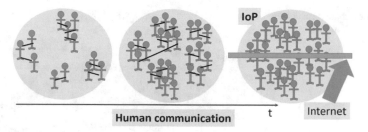

Fig. 1 Human communication along time, ending in the Internet of People (IoP)

scarce, and its dissemination was limited, from one generation to the following one (Fig. 1). Communication in *written* form was almost negligible.

The quantum, gigantic step was the discovery of the press, by Johannes Gutenberg. After 1432 AD, it was now possible to *print* as many copies of whatever knowledge people want. This made whatever knowledge accessible to large audiences, thus spreading human and scientific *knowledge*, and dramatically increasing human communication, in written form.

The knowledge cumulatively gathered in just the last few centuries lead to the spread of science and technology, and to the industrial revolution, as well as to many social transformations. *Education* became relevant in many parts of the world (although for limited subsets of the population). Basic human goals, such as how to get out of poverty, the pursuit of happiness, and the pursuit of wealth, became more visible and humans rush into science-based solutions to build a new society.

In the last decades, the Internet allowed communication among a wide set of humans, and the wide spread of education. Communication among people is carried out, more and more, through the Internet, rather than by direct, face to face communication. *Internet of People (IoP)* became a reality. Today, humans are dependent and addicted to science and technology, and flooded by huge amounts of information, not always reliable. Human society evolution moves very fast, sometimes making humans fragile.

As human society is going through an enormous transformation, one may wonder: is there a *strategy* to move from one model, to another model of society? Unfortunately, the answer seems to be *no*. When shining opportunities emerge, humans usually have no strategy—people just rush for gold. This was the case of 16th century Discoveries performed by the European nations, or the 19th century gold rush towards America's far west, or the industrial revolution, and now the AI. Competitiveness is the driving force.

4 Challenges of IoT-AI Partnership

Unlike humans, all smart objects belonging to a given product are born equal, as far as their hardware part is concerned. Nevertheless, even without AI, constant software

Fig. 2 Key characteristics of a single smart object

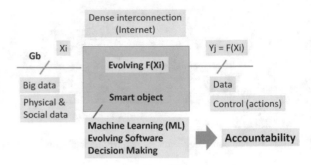

updates make clear that the functionality running on these hardware platforms is constantly changing.

As shown in Fig. 2, each complex smart object has relevant attributes. It receives and processes huge amounts of physical and/or social data, from neighboring objects, or from the Internet, eventually stored in a Cloud environment. It is as densely interconnected, as in a neural network. ML processes constantly examines and classifies data, identifying patterns which may modify internal Data Bases (DB). As a result, the software running in its hardware platform may be constantly evolving due to internal autonomous decisions made by the smart object, using AI techniques. The outcome, in form of an evolving software and functionality, and new data and control variables, may trigger actions, which hopefully are beneficial; however, they can cause harm, if not closely monitored. This is why each smart object must be accountable.

High-performance computing of Big data may require severe high power consumption. Often, large sets of smart objects are working in parallel, in order to achieve fast and accurate results, that may drive ML and decision making. For instance, a recent paper [26] considers model training processes for Natural-Language Processing (NLP), the subfield of AI that focuses on teaching machines to handle human language. They show that advanced techniques are computationally expensive—and highly *energy intensive*. The authors highlight that the carbon footprint required to fuel modern tensor processing hardware is severe, and thus a design constraint. When we think of 24 billion IoT devices, by 2020, the energy required can be awesome [8, 27].

Let us identify some key challenges in the new AI-driven society. As referred, ML and AI modify the software part of smart objects, based on autonomous decisions. Hence, a *first challenge* is (Fig. 3): *how to test an evolving software, being modified along product lifetime?*

Human science is based on the capability *to infer* an analytical formal description of a given characteristic that is present in a huge amount of raw data. Hence, from a set of concrete data, it is possible to abstract some kind of relationship that is susceptible of being described by an equation or equation set.

Let us consider a simple example. From the simple verification that one orange with another orange are two oranges, and one apple with another apple are two

Fig. 3 Key challenges in
IoT—AI partnership

Challenges in IoT – AI partnership

1. Test of an *evolving software*
2. Regulation of *IoP – IoT interactions*
3. Evaluation of *decisions quality*
4. *Data capturing*
5. Evaluation and control of *object's empowerment*
6. *Definition of Laws and Ethics* for IoT, and smart objects *accountability*
7. *Time-to-Market*

apples, and so forth, we can conclude that one plus one (of the same object type) is always two. Then, we can create symbols describing this reality, namely $1 + 1 = 2$. In a more abstract level, we can say $a + b = c$. This is the basics of *arithmetic*.

In another domain, by observing the movement of planets around the Sun, Kepler, Galileo and Newton have derived formal Laws, that carry their names. This is also the way the Universe is being studied, from telescopic observation.

Therefore, the practice of inferring formal relationships, through data observation, has been the way scientists have gone through, and science has advanced. Providing IoT with AI, the goal is also to give these smart objects the ability to learn, to infer, and to make autonomous decisions, based on the knowledge they generate. This not only empowers individual *things*, but also can built an AI-driven *society*, as these objects communicate with each other using the Internet.

So far, the Human society has long been *the sole intelligent society* on this planet. We have the ability to observe, to analyze data, to learn, and to make decisions. Now, the development of AI-driven smart objects leads to the build-up of a *second intelligent society*. Two intelligent societies will simultaneously exist and interact.

Hence, a *second challenge* is *how to regulate IoP—IoT interaction*. The Internet, already crowded with people, will be progressively filled with "*things*". Is the first society (*people*) prepared to build and control a healthy second society (*things*), that will benefit humans, while keeping a sustainable environment in this planet? What will be the *Ethic limits* of IoT and AI, as these two realities converge [2, 3]? How can adequate Ethic values be implemented, regulated by law, and monitored? Like in the area of biotechnology, a significant amount of AI-driven research is performed underground, due to industrial confidentiality. When research results reach the market, it may be too late to control unethical consequences.

Another aspect of IoP-IoT interaction is the fact that smart objects use an extraordinary ability to sense, to analyze Big data [28] in a time frame several orders of magnitude smaller than the human brain can perform. Thus, huge data analysis and high-speed data processing are key advantages of the new smart objects, as compared to the ability of humans. Decisions may be taken much faster than humans do; however, are these decisions better than the ones humans take? Therefore, a *third challenge* is: *how to evaluate decisions' quality?*

A first constraint to the quality of the object decision is the "quality" of the trained *data* provided to these objects as starting point. Data collected by physical means, namely by sensors, can be erroneous if sensors have internal failures or environmentally-induced failures. Social data may also be misleading. Moreover, malicious data is being disseminated through the Internet, and can, maliciously, be introduced in the objects. Hence, a *forth challenge* for achieving sound decisions is *data capturing*. Is the collected data *trustworthy*? Is the data sample *statistically significant*? How and who should evaluate this? May the imported data from an imperfect human society perpetuate its shortcomings, like discrimination? Is there a risk that decisions taken by IoT may be harmful, due to wrong or malicious data, or false information? How *safe* and *reliable* these IoT systems will be? Cybersecurity is already a key concern today [16, 18]. Moreover, similarly to what happens when chips and electronics reliability are analyzed, the identification of false positives and false negatives is relevant. For instance, too much false positives may often halt system performance, reducing system usefulness and customer's satisfaction. Too much false negatives may drive system operation into unsafe behavior. Hence, how to measure the rate of false positives and negatives?

As mentioned before, Artificial Intelligence (AI) tries to mimic the human mind-set. By using Machine Learning (ML) models, raw data is organized in order to derive *order* (or to identify *patterns*) of some interesting kind from even randomly acquired data. This allows smart objects to organize data in more sophisticated data bases, using rules that are inferred by the ML process. If the final conclusions are assessed by humans, the first society retains the control of AI-driven IoT systems. As smart objects move to become, more and more, *autonomous* objects, making (and assessing) their own decisions, how can we be sure that the decisions they make are appropriate? Any decision triggers an action, and actions have *consequences*. Therefore, a *fifth challenge* emerges: *how to evaluate and control object empowerment*?

A key concern is the fact that the AI embedded in these smart objects is created by people, i.e., by the *Natural Intelligence (NI)*. As a byproduct of the existing NI, is not reasonable to assume that AI may be *inferior* to NI in the ability to reach the best possible solutions to problems facing life in this planet?

In fact, being a byproduct of NI, shouldn't AI-driven IoTs be inferior to NI-driven IoPs? If humans do not always make sound decisions, is it reasonable to be expect that human-inspired decisions, made by AI-driven IoTs, designed by humans, will make sound or even *better* decisions?

Moreover, behind every smart object and its artificial intelligence, there is a single or collective "*hidden mind*"—the mind of its designer(s), which may (or may not) target valuable, or good achievements. The outcome of their operation, in the field, will be beneficial to whom? Or, harmful to whom? Often, using NI, the outcomes in human societies are beneficial to a very limited minority, while they exploit the scarce resources of large communities…

This brings to a *sixth challenge*: *the definition of Laws and Ethics in the IoT domain, and object accountability*. Humans face Laws, and Ethics. Adult humans are considered responsible, in face of the Law. Humans are *accountable*. Human behavior is monitored. Should smart objects be facing also Law and Ethics? *Who* should be defining such Ethics and Laws? As mentioned, efforts are under way to define Ethic values for IoT, like the ones carried out by IEEE, the University of Montreal or the European Union [2–4]. Panels of experts try to reach consensus on these issues. However, experts from a given community may be biased. In a recent paper [29], a Harvard Law professor warned that people should not "let industry write the rules for AI". Even unwillingly, they may be protecting corporate interests, not people's interests. AI is not supposed to be solely driven by the goal of maximizing company profit [3]. People (and a sustainable environment in the planet) are more important than companies.

Moreover, how will the design of such objects take into account these Laws, so that they are aware of the Laws, and eventually prevent Law violation? How do we make objects that know what *to do* (mission functionality), and what *to avoid doing*? Who will monitor, in real time, AI-driven IoT systems operation, and the consequences of the decisions and actions they take? Will autonomous smart objects be accountable, as humans are?

Similar to the human society, in which children are not accountable, but, as they grow up, they are progressively allowed, by adults, to take autonomous decisions, the society of *things* should be *progressively accountable*, as the scope of autonomous decisions is progressively enlarged by humans, while retaining control of the process. Hence, *Humans should define the boundaries of smart object autonomy* (either individually, or as part of an object community). This is crucial, in order for humans to keep control of this second intelligent society, making it a useful tool to serve the first intelligent society, Humanity.

So far, the first intelligent society has been "the master", and smart objects have been "the slave". Even the human society has *regulators*, people or Institutions that audit their actions. Shouldn't the AI-driven IoT society have also regulators? Or, should the AI-driven IoT society help humans to decide how AI itself is being progressively introduced? The second society is being developed to enrich people's lives, e.g., alleviating humans from tedious tasks or to perform huge data processing. Nevertheless, in the presence of two competing societies, which one will finally retain the leadership? Is there a risk of a dictatorship of the IoT society over the human society? Or, worst, is there a risk that some humans take control of the IoT society, in order to assume a dictator role over human society?

One way to constrain smart objects decisions, is *to limit the access to external data*. Moreover, *standards* must be established *for* guaranteeing the quality and scope of the *trained data seed,* or initial model with which objects start to "reason" as well as for the scope of the *ML algorithms*. In other words, individuals in human societies act under the Law. The Law establishes the limits of the human freedom. One way to establish this kind of limits in IoT is by limiting the access of each object to the data generated by any other object, or captured through the Internet. The Law also limits the actions each human can take. Similarly, *standards* should limit the scope of the ML engine.

A *seventh challenge* is *Time-to-Market* (*TtM*). Time-to-Market, together with short period product lifetime, push hardware and software development to the edge. New products need to be introduced in the market as soon as possible, prior to products introduced by competitors. Often, products are released without a thorough examination. This can be disastrous for safety-critical applications (automotive, aero spatial, medical applications, to name a few). For instance, it can cause an airplane crash (e.g., due to a flaw in the flight-control system) … Unfortunately, Murphy's Law is not a joke. Should we empower objects allowing that autonomous **critical** decisions will be made by them? Or, should AI and ML data help to regulate and decide the Time to-Market of products and Things?

5 Towards an Intelligent *Strategy* for AI-Driven IoT. Areas of Urgent Research

As referred, all identified challenges (and eventually others we fail to identify), lead to the conclusion that an intelligent *strategy* for AI-driven IoT systems must be developed. Instead of just letting running wild AI-driven IoT systems development, a set of guidelines and actions should be established. As a result, an improvement of the usefulness of IoT to mankind may be achieved.

5.1 Engineering Education

We are aware that two competing worlds already coexist—the *real world*, where humanity prevails, and the *virtual world*, which is assumingly being developed to help people to be happy, and to prosper. The virtual world is populated by hw/sw systems, which tend to be interconnected in arrays of smart objects, using the Internet as a giant platform. The virtual world, which already dominates all aspects of human society (Fig. 4), is rapidly capturing the keen interest of humans. Many people prefer to spend long hours surfing the Internet, touring social networks, posting irrelevant (and often, false) information, playing videogames, rather than to live their lives. Unfortunately, sometimes people look for a refuge in the virtual world, to escape (or forget) an unfortunate real world.

Beyond the social impact, note that *human activity* (e.g., economics, business, investment, government, education, science and so on) is rapidly being transferred and carried out in the virtual world. This may be considered very attractive, as it may be faster, and less expensive. However, such activity may become more difficult to manage and control.

Taking into consideration such impact of virtual world on the real world, it is mandatory that hw/sw systems engineers be prepared to deal with this fact. Hence, beyond the mission subjects of an engineering program, there is a large set of *multi-disciplinary knowledge* that must be acquired, if someone will reach a really "higher" education, that is adequate to the deal with present reality.

Fig. 4 The growing influence of the virtual world

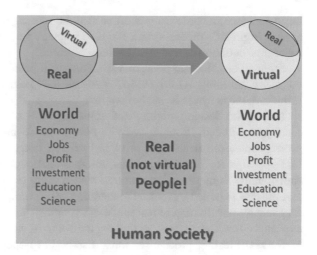

In the past, people realized that engineering education should be complemented with management and business administration skills, so good ideas could be implemented in new products, and these could find a success path to the market and to the creation of wealth. Many engineers took MBA post-graduate courses, in order to be successful management professionals. Many become entrepreneurs, starting new companies.

More recently, as the planet resources are being wasted, and climate change is becoming a nightmare, there has been a growing awareness that a sustainable growth is mandatory. From the engineering education point of view, the concept of *green design* has emerged [30].

Now, the development of AI brought a new reality into stage. Hardware/Software system design has a new dimension, because smart objects are no longer time-invariant, in their architecture. As the ML procedures rearrange data bases, heuristics, algorithms and software code during product lifetime, the operating systems drift away from their original behavior.

Additionally, since new telecommunication technologies, such as 5G technologies [1, 10], are streaming into the market, and AI is continuously being implemented in complex hw/sw systems, making them ever smarter objects in IoT systems, there is a growing need to educate engineers to be aware of the impact that the "*things*" society will have on the human society.

The starting point is that education *professionals*, such as university professors, need to be aware and persuaded that a human society without common principles, Ethics and Law will not lead to a valuable outcome.

Hence, these professionals need to analyze the problem, foresee the consequences and be *convinced* that inserting these themes in a graduate program is a real asset. Moreover, as society is in rapid transformation, Continuing Education (CE) is becoming more and more urgent. Hence, from an engineering education perspective, *formal academic programs* and *CE courses* should be developed and presented to a large audience, in order to guarantee that the message is rapidly spread, not only among engineer's professionals, but also among opinion-makers, decision-makers, managers, economists and social science experts.

From an engineering point of view, education professionals must also acknowledge that building AI-driven IoT systems that help and assist humans clearly influences these systems' *design process*.

In fact, the design variables and constraints are now expanded, and need to be clearly formulated. Problem analysis need to be deepened. Previously, the design process consisted in the design of *the mission functionality*. Then, we moved to DfT—Design for Testability. As a consequence, not only the design of the mission functionality is modified, to accommodate a more testable architecture, but also a test process is added - eventual embedding *self-testing* functionality (e.g., to allow architecture partitioning in test mode of operation) and defining the test stimuli and expected correct responses, so that the hardware of each design copy can be tested.

Software testing procedures must also be developed. At this stage, the design process of smart objects should be considering the ability of monitoring not only *correct operation* of their functional parts, but also *the soundness of the decisions* such smart objects take, during their lifetime, as well as the consequences of *the actions* which are triggered by the decisions taken. For achieving that purpose, Laws and Ethics are to be established for smart objects through the definition of new *standards*. The design process should allow the object, or an external auditor, to verify if these standards are embedded in the architecture of the smart object set. A large avenue of research is opened due to this new perspective. Results of such research should be fueled to academic curricula.

The professor's role on these issues remains essential, to make *students* aware of the challenges of the brave new world we live in, and how to improve the design process, taking into account these additional constraints. The awareness of such challenges, and how to overcome them, needs to be conveyed in graduate and post-graduate courses, and disseminated to industry and to society.

5.2   Design for Accountability (DfA)

Taking into consideration the impact that IoT systems will have in the human society, it is mandatory that smart objects are accountable for the *decisions* they make, and for the *actions* they trigger, since they may lead to people's wellbeing or, alternatively, to catastrophic results. This is why the concept of Design for Accountability (DfA) is introduced.

Accountability deals with being responsible, and being able to show what has been achieved, and how, which allows an external auditor to evaluate and judge the results of a given performance. In this context, by DfA we refer that IoT object and system design must take into consideration the possibility to assess the soundness ML processes, and of any object decision. This requires that the system (or an outside agent) must be able to evaluate the *consequences* of each object decision. This is not a trivial task, particularly, if the final results are not completely known by the designer, which may be the case in various circumstances.

Depending on the *empowerment* given to a smart objects set, and on the domain of application of the IoT system, objects may be granted with the attribute to make *autonomous* decisions or not. If objects can make autonomous decisions, DfA deals with the design process to take into account decision soundness evaluation *after* system operation took place. If objects are not allowed to make autonomous decisions, decisions are first scrutinized by an external auditor, and then implemented or not, depending on the auditor's judgement. Of course, if the application of an IoT system is e.g., autonomous vehicles in urban traffic, objects must be allowed to make autonomous decisions, in real time. In other applications, in which there is time and resources to do so, it may be rewarding to first evaluate decision's soundness, and then applied them, or not.

A first example of object's accountability is antivirus software, in which a software module monitors the legitimacy of another software code to access and modify information in a given system. A second example is when an AI based system is used, to explore space. What is the range of values in the incoming data that is to be taken into consideration? What is the range of values that should be discarded, as outliers? The design engineer can provide an initial guess of that range of values. Yet, if the AI system detects a huge amount of data outside that range, should it be kept or discarded?

As stated, the above mentioned challenges in IoT-AI partnership (Fig. 3) should be seriously taken into account in *engineering education*, since they correspond to the problems new engineers will face. The comments provided in this sub-section are intended to provide some insight on the possible solutions to the identified challenges. Nevertheless, a significant R&D effort needs to be performed, to reach sound solutions.

Design challenges are addressed here from the *engineering* point of view. System designers need to turn out words or sentences into additional engineering requirements and specifications in the design of complex systems constituted by (eventually, among others) smart hw/sw objects, communicating through the Internet.

Prior to assess the soundness of object decisions, it is necessary to guarantee their correct operation, through the implementation of the *hardware test* procedure.

In order to make sure the hardware part is operating correctly, during product lifetime, it is mandatory that no external agent may be allowed to modify the hardware part. This is easy in non-programmable hardware. However, for programmable hardware, like the one using Field-Programmable Gate Arrays (FPGAs), it is mandatory that no external agent is allowed to reprogram the hardware functionality, unless some failure in the configuration memory (caused, e.g., by a Single Event Upset (SEU)) occurs. In such case, reconfiguration is carried out only to reestablish correct functionality.

In terms of the *software test* process, it is another story. As pointed out in [31, 32], the development paradigm of hw/sw systems using ML and AI techniques is completely different from the "traditional" hw/sw systems development. In fact, in the first, the functionality is known, the algorithms are implemented in order to achieve a given functionality, and we know *what* to test, since, functionality is written in a *deductive* way, by writing down the rules as program code.

On the contrary, using ML techniques, rules are inferred from data and requirements are derived *inductively*. As the learning process modifies the rules, the outcome is not entirely known. This makes software test particularly difficult, given the fact that we do not have complete specifications (some of them may be inferred from data), the knowledge of the source code corresponding to inferred specification and yet less some of their critical behaviors. The problem is that, we may not know what to expect, and less *what* or *how* to test the unknown.

Although the ML process may present some opacity, there are some aspects that we know, namely, the *objectives* (or goals) we want the system to reach. Therefore, the decisions taken by smart objects should lead to the achievement of our goals. *Decision's quality* may be defined as its ability to achieve pre-specified goals. Consequently, the design process of such smart systems should consider the ability to monitor, not only the *correct operation* of their functional parts, but also the *soundness of the decisions* such smart objects take, during system operation. In this context, is there room for software self-test and for assessing decisions soundness?

To do so, designers need to establish a set of *metrics*. For instance, we may set (1) the range of values that we consider acceptable in terms of our goals, and (2) the maximum number of "iterations" the ML algorithm should run to reach the goal. *Merit functions* must be established, identifying relevant variables and their weighting factors. To be sure, these variables may now include those corresponding to ethical constraints. The variable set, and the weighting factors may be updated during product lifetime, as the object's learning process progresses.

In traditional hw/sw systems test, it is necessary to identify the input data, the system's functionality and the correct system's response. For the reasons presented in Sect. 4, in smart objects systems design, a key issue is to limit and identify *the data* each object is allowed to access, and to make sure that no malicious data is introduced in the system. The Internet provides an overwhelming set of data and information; however, the system under design is only allowed to access a very limited subset of that data, from other smart objects and/or from humans, navigating also in the Internet. The design should also take into consideration some sort of data evaluation, prior to processing. For instance, is unauthorized data collection taking place? If so, disregard the incoming data.

In terms of testing the software part of the smart object under design, the designer must now take into account that the testing process will include (1) *production* test, and (2) *lifetime* testing. Hence, the ability to test, during product lifetime, must include the ability of the object to test itself (*self-test*), or to be tested by another agent (*external test*), either another higher-hierarchy smart object, or a human. It may be necessary to test in real-time (*online testing*), or periodically (taking into advantage time periods in which the object is idle). In order to test correct operation, correct system responses should be observable. In order to test the soundness of object autonomous decisions, additional software must be embedded in-system.

In order to allow external test, as mentioned before, the external agent must know *what* to test. Hence, a key concept of DfA is that smart objects must *report the soft-ware modifications* introduced by the learning process, and by the decision process that may modify e.g., the internal data bases. This may not be easy to implement, in the design process. However, it makes possible for an external auditor to improve system test, especially software test. Unlike humans, that make changes along their lives and often try to hide their internal feelings and thoughts, smart objects can be designed in such a way that inner modifications are reported to the external world.

System designers must also take into account *design reuse*, and *legacy*. Will the next generation product be smarter than the previous one? This quests for taking advantage of the learning process of the previous generation; again, the *report* of software modifications will save time and money.

In a similar way, reusing design methodologies in a new context can be very rewarding. Software design methodologies, namely an extensive problem analysis, design specifications, architectural design and test, have been reused in hardware (and hw/sw) methodologies with success.

Moreover, hardware methodologies for safety-critical applications, using time or hardware *redundancy*, could be used in software design of complex, smart objects. Time redundancy involves executing a given functionality more than once, to make sure the operation is correct. Hardware redundancy techniques, like Triple Modular Redundancy (TMR) [21] triplicate the hardware module, apply the same input data to the three replicas and compare the outputs of the three modules. If outputs differ, a majority voting process takes place, and the correct output is applied to the following module. Can software redundancy techniques be applied in smart object design? For instance, if the designer does not know beforehand what are the best solutions for a given problem, a set of objects, running in parallel, may be programmed as seeds with different metrics, eventually different variables and/or weighting factors. By using the same input data, it is possible to compare the outputs of different learning and decisions processes, ascertaining what will be the "best" solution to implement in the field, after an experimental period of time.

As referred in Sect. 3, Humans have an inherent feature: *conscience*. The sense of right or wrong guides this mysterious inner moral judge to either praise us when we make good decisions, or find us guilty when we act in an erroneous way. Depending on the circumstances, the trial is made *before* or *after* the decision and action take place. When people face an unknown situation, and have to make fast decisions, usually the trial is made afterwards. On the contrary, if we have time to evaluate and decide, the trial is made in advance, and the moral sense, or the Ethics, are taken into account in the decision process.

Similarly, in smart object design, if acceptable Ethics are assumed as system requirements, and transformed into standards and rules, these ethic rules should be considered in the learning and decision-making processes, leading to more profitable IoT systems. How to do it, is a very interesting theme of research.

Nevertheless, system design needs to be performed in a way that *Ethic rules must be inerasable*, i.e., smart objects must not be allowed to decide without taking Ethics into consideration. As in the human society, where Ethics are changing with time, ethic rules inserted in smart objects should be updatable. Embedded object "conscience" should be mandatory for product homologation, probably as a distinct object, operating as *master*, and should be viewed as the "regulator" in the human society.

Ethics standards will allow each smart object to analyze itself, ascertaining if a given decision or triggered action is (or is not) in compliance to the engraved *principles* embedded in itself, and in its follow objects. As in the human society, the trial may be performed either before, or after decisions and actions take place. Of course, the outcomes of such internal judgement may vary, whether the object itself, a master object or a human auditor will take action if the decisions broke any Law.

These questions led us to this innovative concept—*Design for Accountability (DfA)*. As AI-driven IoT systems are allowed to make more and more autonomous decisions, they should be progressively accountable for their acts.

In order to evaluate the soundness of the decisions taken by the IoT system, often it takes a long time, allowing the cause-effect paradigm to take place. Hence, an Audit process may periodically take place, and be performed either internally, or more reasonably, externally by an independent agent. For instance, in medical diagnosis and subsequent treatment, medical trials and clinical evaluations require many patients, many years, and a significant amount of data. However, is the audit process that leads to progress in medical treatment and in the homologation of pharmaceutical products.

On the contrary, decision and action evaluation in safety-critical IoT systems must be performed as soon as possible, as some actions may be identified as harmful, potentially catastrophic. A collection of forbidden actions for each smart object should be identified, avoiding the object to autonomously make such decisions.

6 Conclusions

Humanity reached a new crossroad, with the advent of the Internet, and of AI-driven IoT systems. A new intelligent society, generated and empowered by humans, is under active development. It is our conviction that the creators must *master* their creation.

Relevant *challenges* in this path have been identified and discussed. How to test the smart object's modified software part, IoP-IoT interactions, the need to evaluate objects decisions quality, data capturing, control of objects empowerment, Ethics and Laws for IoT, smart objects accountability and Time-to-Market are the main challenges for which solutions need to be found.

A *strategy* for intelligent and controlled IoT society is, thus, mandatory, in order to achieve a harmonious, happy outcome. Such strategy should include *engineering education*, and *Design for Accountability (DfA)*. In reality, engineering education should precede a multi-disciplinary education, from academia down to the business environment, as additional actors in other domains need to be aware of the consequences of misusing the emerging AI-driven IoT society. This educational effort should start by engineering education, due to the fact that engineers are the IoT system *designers*, who must develop ways to introduce root solutions for the above mentioned challenges.

As we saw, smart objects monitoring requires the ability of outside agents to be able to verify objects *correct operation* as well as the *soundness* of their *decisions* and triggered *actions*. Embedded functionality in these objects must drive each object to report ongoing transformations. Special care needs to be taken into account, when safety-critical applications are considered.

A fundamental issue is *the insertion of Ethics as standards in IoT* devices and systems, in an inerasable form. Embedded Ethics must be rooted to true Human values, to serve the human society, and act like an inner conscience of these smart objects. No privacy should be given to IoT devices and systems, in the sense that at any moment the inner values, embedded on them, must be accessible and verifiable from outside, eventually corrected if any malicious "values" are introduced in objects. The scope allowed by humans to smart objects to make *autonomous decisions* should be enlarged in a progressive way, as parents do with their children.

As a consequence, it becomes clear that a huge amount of R&D effort lays ahead of us. It must be pursued urgently, in order to obtain a wining partnership between IoT and AI.

References

1. G5: Moving to the next generation in wireless technology. https://www.sciencedaily.com/releases/2015/04/150430082723.htm
2. Ethically Aligned Design—Version II, Request for Input.: IEEE Global Initiative on Ethics of Autonomous and Intelligent Systems, https://standards.ieee.org/industry-connections/ec/autonomous-systems.html
3. Bengio, Y., Université de Montréal.: Montreal Declaration for a Responsible Development of Artificial Intelligence (2018). Available at https://www.montrealdeclaration-responsibleai.com/
4. High-Level Expert Group on Artificial Intelligence set up by the EU.: Ethic Guidelines for Trustworthy AI, European Commission, document made public in 8 Apr 2019. Available at https://ec.europa.eu/digital-single-market/en/news/ethics-guidelines-trustworthy-ai
5. Artificial Inteligence Tutorial. https://www.tutorialspoint.com/artificial_intelligence/
6. Nick McCrea—An Introduction to Machine Learning Theory and Its Applications: A Visual Tutorial with Examples in https://www.toptal.com/machine-learning/machine-learning-theory-an-introductory-primer
7. Xin, Y., Kong, L., Liu, Z., Chen, Y., Li, Y., Zhu, H., Gao, M., Hou, H., Wang, Ch.: Machine learning and deep learning methods for cybersecurity. IEEE Access **6**, 35365–35381 (2018)
8. Nouwens, M., Legarda, H.: China's pursuit of advanced dual-use technologies, Dec 2018. Available at https://www.iiss.org/blogs/analysis/2018/12/emerging-technology-dominance
9. Gubbi, J., Buyya, R., Marusic, S., Palaniswami, M.: Internet of Things (IoT): a vision, architectural elements, and future directions. Future Gener. Comput. Syst. **29**(7), 1645–1660 (2013)
10. Park, T., Abuzainab, N., Saad, W.: Learning how to communicate in the internet of things: finite resources and heterogeneity. IEEE Access: Optim. Emerg. Wirel. Netw.: IoT, 5G Smart Grid Commun. Netw. Spec. Session, IEEE Access **4**, 7063–7073 (2016)

11. Data Management for Artificial Intelligence. https://www.sas.com/en_za/whitepapers/data-management-artificial-intelligence-109860.html
12. Yu Shwe, H., King Jet, T., Han Joo Chong, P.: An IoT-oriented data storage framework in smart city applications. In: 2016 International Conference on Information and Communication Technology Convergence (ICTC), pp. 106–108 (2016)
13. Mohon Ghosh, A., Halder, D., Alamgir Hossain, S.K.: Remote health monitoring system through IoT. In: 2016 5th International Conference on Informatics, Electronics and Vision (ICIEV), pp. 921–926 (2016)
14. Yamauchi, T., Kondo, H., Nii, K.: Automotive low power technology for IoT society. In: 2015 Symposium on VLSI Circuits (VLSI Circuits), pp. T80–T81 (2015)
15. Nelson, G.S.: Getting started with data governance. In: Presented at the Annual Conference of the SAS Global Users Group, Dallas, TX, 28 Apr 2015
16. Ruan, J., et al.: An IoT-based E-business model of intelligent vegetable greenhouses and its key operations management issues. Neural Comput. Appl. 1–16 (2019)
17. He, H., et al.: The security challenges in the IoT enabled cyber-physical systems and opportunities for evolutionary computing & other computational intelligence. In: 2016 IEEE Congress on Evolutionary Computation (CEC), Vancouver, BC, pp. 1015–1021 (2016)
18. Xu, T., Wendt, J.B., Potkonjak, M.: Security of IoT systems: design challenges and opportunities. In: Proceedings of IEEE/ACM International Conference on Computer-Aided Design (ICCAD), pp. 417–423 (2014)
19. Ethically Aligned Design—A Vision for Prioritizing Human Well-being with Autonomous and Intelligent Systems, version 2. IEEE (2018). Available at http://theinstitute.ieee.org/resources/standards/ieee-releases-new-ethical-considerations-for-autonomous-and-intelligent-systems)
20. Mollick, E.: Establishing Moore's Law. IEEE Ann. Hist. Comput. 28(3), 62–75 (2006)
21. Bushnel, M.L., Agrawal, V.D.: Essentials of Electronic Testing for Digital Memory and Mixed-Signal VLSI Circuits. Kluwer Academic Publishers (2000)
22. Will Safety-Critical Design Practices Improve First Silicon Success?. Mentor Graphics White Paper (2017). Available at http://s3.mentor.com/public_documents/whitepaper/resources/mentorpaper_102839.pdf
23. Semião, J., Cabral, R., Cavalaria, H., Santos, M.B., Teixeira, I.C., Teixeira, J.P.: Ultra-low-power strategy for reliable IoE nanoscale integrated circuits. In: Harnessing the Internet of Everything (IoE) for Accelerated Innovation Opportunities. IGI Global (2019)
24. Valdés, M., Freijedo, J., Moure, M.J., Rodríguez-Andina, J.J., Semião, J., Vargas, F., Teixeira, I.C., Teixeira, J.P.: Design and validation of configurable on-line aging sensors in nanometer-scale FPGAs. IEEE Trans. Nanotechnol. 12(4), 508–517 (2013)
25. Charles de Secondat, Baron of Montesquieu "All men are born equal", The Spirit of the Laws, 1748. Available at https://americanart.si.edu/artwork/state-nature-indeed-all-men-are-born-equal-they-cannot-continue-equality-society-makes-them
26. Strubell, E., Ganesh, A., McCallum, A.: Energy and Policy Considerations for Deep Learning in NLP. University of Massachusetts at Amherst (2019)
27. Mavromoustakis, C.X., et al.: Socially oriented edge computing for energy awareness in IoT architectures. IEEE Commun. Mag. 56(7), 139–145 (2018)
28. Big data needs a hardware revolution. Nature 554, 145–146 (2018). https://doi.org/10.1038/d41586-018-01683-1
29. Benkler, Y.: Don't let industry write the rules for AI. Nature 569(161), 2019 (2019). https://doi.org/10.1038/d41586-019-01413-1
30. Irimia-Vladu, M.: Green electronics: biodegradable and biocompatible materials and devices for sustainable future. Chem. Soc. Rev. 43, 588–610 (2014)
31. Cagala, T.: Improving data quality and closing data gaps with machine learning. In: IFC National Bank of Belgium Workshop on Data Needs and Statistics Compilation for Macroprudential Analysis. Brussels, Belgium, 18–19 May 2017
32. Khomh, F., Adams, B., Cheng, J., Fokaefs, M., Antoniol, G.: Software engineering for machine-learning applications—the road ahead. IEEE Comput. Edge, pp. 21–24 (2019) (also in IEEE Software, vol. 35, no. 5, 2018)

AI Architectures for Very Smart Sensors

Peter Malík and Štefan Krištofík

Abstract The chapter describes modern neural network designs and discusses their advantages and disadvantages. The state-of-the-art neural networks are usually too much computationally difficult which limits their use in mobile and IoT applications. However, they can be modified with special design techniques which would make them suitable for mobile or IoT applications with limited computational power. These techniques for designing more efficient neural networks are described in great detail. Using them opens a way to create extremely efficient neural networks for mobile or even IoT applications. Such neural networks make the applications very intelligent which paves the way for very smart sensors.

1 Introduction

Machine learning and especially deep learning are well studied areas with roots in the last century. Nature inspired algorithms with the emphasis on human brain biology produced many great works. For a long time, the main barrier has been the computational complexity. To model and simulate even small artificial neural networks requires significant computational power which limited the size and depth of neural networks for many years. The introduction of convolutional neural networks (CNN) in '80s [14, 17] allowed to share weight parameters and to focus only on a small neighborhood in comparison to fully connected neural networks which significantly reduced the number of parameters. This resulted in a commercially applied 7 layer deep CNN for the recognition of hand written digits and later also hand written text on bank checks [44].

P. Malík (✉) · Š. Krištofík
Institute of Informatics, Slovak Academy of Sciences, Dúbravská Cesta 9,
845 07 Bratislava, Slovakia
e-mail: p.malik@savba.sk

Š. Krištofík
e-mail: stefan.kristofik@savba.sk

© Springer Nature Switzerland AG 2020
G. Mastorakis et al. (eds.), *Convergence of Artificial Intelligence
and the Internet of Things*, Internet of Things,
https://doi.org/10.1007/978-3-030-44907-0_16

More sophisticated applications of neural networks were hindered by computational requirements of more powerful neural network models. The breakthrough came with the work [41] which shows how to efficiently use a modern graphics processing unit (GPU) with many small dedicated computational cores to accelerate the computation of neural networks. It shows 2 orders of magnitude the computational speed improvement which results in the successful training of a 60 million parameters (650,000 neurons) 8 layer deep CNN which won the 1st place in the ImageNet Large Scale Visual Recognition Challenge 2012 (ILSVRC2012) [13] where the general images are classified into 1000 image categories. The winning neural network won with a great margin of 10% absolute or 40% relative accuracy improvement over the 2nd place finisher. This showed to the world the abilities and potential of neural networks and restarted interest in deep neural network research which produced many improvements and new neural network architectures [8, 24, 32, 73, 77, 83]. Highway networks [75] and Residual networks [24] validated that it is possible to design and successfully train more than 1000 layer deep neural networks. The work [23] was the first to propose CNN with classification precision better than that of humans on the ImageNet 1000 class classification dataset. The other superhuman results of carefully designed and trained CNNs applied to specific tasks followed [80].

The classification task is the most basic visual recognition task which has led to many great deep neural network architectures [8, 24, 31, 41, 73, 77, 83]. The advancement is fast-paced due to many accessible datasets including ImageNet [13, 70], CIFAR-10 and CIFAR-100 [40], Birds 200 [82]. The more challenging recognition tasks including visual object detection and semantic segmentation have been rapidly pushed forward by open visual challenges and accessible datasets including the PASCAL Visual Object Classes (VOC) Challenge [15, 16], MS COCO [45], Places [90], Cityscapes dataset [11], Mapillary vistas dataset [60], and EuroCity Persons Dataset [3]. Each challenge requires an evaluation metric and a dataset divided into train, validation and test parts. Using the same dataset and evaluation metric makes it easy to compare different neural network designs and evaluate which new design features lead to more accurate or more computationally efficient neural networks.

Visual object detection has led to many great works. R-CNN [18] became a benchmark with many improvements. Fast R-CNN [19] included more efficient convolutional computation, Faster R-CNN [66] switched the selective search proposal algorithm [81] with the feature proposal neural network embedded within the main neural network, R-FCN [12] further improved computational efficiency by full convolutional approach, and FPN [46] included feature pyramid network before the classification head which effectively transformed single-scale testing into multi-scale testing done in a single step and improved the overall accuracy. Some approaches targeted the more efficient computation [33] and opted for one step detection without feature proposals. This resulted in great works including Single Shot multibox Detector (SSD) [49], You Only Look Once (YOLO) [64], Deconvolutional Single Shot Detector (DSSD) [36], YOLO9000 [65] which successfully detected 9000 different object classes, and RetinaNet [47] with focal loss for dense object detection.

The visual semantic segmentation task has special importance in medical and industrial applications where it is very important to find precise object boundaries. This task also produced many great works including Fully Convolutional Networks (FCN) [52], U-Net [67], DeepLab [6] with fully-connected conditional random fields (CRF), Pyramid Scene Parsing Network (PSPNet) [88], DeepLab-v3 [4] with Atrous Spatial Pyramid Pooling (ASPP) modules, and Global Convolutional Network (GCN) [62] with separable convolutions with large effective field of view.

Visual instance segmentation is one of the most challenging visual tasks due to combined requirements of object detection and semantic segmentation. It must recognize each object instance together with its correct class with pixel level precision. Many great works reuse object detection or semantic segmentation CNN models and modify them for instance segmentation. Mask R-CNN [26] adds new classification branch to faster R-CNN which produces object segmentation maps and MaskLab [5] combines faster R-CNN with semantic and direction features. PANet [51] further improves Mask R-CNN by adding path aggregation network between FPN and Mask R-CNN classification head. Sequential Grouping Networks (SGN) [50] improve the instance segmentation by breaking it into a sequence of easier subtasks while the work [2] proposes the discriminative loss function as a more simple way to handle occlusions.

The interest in efficient computation of CNNs has always been running concurrently with a goal to improve the absolute precision. There are many areas and levels of abstraction where an improvement has a direct impact on computational efficiency. The most beneficial improvements are always those on the highest levels of abstraction. A great example are the works [30, 72, 87] which introduce MobileNet and ShuffleNet. These CNNs significantly reduce the number of parameters utilizing depth-wise separable convolutions while still producing nearly state-of-the-art results. They show a successful run of such CNN on the large core of the Google Pixel 1 phone at 5 FPS while running a complex detection task on the MS COCO dataset [45]. The improvements on the medium level of abstraction can still be significant. Multiple works showed that many parameters do not contribute to most results and can be pruned [20, 21]. This results in significant memory saving. Unfortunately, most computational units are not optimized for the efficient multiplication of spare matrices so the improvement of computation efficacy is not great. One way is to design custom hardware as [22, 37] where several different techniques, including pruning, compression and lookup tables, were combined which resulted in 510 times parameter reduction and efficient computation in a low power custom design implemented in FPGA.

An efficient way to decrease computational complexity is to reduce the precision of all parameters by selecting lower precision format such as half precision, tensor precision, integer or an extreme option—the binary. This way, there is no problem with the spare matrix computation. Even the extreme binary solution with most memory savings has been shown [63, 87] to produce a CNN with relatively high precision. The authors binaryzed all weights and all activations which allowed to use binary computation composed of exclusive-not-or and 1's counting with the result of significantly reduced computational requirements and an option to use no GPUs for inference. However, it was at the cost of significant precision reduction [63]. The combination of binary weights and learnable quantization into few bit-wide activations leads to much better accuracy in comparison [85].

The improvements on the low level of abstraction can still improve computational efficiency by using a good combination of designing framework, inference framework and compatible targeted hardware. Many frameworks for designing neural network models have been created including Caffe [39], Mxnet [7], TensorFlow [1], PyTorch. Some frameworks have been specially developed for designing efficient inference including Caffe2 for efficient inference on mobile devices and TensorRT for efficient inference on embedded devices with NVIDIA GPUs such as Jetson TX2 or AGX Xavier. Targeted hardware can massively vary from standard GPUs through embedded processors with GPUs or other neural network cores, to custom designed FPGAs or fully application specific integrated chips.

Modern hardware has made a significant contribution toward the efficient computation of deep neural networks. A great example is the recent introduction of NVIDIA Tensor Core which is now a standard component of HPC GPUs (Volta architecture for servers and cloud solutions), desktop GPUs (Turing architecture) and autonomous driving embedded devices (AGX Xavier). New smartphones from Apple and Huawei also have embedded hardware for the acceleration of deep neural networks called neural processing units (NPU). Many more companies offer hardware solutions for the acceleration of deep neural networks which open the possibility to use them in very smart sensors. Such sensors can be utilized in many current and emerging IoT-based operations in various fields, such as agriculture, construction, healthcare, aviation, disaster management or transportation. These devices can contribute greatly to the improvement of the management efficiency or the quality of products offered by those operations, e.g., the quality of vegetables produced in IoT-based intelligent greenhouses. However, much research effort is still needed to convince the most influential parties, such as Internet giants, to provide widespread IoT services at affordable prices and speed up building of the infrastructure [68].

The number of IoT-enabled devices is constantly growing and it is predicted that IoT could reshape the business processes in various industries [68]. Moreover, the users' expectations and demands towards reliable, continuous service with low latency from IoT-based applications are also growing. There is an inevitable need to find feasible solutions for providing such level of quality of services. This is especially difficult due to limited resources and energy available on small end user devices. The socially oriented edge computing concept is one of the promising emerging approaches [57]. By offloading energy-consuming computations to other nodes in the cloud that have redundant resources available, the average lifetime of all nodes in the network can be significantly increased. Nodes suitable for offloading are selected based on social characteristics and relations between nodes. This concept has been proved useful for delay-sensitive applications, such as interactive gaming. However, much work is still needed towards the improvement of the cloud (or fog) services and offloading delay reduction [57].

Keeping in mind the limited computational power of the above mentioned devices, this chapter will describe the current trends of designing modern deep CNN models and techniques to reduce computational complexity so the resulting models can be used in mobile or IoT applications with limited hardware resources while still achieving good accuracy and frame-rate.

2 Key Characteristics of Artificial Neural Networks

This section describes few key characteristics of artificial neural networks. Discussed are the advantages and disadvantages and the overall impact on the whole neural network. This section provides basic knowledge and insights into expected impacts of small modifications to any neural network structure.

2.1 Hierarchy of Feature Detectors

The first insights about working of the mammalian visual cortex are from experimental work of David H. Hubel and Torsten Wiesel with an anesthetized cat in 1959. Their findings suggested the existence of hierarchy of feature detectors in the visual cortex. The lower levels correspond to simple feature detectors whose activations are propagated to the middle levels and through middle level activations up to the higher levels where complex feature detectors are created. This way, the complex patterns are recognized by a hierarchical structure comprised of very simple units. Each hierarchical level with multiple units is continually increasing the overall complexity and the ability to recognize more complex patterns. This can be modeled by a multi-layer neural network and visualized by using deconvolutions [84].

2.2 Fully Connected Layers in Neural Networks

The standard multi-layer neural network, e.g., multi-layer perceptron, uses dense interconnection within a layer. This means that each neuron within a layer takes all outputs of the previous layer's neurons as inputs. It is also the reason why this type of a layer is called dense layer or fully connected (FC) layer. Full interconnection between neighboring layers makes this type of a layer very powerful and able to learn very specific features based not only on activation values of the previous layer but also on the spatial characteristic and overall context information. This means that this layer learns not only activation values but also precise spatial coordinates of the whole previous layer. This way, precise context information can be learnt which makes this layer ideal for classification purposes. It is possible to learn to distinguish object characteristics based on the object location in space. Presently, the FC layer is mainly used in the classification part, also called a head, of modern neural networks.

Every powerful advantage usually has its similarly powerful limitation. To be able to precisely map spatial characteristics, one needs a very rigid, unchanging structure. Even minimal change to the structure including modifying resolution or interconnection matrix between neighboring layers has a significant effect on the overall behavior. The FC layer requires fixed dimensionality of input data to work correctly. In other words, the input spatial resolution has to be the same. The inputs with different spatial resolutions have to be resized to the correct resolution by standard algorithms, e.g., linear, cubic or other polynomial interpolation, or region pooling layer has to be used before the FC layer. Another price for the precise spatial mapping ability is a very high parameter count. Each neuron is interconnected with all neurons of the previous layer which represents $O(N_i N_{i-1})$ complexity for each neural layer, where N_i and N_{i-1} are the numbers of neurons of the current and previous layers, respectively.

2.3 Convolutional Layers in Neural Networks

FC layers are not very well suited for image related recognition tasks. The reason lies within the disadvantages described above. We discuss this in more detail.

An image is represented by ordered pixels and each pixel carries not only intensity information of relevant color channels but also the relative position of the captured light source from emitting and or reflecting object surface, placed in the space of the origin. The FC layer is able to distinguish an object placed in different parts of the image. There are some special tasks where this is beneficial; however, this is responsible for wasting resources in most image recognition tasks. We do not need many different neurons specialized for recognizing the same object in different positions. We want the neurons to specialize to different types of objects or different types of object characteristics or states. For this, we need translation invariance, so the same type of an object is recognized by the same neuron which will be activated when the object of interest appears anywhere in the image. There will be no difference between the neuron's activation when the object is placed in the center of the picture or in any of the corners.

The image resolution is usually large. Present mobile devices are capturing the image in 10s of mega pixels resolutions. Such resolution to be processed by an FC layer requires 10s of millions neurons and each has 10s of millions parameters. So there is 100s of thousands billions parameters in the first layer alone. This is a huge unacceptable number. The problem is even worse for memory requirements. Every parameter has to be kept in some memory, and single precision, the standard in computer vision applications, requires 4 bytes. So there is the memory requirement of peta bytes (10^{15} B) for storing parameters of the first layer alone. Again, this is a huge unacceptable number.

These problems were always true and the past hardware limitation was much worse. The solution is a different type of layer with translation invariance and reduced number of parameters. This layer is called convolutional layer. It was introduced by Fukushima in 1980 [17] who had been inspired by the work of David H. Hubel and Torsten Wiesel. One of the first neural networks for recognizing hand-written zip code digits [14] was later improved into the first commercially used neural network with convolutional layers called LeNet-5 applied for the recognition of hand written digits on bank checks in late 1990s [43].

The convolutional layer uses two main principles to reduce the number of parameters. The first is the limited interconnection with the previous layer within close neighborhood and the second is parameter sharing. We discuss both of them in more detail.

The limited interconnection is especially beneficial for image processing. The examples can be found in everyday life. A common user knows that any picture taken by a camera can be improved by applying filters to reduce noise, improve contrast or edges, or make light and color shifts. These filters process the image within the small patches or local neighborhood of the targeted pixel. This results in fast processing because only a small fraction of input data, represented by the whole image, is necessary to calculate one output pixel. The same principle is used in the convolutional layer where each neuron is interconnected with the neuron from the previous layer in the same spatial position and its close neighbors.

The above example with applying filters can also be used to explain the second principle. The mentioned filters are universal and can be used on any image. An image can be small, large, monochromatic, colorful, simple or complex and the results are always predictable. This is also true for any image crop because each filter is represented by a single kernel which is applied sequentially to every input pixel in a convolutional manner. Using the same kernel again and again is an example of kernel sharing and because every kernel is represented by parameters, it is also an example of parameter sharing. The convolutional layer uses the same principle, it even got named after it. Both principles, the local interconnection and parameter sharing together reduce the number of parameters by several orders of magnitude. This allows to use higher input resolution and deeper hierarchical structure with very highly complex filters at its end.

Kernel sharing has an additional benefit. It is translation invariance. Using the same kernel for every spatial position of the input means that the same kernel can recognize the same object or characteristic in the whole input image regardless of its position. The position of the recognized object or characteristic is represented by the relative position of neurons with an elevated activation function. The convolutional layer outputs not only the probability of the recognition of an object of interest but also its relative spatial coordinates—two necessary parts of object detection. Because the kernel recognizes the object of interest efficiently within the whole picture, we can add many more kernels to the same layer and the number of parameters will increase only with linear dependency. This results in very powerful neural networks capable to recognize many different objects or features.

2.4 Vanishing or Exploding Gradient in Connection with Stochastic Gradient Descent

The stochastic gradient descent is the main learning algorithm of modern neural networks. The first use of gradient descent method [69] to automatically learn weights in neural networks was by LeCun in 1989 [43]. The idea is to back-propagate partial derivatives of the loss function through the whole neural network structure in a layer to layer manner. The stochasticity is represented by random selection of input samples. For a long time, the standard nonlinear activation functions were sigmoid and hyperbolic tangent. These functions have many advantages including saturation, smooth non-decreasing character and continuous-differentiability. However, their partial derivative is lower than 1. This directly affects learning by back-propagation, because the gradient may become vanishingly low in the hierarchical structure which stops learning. A very small gradient effectively prevents weights from modifying. A similar mirrored effect is represented by exploding gradient. When the gradient is larger than 1, it can rapidly grow to big values and even a small weight update will make huge modification to weights' values. This also stops learning. For a long time, this represented a major problem which prevented the design of deeper neural networks and limited the neural networks to have only a single digit number of layers.

The efficient solution to this problem is using nonlinear activation functions with partial derivative equal to 1. The most used activation function of modern neural networks is rectified linear unit (ReLU) [59]. This activation function is also very easy to calculate. It has minimal computational requirements. The disadvantage is the positive shift of the mean values which leads to covariance shifts of all layers with ReLUs within the learning process. Continually shifting values are hard to optimize. The solution to combat this is to use batch normalization [38] whenever the ReLU is used. Batch normalization requires additional computations during learning and also during inference. Therefore, many attempts to improve ReLU produced many of its variations including: Leaky ReLU [54], Shifted ReLU [10], Saturated Leaky ReLU, Parametric ReLU [23], Exponential ReLU [10], and Saturated ReLU [72].

Exponential ReLU [10] is a promising ReLU variation which has positive partial derivation equal to 1, negative partial derivation significantly smaller than 1 and the mean values significantly closer to zero in comparison to ReLU. The authors experimentally proved that it does not need batch normalization and is slightly better than a combination of ReLU with batch normalization. The main disadvantage is its higher computational requirements. The exponential function is relatively difficult to calculate.

The selection of the right nonlinear activation function has direct consequences on the final precision, learning duration, learning computational requirements, but also on computational requirements during inference. The inference computational time is very important for applications in IoT where computational resources are limited.

2.5 Shortcut Connections

Modern neural networks have varying internal structures which may be very symmetric [73] or internally highly structured [77]. One of the structure components is the identity connection called the shortcut connection or shortcut. One of the first uses of the shortcut connections was by Ronneberger in 2015 [67] in the U-Net neural network design. It won the ISBI cell tracking challenge for segmentation of neuronal structures in electron microscopic stacks in 2015. The segmentation task is peculiar to the precision on the pixel level. The most important are the border pixels that delineate the object boundary. These are also the most difficult to classify correctly. The segmentation task puts two opposed requirements on the network structure.

The first is the classification requirement. To correctly classify an object, there is a need to incorporate the overall context and as the result many neural networks for the classification task significantly reduce the output resolution to few pixels. The classification task benefits from translation invariance so it can correctly recognize objects placed anywhere in the input image. Few output pixels improve this by incorporating a lot of context information from large effective field of view. The real field of view is much smaller than theoretical field of view but still covers nearly the whole input image [48, 53] in modern neural network designs [8, 24, 32, 72, 77, 83].

The second is the localization requirement. The correct localization is hindered by translation invariance. It needs accurate position information to recognize if the pixel is within or out of the object boundary. Highly reduced output resolution is a problem for precise segmentation. It is almost impossible to correctly restore a precise boundary map from few output pixels alone. However, precise location can be predicted only if the correct object class is already known. The correct class also represents contextual information itself. Many objects have a typical shape and it can help to predict the boundary if a part of the object is missing or occluded. The precise boundary pixel recognition is easier from a higher resolution of early feature maps.

Authors of U-Net [67] used shortcuts to link higher resolution early feature maps to lower resolution higher feature maps to improve the restoration of the output boundary map. The higher feature maps contain a lot of contextual information which improves the classification process. When it is combined with lower feature maps with more spatial information, it is possible to correctly restore the resolution while keeping higher level contextual knowledge. The resulting neural network structure resembles an hourglass and these kind of networks are widely known as hourglass networks. More information about U-Net is given in the section Architectures of modern neural networks.

The shortcut connections have a great benefit in improving the convergence of the learning process. Wherever the shortcuts are used, the flow of information is improved as well. The unhindered flow of information is necessary in a forward process to calculate output predictions, and it is also necessary in a backward process of the gradient back-propagation to modify the weights. This is the main reason why modern very deep neural networks use it regularly. There are special neural network blocks particularly designed to fully utilize shortcut connections. The most known and widely used are residual block [24], residual inception block [79], dense block [32], aggregated residual block [83], dual path block [8], Squeeze-and-Excitation (SE) residual block [28], inverted residual block [72] and the ShuffleNet unit [87] which can be called shuffle residual block. The last two are especially designed for low power mobile devices and IoT. More information about these blocks is given in the section Architectures of modern neural networks.

The introduction of the shortcut connection in a standard building block of neural networks [24] resulted in very deep neural networks. The authors in [24] successfully trained a 1202 layer deep residual network that converges with only slightly worse, about absolute 1.5% output accuracy in comparison to the best 110 layer deep residual network. The authors in [31] found a way to successfully train such deep networks and further improve state-of-the-art. They improve the original 1202 layer deep network by absolute 3% which corresponds to a relative improvement of 27%. Their method introduces an additional shortcut connections that bypass multiple residual blocks and are controlled on stochastic principle. During training, these shortcuts are randomly opened and closed within the whole network to push the network to learn more general characteristics and to be less dependent on one layer. An alternative description is that the method tries to learn an internal ensemble of multiple networks. It has a strong regularization character that is applied on the whole depth of the network. The randomness works similarly as the dropout regularization [74].

2.6 Transfer Learning

The progress in neural network design is very fast paced. Many different factors help to accelerate it further including transfer learning, open accessed scientific works, easily accessible models of trained neural networks, accessible datasets for training neural networks, open challenges in many different scientific fields, social networks, and continually growing amount of digital data produced by companies and common people every year. It is easy to understand how most of them directly influence the progress of neural network design and its practical applications. Transfer learning has a special place there and we describe it in more detail.

Transfer learning represents the ability of any trained neural network to be used for other tasks different from the original task which it was trained for [35]. Given enough expertise, anyone can take a neural network with all its learned weights and specified hyper-parameters, retrain it for the new task with new data and use it. This significantly reduces the time needed to design a neural network solution for a new problem. Sometimes there is no need to retrain it at all, and sometimes minor modifications are needed, such as modifying output dimensionality to accommodate for different number of object classes, and then the retraining is necessary.

Transfer learning also includes reusing parts of trained networks and combining these parts together into a new neural network or simply combining whole networks together into a neural network ensemble. An ensemble of neural networks is a well proven technique used to improve precision at the cost of great computational complexity increase. It is usually used to push absolute results further and be placed high in open challenges leaderboards [13, 45]. However, a very high computational cost limits its practical application.

Reusing neural network parts and combining them into new networks is widely used with many practical applications. Any neural network can be divided into two main parts: the feature extractor also known as the backbone, and the classifier also known as the head. The classification task has usually very simple head composed of very few FC layers and a softmax layer. However, it has a very sophisticated backbone which was trained on huge and diverse datasets. This results in much better output features that are able to describe more abstract characteristics and better generalize [55, 76]. Using the already trained backbone is a very well proven technique to improve precision of more complex tasks and it is known as pre-training [55, 76].

More complex tasks regularly use parts of the proven design of less complex tasks. An example is the instance segmentation, a very difficult image recognition task, which combines the semantic segmentation and detection tasks into one. Many neural network designs for instance segmentation [5, 26, 51] reuse parts of the proven classification head design of neural networks for detection [66] or semantic segmentation [4, 6, 52] tasks.

3 Architectures of Modern Neural Networks

This section depicts selected neural network architectures that significantly improved state-of-the-art. The internal architectures are briefly described with all their advantages. The novel ideas and all major architecture improvements are described in more detail. Some experimental achievements are also included.

3.1 LeNet-5

LeNet-5 is not a modern neural network architecture. However, it has its place here due to several historical reasons. LeNet-5 is composed of 7 layers, 5 convolutional layers and 2 fully connected layers [44], see Fig. 1. It is one of the first convolutional neural networks successfully applied and commercially used for image recognition tasks in the real world. Also, it is one of the first architectures showing that a deeper neural network architecture is more powerful than a wider network architecture. It was specially designed for image recognition tasks and with very limited computational resources in mind. This is similar to very smart sensors today, except LeNet-5 ran on high performance computational hardware and the present smart sensors are executed on small low-power computational hardware. Huge computational saving is accomplished by using convolutional layers due to local connectivity and parameter sharing. The additional benefit is translation invariance as the authors originally intended.

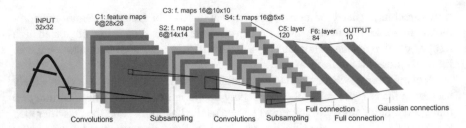

Fig. 1 Architecture of LeNet-5 [44]. Each plane is a feature map with a single shared kernel

3.2 AlexNet

AlexNet is the neural network that restarted interest in supervised learning of deep neural networks and started a new trend of applying deep neural networks to practical problems of the real world. The main reason behind this is the novel use of graphics processing units (GPUs) to calculate all necessary processes of the forward and backward information flow. Krizhevsky et al. used two high performance desktop GPUs GTX 580 3 GB to train a deep neural network with 60 million parameters [41]. This was novel work which included writing a new highly-optimized GPU implementation code of the 2D convolution and all the other operations necessary for training convolutional neural networks. They used the CUDA environment to write the code and made the source code publicly available. This started the interest in using high performance GPUs to train neural networks. This trend has been monitored by GPU designers and the result is that the modern GPU architecture has specialized computational cores called tensor cores to efficiently compute tensor multiplication and addition operations [56] which are the most used operations in neural network computations.

This trend continues also for low power devices. Only recently, a new generation of processors for mobile devices from Apple and Huawei with special computation cores for efficient computation of neural networks has been presented and introduced to new mobile phones. Apple's A11 Bionic and A12 Bionic and Huawei's Kirin 970 and Kirin 980 mobile processor chips are embedded with Neural network Processing Units (NPUs). This trend will definitely continue which will bring more hybrid hardware architectures with capabilities of efficient neural network computations for IoT.

AlexNet became a standard benchmark and it is included in most of neural network frameworks such as Caffe [39], TensorFlow [1], Mxnet [7], PyTorch and others. It is composed of 5 convolutional layers, 3 pooling layers, 3 FC layers and a softmax layer, see Fig. 2. It was originally designed to compete in the ImageNet Large-Scale Visual recognition Challenge (ILSVRC) [13] where it won the 1st place in 2012 by a large margin. AlexNet ensemble reached 15.3% top-5 error rate and the second best entry which uses classical computer vision design of classifiers with Fisher Vectors [71] reached 26.2% top-5 error rate.

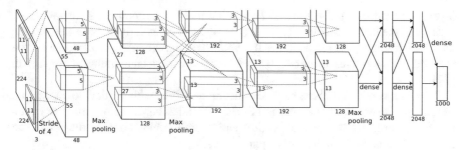

Fig. 2 Architecture of AlexNet [41]. Resolutions of feature maps and kernels are placed near corresponding edges. Numbers of feature maps are placed under layers

AlexNet is significant in two additional aspects. The first is a novel use of ReLU activation function [59] in a very deep convolutional neural network for improving the learning process. The work presents a clear advantage of using ReLU which reduces the time necessary to reach 25% training error rate to 1/7 in comparison to hyperbolic tangent. The authors also added dropout regularization [74] to reduce overfitting. They only used it before the first two FC layers. Dropout, first introduced in [29], reduces complex co-adaptations of neurons by introducing stochastic behavior during the learning process. Dropout randomly keeps or disables connections of the selected layer with a probability specified beforehand. This forces the network to learn more robust features because each neuron cannot rely on a particular neuron from the previous layer alone. Instead, each neuron has to find multiple solutions by combining more neurons from the previous layer.

The second is the use of grouped convolutions also known as aggregated convolutions [83]. The authors use it due to memory and cross-GPU communication limitations. To improve communications, they spread the network across two GPUs by dividing the layers into two groups while limiting some interconnections to communicate only within the part of the network stored in single GPU memory, see Fig. 2. This means that some of the groups are interconnected with only one of the two groups from the previous layer. The group convolutions are proven to be more efficient in comparison to a basic convolution [83] which will be discussed in more detail later.

3.3 VGGNet

VGG neural network is introduced by Simonyan et al. [73]. The authors introduce 5 different versions ranging from a 11 layer deep neural network with 133 million parameters to a 19 layer deep neural network with 144 million parameters. The best of them, the 16 layer deep neural network with 138 million parameters, became a standard benchmark and is included in many neural network frameworks. It is composed of 5 blocks of convolutional layers followed by a maxpooling layer, 3 FC layers and a softmax layer. The first 2 convolutional blocks comprise 2 convolutional layers and the last 3 convolutional blocks comprise 4 convolutional layers. It was the best performing single model in the ILSVRC 2013 classification challenge where it reached 7.0% top-5 test error rate, 0.9% better in comparison to a single GoogLeNet [77].

VGGNet is one of the first very deep neural networks with the double digit depth of trainable layers (convolutional and FC layers). It has a symmetric structure which improves hardware resource allocation and overall computational efficiency. The disadvantage is the high number of parameters.

VGGNet is special in using only 3×3 convolutions or 1×1 convolutions and 1×1 convolutions are used by only one VGGNet version. The 3×3 convolutions have very small receptive field of only 9 pixels. However, two 3×3 convolutions in a sequence have an effective receptive field of $5 \times 5 = 25$ pixels. Three 3×3 convolutions in a sequence have an effective receptive field of $7 \times 7 = 49$ pixels. The advantage is using more nonlinear activation functions (ReLUs) placed after each convolution which makes the decision function more discriminative. The second advantage is the reduced number of parameters. The sequence of three 3×3 convolutions has $3 \times (3 \times 3 \times C \times C) = 27C^2$ parameters assuming all layers have C input channels and C output channels. A single 7×7 convolution with the same effective receptive field and input and output channels has $7 \times 7 \times C \times C = 49C^2$ parameters which is 81% more parameters. This can also be seen as a separable convolution which uses computational decomposition into more layers. Using only 3×3 or 1×1 convolutions became a standard in all modern neural networks.

3.4 GoogLeNet

GoogLeNet has been introduced by Szegedy et al. [77] in nearly the same time as VGGNet. It is a very successful neural network with slightly less than 6.8 million parameters which is approximately 12 times less than AlexNet and yet the ensemble of 7 GoogLeNets won IVLSRC 2013 with 6.7% top-5 error rate. It is composed of 22 trainable layers which are internally highly irregular.

GoogLeNet has been designed with low parameter count as a goal. To reach it, the internal structure has to be designed more efficiently by being more sparsely connected. The authors designed a new neural network building block called inception module, see Fig. 3. It uses multiple parallel dataflow paths represented by different convolutions. The authors use a set of convolutions with the different receptive field that are applied to the previous layer in parallel. The convolution with a larger receptive field has fewer filters; it is thinner to help reduce the number of parameters. To reduce parameters further, the number of channels from the previous layer has to be reduced. This is more critical for convolutions with a larger receptive field. The authors solved it by introducing an additional 1×1 convolution into the dataflow path placed before the convolution with a larger receptive field. They chose to use multiple 1×1 convolutions to better control the number of input channels. Convolutions with a larger receptive field have more reduced number of input channels.

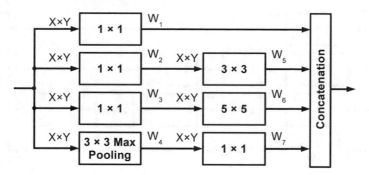

Fig. 3 Inception module

The resulting inception block has 4 parallel dataflow paths. One is single layer deep and contains a 1×1 convolution. Other three are two layer deep. Two of them start with 1×1 convolutions to reduce the input dimension and continue with 3×3 or 5×5 convolutions, respectively. The last dataflow path starts with a 3×3 max-pooling layer and continues with a 1×1 convolution. All four dataflow paths are concatenated at the end of the inception block.

This design has several advantages. The first is a significant parameters count reduction and therefore a strong reduction of computational requirements. The second is the different effective depth of the overall neural network structure. There is a lot of dataflow paths with the effective depth of 22 trainable layers which is important for stronger recognition abilities. However, there are faster paths with half the depth to back propagate gradients which is important for better and faster learning. The authors also experimented with a single linear classification layer placed before the final softmax layer which not only reduced parameters but improved the results for the ImageNet challenge.

GoogLeNet has been improved several times [78, 79]. The first improved version with Inception-v2 [78] utilizes separable convolutions, see Figs. 4, 5 and 6. Similarly as VGGNet, it replaces a 5×5 convolution by a sequence of two 3×3 convolutions.

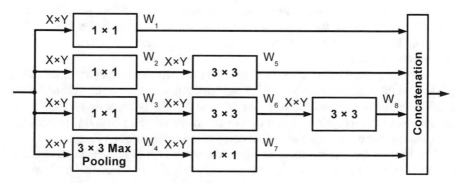

Fig. 4 Inception-v2 block type A

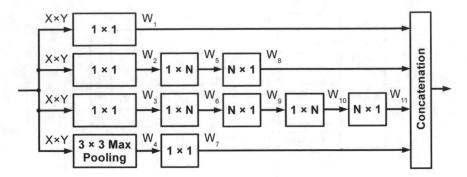

Fig. 5 Inception-v2 block type B

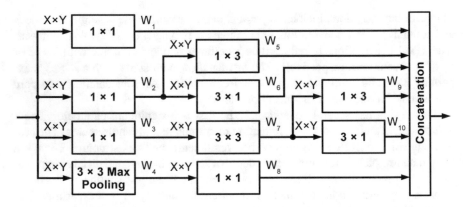

Fig. 6 Inception-v2 block type C

However, Inception-v2 goes further. It replaces an N × N convolution by a sequence of an N × 1 convolution and a 1 × N convolution which has the same effective receptive field. This is again a computational decomposition into more layers known as factorization of an N × N convolution into a sequence of two N × 1 and 1 × N convolutions. Inception-v2 uses many different versions of inception blocks in the different depths of the neural network, e.g., in the middle section N = 7. The inception blocks are optimized for the feature map grid size and position within the neural network structure.

Further improved Inception-v2 with auxiliary classifiers with batch normalization [38] called Inception-v3 reached 4.2% top-5 error rate with a single model and 3.58% top-5 error rate with an ensemble of 4 models and won the 1st place in ILSVRC 2014. GoogLeNet with Inception-v3 is a deeper and more complex neural network with 3.2 times higher computational cost in comparison to the original GoogLeNet.

The next major improvement of Inception block [79] is introducing a shortcut connection to the Inception block in the form of residual connections [24]. It is called Inception-ResNet block and again it has several variants, see Figs. 7 and 8. The single

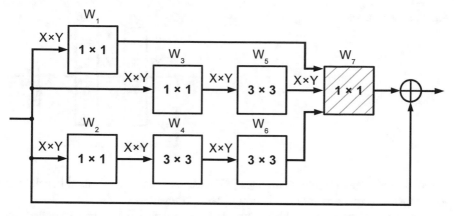

Fig. 7 Inception-ResNet block type A. The diagonally hatched texture indicates layers with no non-linearities

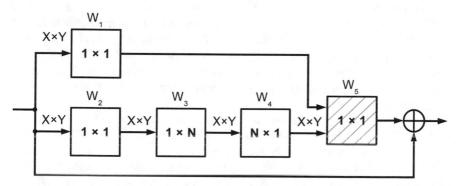

Fig. 8 Inception-ResNet block type B and C. The diagonally hatched texture indicates layers with no non-linearities

model and an ensemble of 4 GoogLeNets with Inception-ResNet-v2 reached 3.7% and 3.1% top-5 error rate on the ILSVRC 2012 validation dataset, respectively.

3.5 ResNet

ResNet has been introduced by He et al. [24]. It is a very successful neural network with many variations in the number of trainable layers, layers' width and overall number of parameters. A 152 layer deep ResNet won the 1st place in ILSVRC 2015 in both detection and localization tasks. It reached 4.49% top-5 error rate of a single model and 3.59% top-5 error rate of an ensemble of 6 models which won the 1st place in the ILSVRC 2015 main classification task. It also won the 1st place in COCO 2015

Fig. 9 Residual block

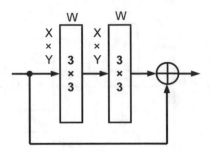

detection and segmentation challenges [45]. ResNet became a standard benchmark and is included in many modern neural network frameworks.

The main advantages of ResNet are its basic novel building blocks called the residual block and bottleneck residual block. The residual block is a combination of a shortcut connection and a sequence of two 3×3 convolutions, see Fig. 9. The shortcut connection and the second 3×3 convolution outputs are added at the end of the block. Using addition to combine two pathways is the characteristic feature of residual blocks. It has several advantages. The first is the same number of channels at the block input and output which makes it easier to combine multiple residual blocks into a sequence. This also reduces the number of hyper-parameters. The residual blocks have a small number of hyper-parameters in general which makes it easier to design new neural networks.

The second advantage is the shortcut connection which improves both forward and backward data flow. The authors proved it by a successful training of a 1202 layer deep neural network that is only slightly worse in comparison to an optimal 110 layer deep network trained on the CIFAR-10 dataset [40]. The successful training of the 1202 layer deep network is partly done due to introducing batch normalization [38] after each convolution and before activation function. Batch normalization significantly speeds up training and combats overfitting.

The third advantage is the internal structure which facilitates filter reuse. The residual block output is represented by the addition of the block input and a small few layer deep neural network output which simplifies the learning process. The small network only has to learn a residual function which slightly improves the output of the previous residual block or layer. This is the reason why the block is named residual block. Learning a residual function is easier in comparison to learning the whole improved function. A good example is an extension of the classification set. If there is a neural network trained on the original classification set and we want to add a new block to accommodate increased complexity of the classification task with higher number of objects under interest, the residual function has to learn to recognize only the new object types that have been added. The original object types are recognized by the original network and propagated to the output by the shortcut connection. The residual function just has to be zero when the original object types are recognized.

The bottleneck residual block introduces a dimension reduction to reduce the number of parameters similarly as GoogLeNet [77]. A sequence of two 3×3

Fig. 10 Bottleneck residual block

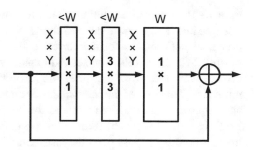

convolutions is replaced by a three layer deep sequence of 1×1, 3×3 and 1×1 convolutions, see Fig. 10. The first 1×1 convolution reduces the number of input channels. The last 1×1 convolution restores the number of channels to match it with the input. This allows to build deeper neural networks while keeping the number of parameters similar.

Improved ResNet is introduced in [25]. It introduces full pre-activated residual blocks. The original residual blocks use final ReLU activation function placed after the addition. This means that the shortcut connections of two following blocks are separated by a nonlinear activation function. Any nonlinearity negatively influences the information flow within the network. Placing the activation function with the batch normalization to the residual path before the first convolutional layer makes the shortcut connection clear and purely linear. This type of shortcut connection is called identity connection because there is no element that changes its input value. This results in greatly improved information flow in both directions that positively influences learning and its speed. The effect is verified with a 1001 layer deep ResNet with pre-activated residual blocks that reaches better precision than any less deep ResNet and much better accuracy than any original ResNet on the CIFAR-10 dataset. The new 200 layer deep ResNet is introduced which reaches a single model single crop 4.8% top-5 error rate on the ILSVRC 2012 validation dataset.

The inverted residual block is introduced by Sandler et al. [72]. The idea is to propagate a reduced number of channels through the identity connections and let the channel expansion be kept only within the residual path. This significantly reduces the number of parameters. It will be discussed in more detail in the section Computational complexity reducing techniques of modern neural networks.

3.6 U-Net

U-Net is introduced by Ronneberger et al. [67]. It won the ISBI cell tracking challenge for segmentation of neuronal structures in electron microscopic stacks in 2015. The network is specially designed for the visual segmentation task with a focus on the accurate border pixel recognition. Therefore, the network needs the output with high resolution. A successful segmentation task requires correct classification which is improved by reduced resolution of later feature maps with a lot of included contextual information. The authors divided the U-Net into 2 parts. The first contracting part processes data mainly for the correct classification and it has an output resolution of only 28 × 28 pixels. The second expansive part with nearly identical but order-inverted structure slowly restores the resolution step by step to nearly the original input size. This makes the internal structure look like an hourglass and many similar and improved versions are known as the hourglass type of neural networks.

To improve the localization precision of restored feature maps, authors link the first restored layer data from the last feature map with the similar resolution of the contracting part. Two different parts of the network are connected by identity connections. There are 4 identity connections in total as there are 4 maxpooling operations each reducing the resolution by a half and similarly 4 upsampling convolutions [52] each restoring the resolution to the double size.

3.7 DenseNet

Densely connected convolutional neural network (DenseNet) is introduced by Huang et al. [32]. It proposes a new type of neural network building block called dense block which is composed of a sequence of convolutional layers where each layer is connected with all previous feature maps within the dense block, see Fig. 11. This creates a densely interconnected structure with many identity connections that link the layer with feature maps on many different depths. The identity connections greatly improve the convergence rate and the overall learning speed. They also improve feature reuse similarly as within the residual blocks. However, the dense block has other greatly important advantage which strongly encourages creating new filters [8]. The access to many feature maps from different depths allows direct filter reuse and their new combination within a layer. The result is a new filter set that better assists the following layers. Because the following layers are connected with all previous feature maps within the block, there is no need to make similar feature maps and each layer is pushed to create new more advanced filters that better suit the main recognition task. This is partially confirmed by its overall number of parameters.

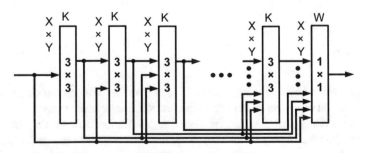

Fig. 11 Dense block

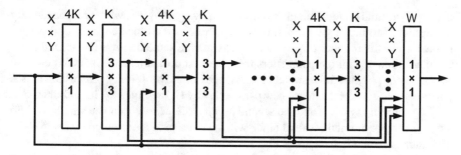

Fig. 12 Bottleneck dense block

DenseNets need less than half the number of parameters in comparison to ResNets with the same precision evaluated on the ILSVRC 2012 validation dataset [32].

The dense block introduces a new parameter k called the growth rate. It is a number that represents how many new filters are created by a layer. Due to the direct access to all previous filters within the block which promotes filter reuse, the growth rate with small value is sufficient. A typical value is 12, 24 or 32 for ImageNet optimized networks. The lth layer of the dense block has $k_0 + k \times (l - 1)$ input feature maps. Dense blocks are separated by a 1×1 convolution followed by a pooling layer reducing the spatial resolution.

The authors also introduced a dense block with bottleneck layers, see Fig. 12. Inspired by the bottleneck residual block, they transformed each convolution layer within a dense block into a sequence of 1×1 and 3×3 convolutions. The 1×1 convolution reduces the number of input channels to 4 times of the growth rate. This allows to use higher growth rates, improves the overall efficiency and reduces the number of parameters. A 100 layer deep DenseNet with 12 growth rate has 7 million parameters in comparison to 0.8 million parameters of a 100 layer deep DenseNet with bottleneck layers, 12 growth rate and only slightly worse precision. Both networks are evaluated on the CIFAR-10 and CIFAR-100 datasets [40]. Memory efficient implementation of DenseNet is presented in [61].

3.8 PolyNet

PolyNet is introduced by Zhang et al. [86]. It proposes a new PolyInception family of neural network building blocks that can be flexibly inserted into a neural network as a replacement for any neural network part. The advantage of the new modules is in the increased internal structure diversity. The rising neural network depth results in increasingly smaller returns. The growing neural network width results in the quadratic rise of the number of parameters. The authors explore the diversity of neural network structure as a new dimension for increasing the performance of neural networks. They generalize the inception residual blocks [79] via various polynomial compositions. If the residual path of the residual block is denoted as F, the second order polynomial residual block is represented by $I + F + F^2$, where I is the identity connection, see Fig. 13. This significantly increases the block capacity and performance while the number of parameters remains the same because the second order element is reusing all the parameters from the first order element. A more powerful second order polynomial residual block is represented by $I + F + GF$ which reuses parameters of F and introduces new parameters by G, see Fig. 14. The authors called the first version as *poly-2* and the second as *mpoly-2*. *Mpoly-2* can be simplified into a first order polynomial residual block with an extra path represented by $I + F + G$. The authors called it 2-*way*, see Fig. 15.

The proposed PolyInception family is a general scheme which includes higher order blocks. *Poly-3* is $I + F + F^2 + F^3$, *mpoly-3* is $I + F + GF + FGH$ and 3-way is $I + F + G + H$. Experimental results verified that the best precision is achieved by using a mixed solution of all third order blocks in the interleaving repeated pattern: 3-*way*, *mpoly-3*, *poly-3*. The proposed very deep PolyNet reaches a single model single crop 4.25% top-5 error rate on the ILSVRC 2012 validation dataset and a single model multi crop 3.45% top-5 error rate on the ILSVRC 2012 test dataset.

Fig. 13 Poly-2 block

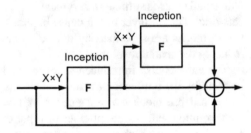

Fig. 14 Mpoly-2 block

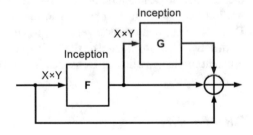

Fig. 15 2-way block

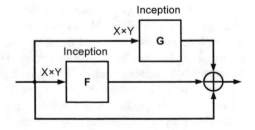

Fig. 16 ResNeXt block. The horizontally hatched texture indicates grouped convolutions

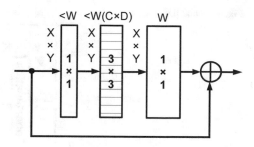

3.9 ResNeXt

ResNeXt is introduced by Xie et al. [83]. It introduces a new type of the aggregated residual unit that uses grouped convolutions, see Fig. 16. The grouped convolution is a set of low-dimensional 3×3 convolutions instead of one 3×3 convolution. The idea is to use low-dimensional embedding of higher number of channels into few number of channels, then compute a set of 3×3 convolutions and after that restore the number of channels by concatenating the results of the whole set. The number of 3×3 convolutions within a set represents a new dimension of the layer and the authors called it cardinality. The higher cardinality is represented by more 3×3 convolutions with fewer channels that are more efficient to calculate because they have much less parameters in total. Therefore, the number of all filters within a 3×3 convolutional set can be much higher while keeping the overall number of parameters the same. Higher cardinality means wider layer with the similar number of parameters. A wider layer is more representative which is the reason why the aggregated residual block is more powerful in comparison to a standard residual block. The authors call it ResNeXt block. They experimentally verified that embedding into 4 channels is the most efficient.

A new 101 layer deep ResNeXt is introduced which reaches a single model single crop 4.4% top-5 error rate on the ILSVRC 2012 validation dataset. This is 0.4% better result in comparison to a 200 layer deep ResNet. Both networks have similar number of parameters and computational complexity.

3.10 DualPathNet

Dual path neural network (DPN) is introduced by Chen et al. [8]. The authors propose a new neural network building block that utilizes both residual connections and dense connections. They wanted to combine both strong advantages of ResNets' feature reuse and DenseNets' new features creation into a single neural network structure. This also eliminates ResNet and DenseNet weaknesses. ResNets have a great information sharing strategy due to residual connections which makes it difficult to explore new features. DenseNets have a great new feature exploring

Fig. 17 Dual path block

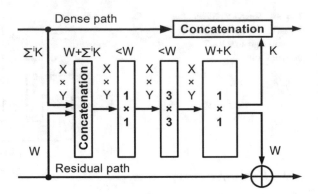

potential which leads to redundancy because the layers are densely interconnected only within a dense block and there is no direct way to access earlier feature maps.

The DPN block is composed of a bottleneck styled sequence of 1×1, 3×3 and 1×1 convolutions where the last 1×1 convolutional results are split into two parts, see Fig. 17. The first part is element-wisely added to the residual path and the second part is concatenated with the densely connected path. The middle 3×3 convolutional layer uses grouped convolutions as ResNeXt [83] to improve learning capacity. The result is a more efficient neural network structure.

92 layer deep DPN-92 (32×3d) with 6.5 GFLOPs computational complexity reaches a single model single crop 5.4% top-5 error rate on the ILSVRC 2012 validation dataset. In comparison, 101 layer deep ResNeXt101 (32×4d) with 8.0 GFLOPs reaches a single model single crop 5.6% top-5 error rate only. DPN-131 (40×4d) with 16 GFLOPs reaches a single model single crop 5.12% error rate. In comparison, ResNeXt101 (64×4d) with 15.5 GFLOPs reaches a single model single crop 5.3% top-5 error rate only.

4   Classification Accuracy of Modern Neural Networks

The most common way to evaluate the quality of neural networks found in recent literature is to compare their accuracy for the image classification task on the ImageNet validation dataset. Such results are readily available in most papers. Following this scenario, in Table 1 we summarize the error rates (in %) achieved by various modern networks described in the previous section.

More precisely, we report error rates for AlexNet [41], VGG [73], GoogLeNet [77] with its improvements Batch-Normalized Inception v1 [38], Inception v2 and v3 [78], Inception v4 and Inception-ResNet v2 [78], ResNet [24] and its improvement with pre-activated residual blocks [25], DenseNet [32], PolyNet [86], ResNeXt [83] and DualPathNet [8]. The results are sorted by the top-5 error rate and by top-1 if top-5 data is not available in a descending manner (less rate is better).

Table 1 ImageNet validation error rates for various modern neural networks. *BN:* batch-normalized; *BOT:* bottleneck architecture; *PS:* projection shortcuts; *C:* cardinality; *D:* bottleneck width; ① reported by [86]; ② reported by [83]; ③ reported by [25]; ④ reported by [8]

Network	Type	#	Image size	Crop size	# crp	top-1 error	top-5 error	Additional information
AlexNet	–	1	256	224	10	40.70	18.20	60 M
AlexNet	–	5	256	224	10	38.10	16.40	–
GoogLeNet	Incep. v1	1	256–352	224	1	–	10.07	6.8 M
GoogLeNet	Incep. v1	1	256–352	224	10	–	9.15	6.8 M
ResNet	18	1	224–640	224	10	27.88	–	–
VGG	19	1	256	224	10	27.30	9.00	144 M
VGG	16	1	256	224	10	27.00	8.80	138 M
VGG	19	1	224–288	224	30	26.90	8.70	144 M
VGG	16	1	224–288	224	30	26.60	8.60	138 M
VGG	16	1	352–416	224	30	26.50	8.60	138 M
VGG	19	1	352–416	224	30	26.70	8.60	144 M
VGG	16	1	384	224	10	25.60	8.10	138 M
GoogLeNet	Incep. v1	7	256–352	224	1	–	8.09	–
VGG	19	1	384	224	10	25.50	8.00	144 M
GoogLeNet	Incep. v1	1	256–352	224	144	–	7.89	6.8 M
Inception	BN-v1	1	>224	224	1	25.20	7.82	13.6 M
ResNet	34	1	224–640	224	10	25.03	7.76	PS
DenseNet	121	1	224–640	224	1	25.02	7.71	–
GoogLeNet	Incep. v1	7	256–352	224	10	–	7.62	–
VGG	16	1	256–512	224	30	24.80	7.50	138 M
VGG	16	1	256–512	224	50	24.60	7.50	138 M
VGG	19	1	256–512	224	30	24.80	7.50	144 M
VGG	19	1	256–512	224	50	24.60	7.40	144 M
VGG	16	1	256–512	224	150	24.40	7.20	138 M
VGG	–	2	256–512	224	50	23.90	7.20	–
VGG	19	1	256–512	224	150	24.40	7.10	144 M
VGG	–	2	256–512	224	30	24.00	7.10	–
DenseNet	169	1	224–640	224	1	23.80	6.85	28.9 M
VGG	–	2	256–512	224	150	23.70	6.80	–
ResNeXt	50	1	256	224	1	23.00	–	$C \times D = 2 \times 40d$
ResNet	50	1	224–640	224	10	22.85	6.71	25.5 M ②, PS
GoogLeNet	Incep. v1	7	256–352	224	144	–	6.67	–

(continued)

Table 1 (continued)

Network	Type	#	Image size	Crop size	# crp	top-1 error	top-5 error	Additional information
DenseNet	121	1	224–640	224	10	23.61	6.66	–
ResNeXt	50	1	256	224	1	22.60	–	C × D = 4 × 24d
DenseNet	201	1	224–640	224	1	22.58	6.34	–
ResNeXt	50	1	256	224	1	22.30	–	C × D = 8 × 14d
ResNeXt	50	1	256	224	1	22.20	–	C × D = 32 × 4d
ResNet	152	1	–	224	1	22.16	6.16	①
DenseNet	264	1	224–640	224	1	22.15	6.12	–
ResNet	101	1	224–640	224	10	21.75	6.05	PS
ResNet	200	1	–	320	1	21.80	6.00	③
DenseNet	169	1	224–640	224	10	22.08	5.92	–
Inception	BN-v1	1	>224	224	144	21.99	5.82	13.6 M
ResNeXt	101	1	256	224	1	21.70	–	C × D = 2 × 40d
Inception	v2	1	299	224	1	21.60	5.80	–
ResNet	152	1	224–640	224	10	21.43	5.71	PS
Inception	v3	1	299	224	1	21.20	5.60	<25 M
ResNet	34	1	224–640	224	10	21.53	5.60	BOT
ResNeXt	101	1	256	224	1	21.40	–	C × D = 4 × 24d
ResNeXt	101	1	256	224	1	21.30	–	C × D = 8 × 14d
ResNeXt	101	1	256	224	1	21.20	5.60	44.3 M ④, C × D = 32 × 4d
ResNet	152	1	–	224	1	20.93	5.54	①
DenseNet	201	1	224–640	224	10	21.46	5.54	–
ResNet	152	1	–	320	1	21.30	5.50	③
ResNet pre-act	152	1	256–352	320	1	21.10	5.50	–
ResNeXt	101	1	256	224	1	20.70	5.50	C × D = 2 × 64d
DualPathNet	DPN-92	1	–	224	1	20.70	5.40	37.8 M
ResNeXt	101	1	256	224	1	20.40	5.30	83.7 M ④, C × D = 4 × 64d

(continued)

Table 1 (continued)

Network	Type	#	Image size	Crop size	# crp	top-1 error	top-5 error	Additional information
DenseNet	264	1	224–640	224	10	20.80	5.29	–
ResNet	50	1	224–640	224	10	20.74	5.25	25.5 M ②, BOT
DualPathNet	DPN-98	1	–	224	1	20.20	5.20	61.7 M
DualPathNet	DPN-131	1	–	224	1	19.93	5.12	79.5 M
Inception-ResNet	v2	1	–	224	1	20.05	5.05	①
Inception	v4	1	299	224	1	20.00	5.00	–
Inception-ResNet	v2	1	–	224	1	19.90	4.90	133 M ①
ResNet	269	1	–	224	1	19.78	4.89	①
Inception	BN-v1	6	>224	224	144	20.10	4.82	–
ResNet pre-act	200	1	256–352	320	1	20.10	4.80	–
ResNet	500	1	–	224	1	19.66	4.78	①
DualPathNet	DPN-92	1	–	320	1	19.30	4.70	37.8 M
ResNet	101	1	224–640	224	10	19.87	4.60	BOT
ResNet	152	1	224–640	224	10	19.38	4.49	BOT
Inception	v3	1	299	224	12	19.47	4.48	<25 M
ResNeXt	101	1	256	320	1	19.10	4.40	83.7 M ④, $C \times D = 4 \times 64d$
DualPathNet	DPN-98	1	–	320	1	18.90	4.40	61.7 M
PolyNet	–	1	299	224	1	18.71	4.25	92 M
DualPathNet	DPN-131	1	–	320	1	18.62	4.23	79.5 M
Inception	v3	1	299	224	144	18.77	4.20	<25 M
Inception	v4	1	299	224	12	18.70	4.20	–
Inception-ResNet	v2	1	–	224	12	18.70	4.10	133 M ①
ResNet	152	1	–	224	10	18.50	3.97	①
Inception	v4	1	299	224	144	17.70	3.80	–
Inception-ResNet	v2	1	–	224	144	17.80	3.70	133 M ①
ResNet	500	1	–	224	10	17.59	3.63	①
Inception	v3	4	299	224	144	17.20	3.58	–
ResNet	269	1	–	224	10	17.54	3.55	①
PolyNet	–	1	299	224	10	17.36	3.45	92 M
Inception	v4	4	299	224	144	16.50	3.10	–
Inception-ResNet	v2	4	–	224	144	16.50	3.10	133 M ①

In Table 1, network names are listed in the first column. Those are the names by which the networks could be best recognized in the community. The second column further specifies the particular type of the network. This is typically specified by the number of layers or the network version number. The third column specifies whether the result was achieved by an ensemble of networks (value > 1) or by a single network (value = 1). The fourth column lists the sizes of the shorter side of test images used during test time. In cases when multiple sizes were used, we provide the interval ranges. The fifth column specifies the test image square crop size. The number of test crops used per test image is in the sixth column. The seventh and eight columns provide the actual top-1 and top-5 error rates, respectively. The last column provides complementary information about the network, such as the number of learnable parameters in millions (M), also with references in case this number was estimated by other teams, additional information about the network architecture, or a reference in cases when the results were reported by other teams by training the network from scratch themselves. For ResNeXt networks, the architecture is denoted in the following format: $C \times D$, where 'C' is cardinality and 'D' is bottleneck width.

A few trends can be observed from Table 1. Currently the best performing network types in terms of raw classification accuracy seem to be very deep ResNet variants, Inception networks and their combinations. These often use multi-crop evaluation and an ensemble of networks at times. However, the improved accuracy comes at the cost of fairly high number of parameters.

It is very hard to do a fair comparison of computational efficiencies of neural network architectures. Ideally, the trained models should be compared on the same accuracy metric (top-1 or top-5 error rate) and then the computational efficiency be represented by the time the model needs to classify the same amount of test images on the same hardware setup. This is impossible, because nearly each model has different accuracy, see Table 1, and every research group uses different hardware setup. However, there are a few ways to approximately compare computational complexity. They are based on the evaluation of the number of parameters or the number of all multiplication and accumulate operations (in general, the number of floating-point operations in millions, i.e., MFLOPS) with respect to the accuracy (top-1 or top-5 error rate). The lower number of parameters or floating-point operations indicate the more computationally efficient model when the accuracies of models under comparison are close. This is only an approximate evaluation because the complexity of the internal model dataflow can significantly influence the real computational duration. Very symmetric models with large layers, e.g., VGG, are computed faster in comparison to highly asymmetric models with large number of layers, e.g., Inception family, even if the numbers of parameters are similar. The difficulty of improving accuracy is monotonically rising with the accuracy value. Therefore, a model with higher accuracy and lower parameter count (or lower MFLOPS) can be evaluated as being more computationally efficient. Based on this, we can see from Table 1 that the DualPath models and models from Inception family are computationally more efficient.

5 Computational Complexity Reducing Techniques of Modern Neural Networks

Modern neural network architectures described in the previous section have been designed to reach high precision and to push the state-of-the-art further. The disadvantage is a high computational cost which is too high for most practical applications. Many applications are limited by available computational resources, e.g., mobile applications, limited by maximal latency, e.g., control industrial applications, or both, e.g., autonomous driving.

There are several ways to improve the computational efficiency of neural networks. The first is modifying the neural network structure. The second is reducing the precision of weights, activations or both. The third is modifying the computational hardware to be more optimized for neural networks.

5.1 Reducing the Depth and Width of Neural Networks

Reducing the depth and width of neural networks are the standard techniques used to reduce the number of parameters and therefore the computational complexity. Reducing the number of filters within a layer is more efficient because the number of parameters is growing with quadratic dependency on the linearly increased number of filters. Reducing the number of layers lowers parameters linearly. These techniques have a limitation because the reduced network loses some of its accuracy and a very thin network may not have enough capacity to reach the required performance.

There are many reduced networks that are successfully applied in many applications, e.g., ResNet-18, ResNet-36, ResNet-50.

5.2 Reducing the Resolution of the Input

Reducing the input resolution keeps the parameters the same but it has a very strong effect on the computational complexity which is directly affected by the number of pixels. This technique has a limitation because it affects the accuracy and some architectures may have internal limits, e.g., FC layers may have to be modified. The best suited architectures are fully convolutional neural networks [52], because they produce output resolution maps directly proportional to the input resolution and work with different input resolutions without any modification.

5.3 More Efficient Building Blocks of Neural Networks

Using more efficient building blocks reduces the number of parameters while keeping network capacity the same. Great examples are the bottleneck styled blocks [24], blocks with grouped convolutions [83], PolyInception modules [86] or DPN blocks [8].

5.4 Separable Convolution

Using separable convolutions also known as spatially separable convolutions is another way to reduce the number of parameters while keeping network capacity the same. Using a sequence of 3×3 convolutions instead of a convolution with a larger receptive field is more efficient [73]. Similarly, using a sequence of 2 one-dimensional convolutions with the second rotated by $90°$ is more efficient than a two-dimensional convolution [78].

5.5 Depth-Wise Separable Convolution

Depth-wise separable convolution presented in [9] is a more general version of separable convolutions where the depth-wise convolution is followed by a point-wise convolution. The depth-wise convolution is a spatial convolution performed independently over each input channel which is very computationally efficient, see Fig. 18. The point-wise convolution, known from the bottleneck structure [24], projects the

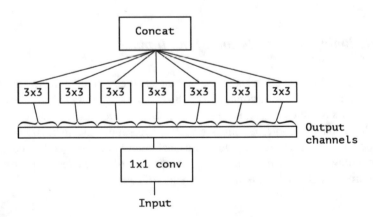

Fig. 18 Point-wise convolution and depth-wise convolution shown on the extreme version of Inception module [9]. The depth-wise separable convolution differs only in the execution order where the point-wise convolutions follows the depth-wise convolution

channels' outputs of the depth-wise convolution onto new channel space. The point-wise convolution is more computationally intensive in comparison to depth-wise convolution due to a high number of channels in later layers. The depth-wise separable convolution is $1/N + 1/D_K^2$ times less computationally intensive in comparison to the standard $D_K \times D_K$ convolution with N output channels. The standard 3×3 convolution is between 8 and 9 times more computationally expensive. Using depth-wise separable convolution leads to a much more efficient neural network. The real parameter reduction is about 4 times because modern networks use many 1×1 convolutions in bottleneck structures.

The depth-wise separable convolutions slightly reduce the overall precision [30] when the layer width is kept the same. The reduced parameters can be used to increase the layer width which increases the capacity and overall performance [9].

5.6 Pruning

Pruning is a technique used to reduce the number of parameters by discarding parameters with the least impact on the overall precision [27]. The parameters with values close to zero are thrown away. This results in a sparse matrix of parameters. The actually achieved computational reduction is much less than theoretical because the most of computational hardware is optimized for an efficient computation of dense matrices with very limited support for sparse matrices.

Pruning alone reduces the output accuracy. Most of neural networks have many parameters and the capacity to successfully perform the task even after significant pruning. To recover the lost accuracy, a neural network has to be retrained with the remaining parameters [20].

The most efficient pruning is done within many iterations where the small pruning is followed by weight retraining. The iterative process limits the damage caused by reducing the parameters. Retraining recovers the lost accuracy more easily when the loss is small. Many small iterations can discover minimal number of parameters without the loss of accuracy. Iterative pruning is from 5 to 9 times more efficient without the loss of accuracy in comparison to one-step aggressive pruning [20]. The most important parameters for keeping are found by greedy search during each iteration.

5.7 Compression of Parameters

The compression of parameters is a technique used to reduce the memory space taken up by learned weights. Any data compression scheme can be used, e.g., Huffman coding. It is usually used after pruning. Pruning creates sparse matrices with a lot of zero values which take up the same amount of space as non-zero parameters. The compression of parameters primarily minimizes the space taken by zero values.

More efficient compression techniques optimize also the non-zero parameters by introducing a parameter sharing scheme. Weights with very close values are substituted by one common value placed within a lookup table and the encoded position within the lookup table is used as a substituting index for the original values [21]. Substituting indexes are encoded by Huffman coding into few bits which results in large memory savings. The number of different valued parameters is significantly reduced which makes it possible to design an efficient scheme that uses lookup tables to compute only non-zero parameters and skips the rest. The result is an efficient but highly complex computation scheme. To use it without the loss of accuracy, retraining is necessary to compensate for the reduced number of parameters' values. The best results are produced by many iterations with small steps [21].

5.8   Reduced Floating-Point Precision

The state-of-the-art neural networks use 32 bit floating-point precision known as single precision for storing the weights and also for all necessary computations. This leads to best results. Modern neural networks have many parameters which makes them flexible to reduced precision schemes. The key idea is to retrain a neural network with reduced parameters which recovers the most of lost accuracy. If only the weights are transformed into lower floating-point precision, the memory requirements are reduced but computational complexity remains the same. Modern professional GPUs have support for 16 bit floating-point precision known as half-precision with quadrupled efficiency in comparison to single precision. Transforming all weights into half precision and using half precision in all computations leads to 4 times improved computational speed on the same hardware. It is achieved only with the full support for half-precision operations. Some modern GPUs have limited support for half precision meaning that the half precision operations will be computed at the same or even slower rate in comparison to single precision operations, e.g., some desktop or mobile GPUs.

Google introduced tensor precision which also uses 16 bits to encode the values but uses the same 8-bit exponent encoding as single precision and therefore it uses only a 7-bit wide mantissa encoding in comparison to a 10-bit wide mantissa encoding of half precision. Tensor precision is used in Google's Tensor Processing Units (TPUs) hardware accelerators of neural network computing.

5.9   Quantification of Weights into Integer or Binary Values

Significant memory saving and computation complexity reduction can be achieved by using a more aggressive quantization scheme for neural network weights. Any quantization scheme, e.g., uniform [34, 89] or logarithmic [58], can be used. Memory saving is straightforward, the weights just occupy less memory. The computational complexity reduction becomes significant with shorter bit-widths where the multiplications are transformed into series of additions and shifts. The shift with a floating-point value is represented by increasing or decreasing the exponent. The extreme quantization into a single bit omits the multiplication altogether. Binary weights are usually using $\{-1, 1\}$ values [63, 85]. Any quantization leads to accuracy loss. However, retraining of weights after quantization can recover almost all lost accuracy even when the weights are represented by very few bits [85]. The incremental quantization scheme has better results [89]. Training with quantized weights is not trivial because the quantization function is often non-differentiable and gradients have to be approximated.

The full potential of only weights quantization into integers can be utilized in custom designed hardware, e.g., FPGA. Modern professional GPUs have support for efficient integer multiplications.

5.10 Quantification of Both Weights and Activations into Integer or Binary Values

Quantization of activations requires the quantization scheme to be embedded within the neural network because it has to work on-line with any input data. The quantization of activations degrades the accuracy more intensely than the weights quantization because the approximation of input distribution is much more difficult. Therefore, the quantization of activations is done together with the weights quantization. This leads to very high memory savings and computational complexity reduction. Usually, the quantization of activations is done after the first layer where there is enough information transformed into multiple channels which limits the accuracy loss. The extreme quantization into binary values of both activations and weights transforms all mathematical operations within a layer into a combination of an exclusive-not-or known as *xnor* and 1's counter within a string known as popcounting or *popcnt* [63]. It offers 58 times computational speed-up on CPUs and an option of CPU-only inference. The cost is a significant accuracy reduction.

When the quantizations of both the activations and weights are learned jointly with a neural network, the accuracy loss is much smaller [85]. The proposed method allows arbitrary bit-precision of activations and weights and is easy to train. Another method to reach high precision results with both the binary activations and weights is proposed in [91]. The authors proposed a novel network decomposition scheme called Group-Net where the network is divided into full precision groups and each group is approximated by a set of homogeneous binary branches. The effective connections between groups are learned to improve representational capability. The trade-off between the overall accuracy and computation complexity is controlled by the number of binary branches called bases. Group-Nets with 5 bases reached very high accuracy close to full precision models.

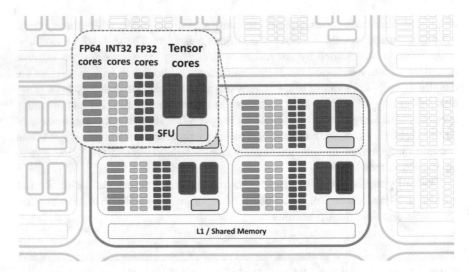

Fig. 19 Simplified diagram of the Volta SM architecture. The NVIDIA Tesla V100 uses 80 SMs [56]

5.11 More Efficient Computational Hardware

The progress of modern computational devices is fast and it slowly converges with the neural network requirements. Modern GPUs include tensor cores for efficient computations of neural networks [56] with varying precision, see Fig. 19. New mobile processors such as Apple's A11 Bionic and A12 Bionic and Huawei's Kirin 970 and Kirin 980 include neural processing units. Many small processors have an embedded GPU that can be used to accelerate neural network computations.

5.12 Custom Hardware Design

Custom hardware design is the best way to maximize the computational efficiency of neural network inference [22]. The best platform is Field Programmable Gate Array (FPGA) due to its inherent ability to repeatedly modify its own internal structure. It can simply be used to accelerate different neural networks after reprogramming. The disadvantage is the necessity to have knowledge to design computational structures in hardware.

6 Neural Networks for Mobile Devices and IoT

This section depicts few modern neural network architectures oriented specifically for mobile and IoT applications. The selected architectures significantly pushed state-of-the-art in their application domain. The internal structure and advantages are described in more detail. Some interesting results are also pointed out.

6.1 SqueezeNet

The SqueezeNet neural network is introduced by Iandola et al. [37]. It is an AlexNet-level accuracy network with $50 \times$ fewer parameters and the model size of 0.48 MB. The original AlexNet has 80.3% top-5 ImageNet accuracy and the model size of 210 MB. SqueezeNet represents a 510 times model size reduction. This very impressive reduction is achieved by combining several techniques described in the previous section.

The network reduces the number of 3×3 convolutions where some are reduced into 1×1 convolutions. Some 3×3 convolutions are transformed into a sequence of 1×1 and a mix of 1×1 and 3×3 convolutions. The first reduces the number of input channels so the second mixed layer has less parameters. The network is downsampled later which keeps information mainly in feature maps and allows for less wide layers with less parameters. The authors call this sequence a fire module, see Figs. 20 and 21. This results in a deeper less wide neural network with more efficient building blocks.

Pruning reduced some layer interconnections. Mostly, the interconnections of 3×3 convolutions are reduced. This results in 4.8 MB model size and 50 times parameter reduction while still using single precision.

The parameters are quantized into 6 bits using deep compression [21] which combines repeated quantization with retraining, Huffman coding and lookup tables.

Fig. 20 Fire module

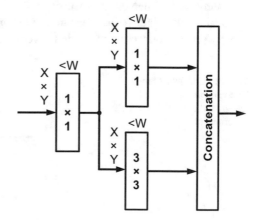

Fig. 21 Residual fire
module

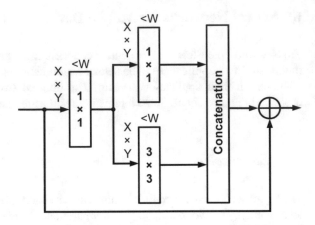

This results in a 510 times model size reduction while keeping 80.3% top-5 ImageNet
accuracy.

6.2 MobileNet

MobileNet neural network is introduced by Howard et al. [30]. It is oriented for
mobile applications. It uses depth-wise separable convolutions [9] which reduces
the number of parameters multiple times while the ImageNet accuracy drop is only
1% in comparison to the standard convolution, see Fig. 22. The network is also thinner
and uses reduced input resolution. This results in a network with about 4% better
accuracy in comparison to AlexNet while it is also 45 times smaller in model size and
9.4 times less computationally intensive. Using different multiplier to the number
of filters within layers and different input resolution, many variations of MobileNet
are created. The variations differ in the accuracy, model size and computational
complexity.

Improved MobileNet-v2 is introduced by Sandler et al. [72]. It uses inverted
residual blocks which reduce the number of parameters by connecting thinner layers
as opposed to connecting expanded layers, see Fig. 23. This directly impacts the

Fig. 22 MobileNet block.
The vertically hatched
texture indicates depth-wise
separable convolutions

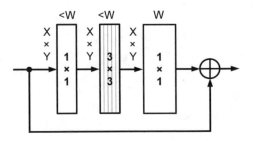

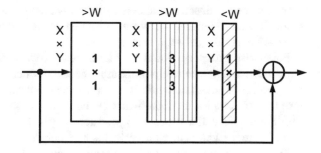

Fig. 23 Inverted residual block. The vertically hatched texture indicates depth-wise separable convolutions and the diagonally hatched texture indicates layers with no non-linearities

width of point-wise convolutions in the bottleneck residual block. These thinner layers do not use any non-linearity which improves the overall accuracy and training speed. The inverted bottleneck residual block starts by a point-wise expanding convolution followed by a depth-wise separable convolution and ending with a linear point-wise convolution. The expanding layer uses higher-dimensional embedding before the point-wise convolution which limits the information loss and motivates to create new filters. The resulting neural network has lower number of parameters and computational complexity while keeping the accuracy the same as the original.

6.3 ShuffleNet

ShuffleNet neural network is introduced by Zhang et al. [87]. It is oriented for mobile and IoT applications. ShuffleNet improves MobileNet [30] and specifically targets applications with very strict computational requirements, i.e., under 150 MFLOPs.

MobileNet reduces the number of parameters by depth-wise separable convolutions. However, the point-wise convolutions are not optimized and they consume a lot of parameters. ShuffleNet introduces point-wise group convolutions to reduce the complexity of point-wise convolutions, see Fig. 24. The simple division of channels within all 1×1 convolutions into several groups leads to communication only within a group which limits data flow between the groups and results in weak feature representations. To remedy this, ShuffleNet introduces a shuffle layer which changes the order of all channels within a layer so the channel group in the next layer is

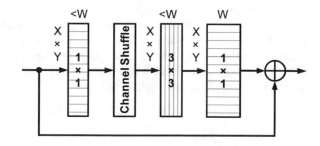

Fig. 24 ShuffleNet residual block. The horizontally hatched texture indicates grouped convolutions and the vertically hatched texture indicates depth-wise separable convolutions

interconnected with some channels from all channel groups from the previous layer. A shuffle layer is placed before all grouped convolutions. The result is a greatly improved data flow between channels.

Point-wise group convolutions significantly reduce the number of parameters while the capacity is kept approximately the same which is important for representational power of neural networks. Overly thin neural networks rapidly lose the ability to represent complex characteristics which is in turn mirrored in the fast accuracy drop. ShuffleNet is able to produce good accuracy in the 10–150 MFLOPs complexity range where it significantly outperforms MobileNets.

ShuffleNet is computationally efficient because the cuda-convnet CNN library [42] supports the random sparse convolution layer which is equivalent to the random channel shuffle followed by a group convolutional layer. Due to low computational complexity, ShuffleNet can be run on small ARM based processors with an acceptable frame-rate and can be utilized for mobile and IoT applications.

6.4 LQ-Nets

LQ-Nets are neural networks with quantized weights and activations with novel learnable quantizers proposed by Zhang et al. [85]. Many quantization schemes are using uniform [34, 89] or logarithmic quantizers [58] which are not optimal for all neural network layers. The distributions of weights and activations differ a lot within a neural network [85] and using optimal quantizers for each layer activations and filters leads to much better accuracy. The key is to minimize the quantization errors. This is accomplished by jointly training the network and quantizers together which allows to find optimal quantization levels and recover some of the accuracy lost due to the quantization by retraining the weights. The effect is magnified by the iterative process of learning. The neural network weights are learned by the standard back-propagation and the quantizers are learned by the new proposed algorithm called Quantization Error Minimization (QEM) in the forward pass. Using many quantizers for layers and filters introduces extra parameters, but their number is negligible compared to the large volume of network weights. The QEM algorithm is universal and can be used for any neural network. It is designed to maximize computational efficiency benefits of bitwise operations *xnor* and *popcnt*. The numbers of bits for the weights and activations are set separately and can be chosen by a user arbitrarily.

LQ-Nets quantized by the QEM algorithm significantly improve state-of-the-art neural networks with limited precision weights and activations. They have significantly higher accuracy when compared with the other quantization schemes using the same number of bits. There is only 1–2% absolute accuracy drop in comparison to the single precision neural network counterparts when 3 or 4 bits are used for the quantization of both weights and activations.

6.5 GroupNet

GroupNet is introduced by Zhuang et al. [91]. It is a binary neural network with binary weights and activations that approximates the single precision by multiple binary branches organized into groups. The direct binarization leads to high error from quantization. The lower error is produced by layer-wise binary decomposition where each single precision convolution is decomposed into a linear combination of homogeneous binary bases represented by a group of binary branches. This layer-wise decomposition may still incur severe quantization errors and bring large gradient deviation due to the aggregation of multiple branches after each layer. The authors introduced a flexible decomposition strategy called group-wise binary decomposition. A group represents a whole part of the original neural network. The number of groups is set at the beginning. The original neural network is divided into that many groups where each group should be at least one residual block long. These groups are decomposed into a linear combination of binary branches with the same structure as the part of the original network, see Fig. 25.

There is a high number of possibilities how to divide and decompose a network optimally. The authors solved this by proposing a strategy that allows to learn the network internal decomposition dynamically. They introduced a fusion gate as a soft connection with learnable parameters between adjacent blocks, see Fig. 26. Its function can be described as gated adder which learns to better utilize binary branches or their aggregation into a better result for the next block. The effective connections between blocks are learned to improve representational capability. The trade-off between the overall accuracy and computational complexity is controlled by the number of binary branches called bases. GroupNets with a shortcut bypassing every binary convolution and 5 bases reached very high accuracy close to full precision models. They can be computed efficiently even on CPUs and are oriented for mobile or IoT applications.

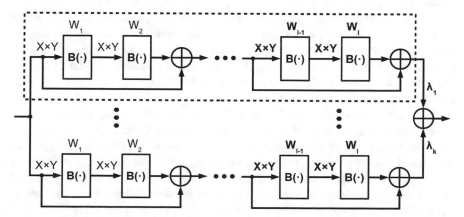

Fig. 25 Group-wise binary decomposition. A whole group composed of several blocks is decomposed into binary branches. B(•) represents a binarized convolution

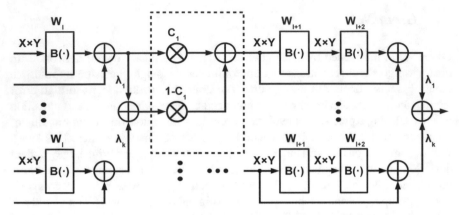

Fig. 26 Fusion strategy represented by a fusion gate marked by a dashed line

7 Classification Accuracy of Neural Networks for Mobile and IoT

To compare the quality of networks suited for mobile devices and IoT applications, we follow the same procedure as with modern networks in section Architectures of modern neural networks. Table 2 summarizes the error rates (in %) achieved by various networks described in the previous section on the ImageNet validation dataset for the image classification task.

More precisely, we report error rates for SqueezeNet [37], MobileNet v1 [30], MobileNet v2 [72], ShuffleNet [87], LQ-Net [85] and GroupNet [91]. The results are sorted by the top-1 error rate in a descending manner (less rate is better).

Table 2 was constructed in a similar fashion as Table 1, listing the names, types, and top-1 and top-5 error rates. In case of LQ-Nets and Group-Nets, the type column specifies the base network to which the proposed principles were applied. In the last column, we provide complementary information, including the number of learnable parameters in millions (M), complexity in terms of millions of floating-point operations (MFLOPs) and further network architecture details, such as the number of groups for ShuffleNets (denoted as g), the bit-widths of weights and activations for LQ-Nets and Group-Nets (denoted as w/a, where 'w' is weight bit-width and 'a' is activation bit-width), and the number of bases for Group-Nets (denoted as b).

A few conclusions can be drawn from Table 2. Clearly, in terms of raw classification accuracy, these types of networks lag behind larger models (see Table 1) by few percent, but still offer fairly high classification power. However, these networks have an advantage of much smaller parameter counts. Among the best performing networks, one can find a varying range of network architectures, currently mainly dominated by LQ-Nets with ResNet-50 as a base network and also some variants of Group-Nets with again ResNet-50 as a base network, a special variant of ShuffleNet with squeeze and excitation blocks and some MobileNet v2 variants. This further

Table 2 ImageNet validation error rates for various neural networks for mobile and IoT applications. *SE*: squeeze & excitation; *GBD*: group-wise binary decomposition; *g*: number of groups; *b*: number of bases; *w*: weight bit-width; *a*: activation bit-width

Network	Type	top-1 error	top-5 error	Additional information
ShuffleNet	0.25x	55.00	–	13 MFLOPs, g = 3
ShuffleNet	0.25x	54.20	–	13 MFLOPs, g = 4
ShuffleNet	0.25x	52.70	–	13 MFLOPs, g = 8
MobileNet v1	0.25 MNet-224	49.40	–	0.5 M, 41 MFLOPs
Group-Net	ResNet-18	44.40	21.40	b = 1
LQ-Net	AlexNet	44.30	21.20	w/a = 1/2
ShuffleNet	0.5x	43.20	–	38 MFLOPs, g = 3
Group-Net	AlexNet	42.70	19.90	GBD v1, w/a = 1/2
LQ-Net	AlexNet	42.60	19.90	w/a = 2/2
SqueezeNet	32 bit	42.50	19.70	1.25 M, 833 MFLOPs
SqueezeNet	8 bit	42.50	19.70	0.165 M, 833 MFLOPs
SqueezeNet	6 bit	42.50	19.70	0.118 M, 833 MFLOPs
ShuffleNet	0.5x	42.30	–	38 MFLOPs, g = 8
Group-Net	AlexNet	42.20	19.10	w/a = 1/2
ShuffleNet	0.5x	41.60	–	38 MFLOPs, g = 4
SqueezeNet	–	41.20	18.00	complex bypass
Group-Net	ResNet-18	40.80	17.70	b = 5, GBD v3
MobileNet v1	0.5 MNet-160	39.80	–	1.32 M
SqueezeNet	–	39.60	17.50	simple bypass
LQ-Net	AlexNet	39.50	17.30	w/a = 2/32
Group-Net	ResNet-18	38.50	16.80	b = 1, w/a = 1/4
Group-Net	ResNet-18	37.80	15.90	b = 5, GBD v2
Group-Net	ResNet-18	37.50	15.80	b = 3
LQ-Net	ResNet-18	37.40	15.70	w/a = 1/2
Group-Net	ResNet-18	37.00	15.20	b = 5, GBD v1
MobileNet v1	0.5 MNet-224	36.30	–	1.3 M, 149 MFLOPs
Group-Net	ResNet-18	35.80	14.40	b = 4
MobileNet v1	1.0 MNet-128	35.60	–	4.2 M
Group-Net	ResNet-18	35.50	15.00	b = 5, GBD v3, w/a = 1/4
Group-Net	ResNet-18	35.20	14.30	b = 5
LQ-Net	ResNet-18	35.10	14.10	w/a = 2/2
MobileNet v1	Shallow MNet	34.70	–	2.9 M
LQ-Net	GoogleNet	34.40	13.60	w/a = 1/2

(continued)

Table 2 (continued)

Network	Type	top-1 error	top-5 error	Additional information
Group-Net	ResNet-18	33.70	13.40	b = 4, shortcut bypass
LQ-Net	ResNet-34	33.40	13.10	w/a = 1/2
Group-Net	ResNet-18	33.00	12.50	b = 5, shortcut bypass
LQ-Net	VGG	32.90	12.40	w/a = 1/2
MobileNet v1	1.0 MNet-160	32.80	–	4.2 M
ShuffleNet	1x	32.80	–	140 MFLOPs, g = 4
ShuffleNet	1x	32.60	–	140 MFLOPs, g = 3
Group-Net	ResNet-18	32.50	12.00	b = 8
ShuffleNet	1x	32.40	–	140 MFLOPs, g = 8
LQ-Net	ResNet-18	32.00	12.00	w/a = 2/32
LQ-Net	ResNet-18	31.80	12.10	w/a = 3/3
LQ-Net	GoogleNet	31.80	11.90	w/a = 2/2
Group-Net	ResNet-18	31.70	12.10	b = 5, GBD v2, w/a = 1/4
MobileNet v1	0.75 MNet-224	31.60	–	2.6 M, 325 MFLOPs
Group-Net	ResNet-34	31.50	12.00	b = 5
Group-Net	ResNet-18	31.50	11.30	b = 3, w/a = 1/4
LQ-Net	ResNet-50	31.30	11.60	w/a = 1/2
LQ-Net	VGG	31.20	11.40	w/a = 2/2
MobileNet v1	1.0 MNet-192	30.90	–	4.2 M
Group-Net	ResNet-18	30.80	11.50	b = 5, GBD v1, w/a = 1/4
LQ-Net	ResNet-18	30.70	11.20	w/a = 3/32
LQ-Net	ResNet-18	30.70	11.20	w/a = 4/4
Group-Net	ResNet-50	30.50	10.80	b = 5
LQ-Net	DenseNet-121	30.40	10.90	w/a = 2/2
Group-Net	ResNet-18	30.40	11.00	b = 5, w/a = 1/2
LQ-Net	ResNet-34	30.20	10.90	w/a = 2/2
LQ-Net	ResNet-18	30.00	10.90	w/a = 4/32
Group-Net	ResNet-18	29.90	10.50	b = 5, w/a = 1/4
Group-Net	ResNet-18	29.60	10.20	b = 5, w/a = 1/32
Group-Net	ResNet-18	29.50	10.70	b = 5, shortcut bypass
MobileNet v1	1.0 MNet-224	29.40	–	4.2 M, 569 MFLOPs
ShuffleNet	1.5x	28.50	–	292 MFLOPs, g = 3
LQ-Net	ResNet-50	28.50	9.70	w/a = 2/2

(continued)

Table 2 (continued)

Network	Type	top-1 error	top-5 error	Additional information
MobileNet v1	Conv MNet	28.30	–	29.3 M
Group-Net	ResNet-34	28.20	9.60	b = 8
LQ-Net	ResNet-34	28.10	9.80	w/a = 3/3
MobileNet v2	–	28.00	–	3.4 M
Group-Net	ResNet-50	27.20	9.50	b = 8
ShuffleNet	2x	26.30	–	524 MFLOPs, g = 3
LQ-Net	ResNet-50	25.80	8.40	w/a = 3/3
Group-Net	ResNet-50	25.50	8.50	b = 5, w/a = 1/2
MobileNet v2	1.4	25.30	–	6.9 M
LQ-Net	ResNet-50	24.90	7.70	w/a = 2/32
LQ-Net	ResNet-50	24.90	7.60	w/a = 4/4
ShuffleNet	2x (with SE)	24.70	–	527 MFLOPs, g = 3
Group-Net	ResNet-50	24.00	7.30	b = 5, w/a = 1/4
LQ-Net	ResNet-50	23.60	6.90	w/a = 4/32

proves that the bit-width reduction of weights and activations in well-established networks such as ResNets, is a promising research direction.

The computational efficiency of mobile and IoT neural network architectures can be compared by using the same evaluation techniques as presented in section Classification accuracy of modern neural networks described after Table 1. Based on those evaluation techniques, we can see that ShuffleNet is a much more computationally efficient architecture in comparison to MobileNet v1. We can see this in the 140 MFLOPs range where ShuffleNet with 140 MFLOPs reaches 32.40% top-1 error rate and MobileNet v1 with 149 MFLOPS reaches only 36.2% top-1 error rate. It is also valid in the 300 MFLOP range where ShuffleNet with 292 MFLOPs reaches 28.5% top-1 error rate while MobileNet v1 with slightly worse top-1 error rate 29.4% requires 569 MFLOPs. MobileNet v2 models are much more computationally efficient in comparison to MobileNet v1. We can see that MobileNet v2 with 3.4 million parameters reaches 28% top-1 error rate while MobileNet v1 with 4.2 million parameters reaches only 29.4% top-1 error rate. Data in Table 2 suggest that the difference in a computational efficiency between ShuffleNet and MobileNet v2 is small and which of them is better may be dependent on the application and targeted computational device.

The quantized and binaryzed models including LQ-Nets and Group-Nets can be compared with floating-point models including ShuffleNets and MobileNets v2 on the same computational device only. Binaryzed models have the theoretical potential to accelerate the computation 64 times on CPUs; however, the true acceleration value is much smaller. Floating-point models can further reduce this advantage by using efficient GPUs or neural processing cores with combination of 16bit floating-point inference or inference using quantized weights. Both of these inferences are supported and efficiently accelerated by modern GPUs.

8 Conclusion

The chapter described modern neural network designs and discussed their advantages and disadvantages. Many modern techniques used to design more efficient neural networks were described in detail. By using these techniques, it is possible to create extremely efficient neural networks for mobile or even IoT applications by designing them from scratch or simply by modifying existing state-of-the-art models. Such models will make the applications very intelligent which opens a way for very smart sensors.

The designers of computational hardware are aware of neural network popularity and their massive potential. They are working on much more efficient computational cores for the acceleration of neutral network computing. Some results of their efforts can already be seen in modern computational devices including GPUs, mobile processors and special computational boards for autonomous driving and drones. This trend will continue and we will see much more highly efficient devices for neural networks in the future.

Designers of the upcoming neural network designs can focus more on highly computationally efficient neural network architectures instead of bending neural network designs for the current computational devices. History has many similar examples. The current computational devices always use highly optimized computational cores for decoding audio, video or encrypted data because it is much more energy efficient. To calculate it on a general computational core would waste power, generate more heat and limit battery life. This is exactly why there is no doubt we will see many more highly efficient computational devices for neural networks emerging in the future for all kinds of power, size or weight specifications.

Future research will continue to focus on the design of more efficient neural network architectures. MobileNets and ShuffleNets are early examples of neural networks specially designed for the very limited computational budget. Many state-of-the-art architectures are far from optimal. They have high capacity, i.e., the amount of all learnable weights, a large portion of which is used to overfit the training set. It can be seen in the high difference between accuracy on the testing and training sets. Few of the solutions how to improve it include the more efficient architectures, larger and more diverse training sets and better regularization techniques.

Another research area with high potential includes efficient quantization techniques and neural networks with low bit-width computations. Binaryzed neural networks have small memory requirements and also small requirements of computational resources and they can be potentially computed on ARM processing cores without any GPU. IoT applications can benefit from them greatly.

This chapter entirely focused on the feedforward neural network architectures. However, combining the benefits of the feedforward architectures with the advantages of recurrent architectures that utilize feedback connections can lead to novel neural network architectures and new design techniques that can push the state-of-the-art further.

Acknowledgements This work has been supported by Slovak national project VEGA 2/0155/19.

References

1. Abadi, M., Barham, P., Chen, J., Chen, Z., Davis, A., Dean, J., Devin, M., Ghemawat, S., Irving, G., Isard, M., et al.: TensorFlow: a system for large-scale machine learning. In: OSDI **16**(2016), pp. 265–283 (2016)
2. Brabandere, B., Neven, D., Gool, L.: Semantic instance segmentation with a discriminative loss function. In: CoRR (2017). arXiv:1708.02551v1
3. Braun, M., Krebs, S., Flohr, F., Gavrila, D.M.: The EuroCity persons dataset: a novel benchmark for object detection. In: CoRR (2018). arXiv:1805.07193v2
4. Chen, L.-C., Papandreou, G., Schroff, F., Adam, H.: Rethinking atrous convolution for semantic image segmentation. In: CoRR preprint (2017). arXiv:1706.05587
5. Chen, L.-C., Hermans, A., Papandreou, G., Schroff, F., Wang, P., Adam, H.: MaskLab: instance segmentation by refining object detection with semantic and direction features. In: CVPR, pp. 4013–4022 (2018). arXiv:1712.04837v1
6. Chen, L.C., Papandreou, G., Kokkinos, I., Murphy, K., Yuille, A.L.: Deeplab: semantic image segmentation with deep convolutional nets, atrous convolution, and fully connected crfs. IEEE TPAMI **40**(4), pp. 834–848 (2018)
7. Chen, T., Li, M., Li, Y., Lin, M., Wang, N., Wang, M., Xiao, T., Xu, B., Zhang, Ch., Zhang, Z.: Mxnet: a flexible and efficient machine learning library for heterogeneous distributed systems. In: CoRR (2015). arXiv:1512.01274
8. Chen, Y., Li, J., Xiao, H., Jin, X., Yan, S., Feng, J.: Dual path networks. In: NIPS (2017) arXiv:1707.01629v2
9. Chollet, F.: Xception: deep learning with depthwise separable convolutions. In: CVPR, pp. 1251–1258 (2017). arXiv:1610.02357v3
10. Clevert, D., Unterthiner, T., Hochreiter, S.: Fast and accurate deep network learning by exponential linear units (ELUs). In: ICLR (2016). arXiv:1511.07289v5
11. Cordts, M., Omran, M., Ramos, S., Rehfeld, T., Enzweiler, M., Benenson, R., Franke, U., Roth, S., Schiele. B.: The cityscapes dataset for semantic urban scene understanding. In: CVPR (2016)
12. Dai, J., Li, Y., He, K., Sun, J.: R-FCN: object detection via region-based fully convolutional networks. In: NIPS, pp. 379–387 (2016). arXiv:1605.06409v2
13. Deng, J., Dong, W., Socher, R., Li, L.J., Li, K., Fei-Fei, L.: Imagenet: a large-scale hierarchical image database. In: CVPR (2009)
14. Denker, J.S., Gardner, W.R., Graf, H.P, Henderson, D., Howard, R.E., Hubbard, W., Jackel, L.D., Baird, H.S., Guyon I.: Neural Network Recognizer for Hand-Written Zip Code Digits. AT&T Bell Laboratories (1989)
15. Everingham, M., Van Gool, L., Williams, C.K., Winn, J., Zisserman, A.: The PASCAL Visual Object Classes (VOC) challenge. In: IJCV **88**(2), pp. 303–338 (2010)
16. Everingham, M., Eslami, S.M.A., Gool, L.V., Williams, Ch.K.I., Winn, J., Zisserma, A.: The pascal visual object classes challenge a retrospective. In: IJCV **111**(1), pp. 98–136 (2014)
17. Fukushima, K.: Neocognitron: a self-organizing neural network model for a mechanism of pattern recognition unaffected by shift in position. In: Biol. Cybern. **36**(4), pp. 193–202 (1980). https://doi.org/10.1007/bf00344251

18. Girshick, R., Donahue, J., Darrell, T., Malik, J.: Rich feature hierarchies for accurate object detection and semantic segmentation. In: CVPR (2014). arXiv:1311.2524v5
19. Girshick, R.: Fast R-CNN. In: ICCV (2015)
20. Han, S., Pool, J., Tran, J., Dally, W.: Learning both weights and connections for efficient neural network. In: Proceedings of Advanced in NIPS, Montreal, Canada, pp. 1135–1143 (2015). arXiv:1506.02626v3
21. Han, S., Mao, H., Dally, W.J.: Deep compression: compressing deep neural networks with pruning, trained quantization and Huffman coding. In: ICLR (2016). arXiv:1510.00149v5
22. Han, S., Liu, X., Mao, H., Pu, J., Pedram, A., Horowitz, M.A., Dally, W.J.: EIE: efficient inference engine on compressed deep neural network (2016). arXiv:1602.01528v1
23. He, K., Zhang, X., Ren, S., Sun, J.: Delving deep into rectifiers: surpassing human-level performance on ImageNet classification. In: ICCV (2015). arXiv:1502.01852v1
24. He, K., Zhang, X., Ren, S., Sun, J.: Deep residual learning for image recognition. In: CVPR (2016), pp. 770–778. arXiv:1512.03385v1
25. He, K., Zhang, X., Ren, S., Sun, J.: Identity mappings in deep residual networks. In: ECCV (2016), pp. 630–645. arXiv:1603.05027v3
26. He, K., Gkioxari, G., Dollar, P., Girshick, R.: Mask R-CNN. In: ICCV (2017). arXiv:1703.06870v3
27. He, Y., Zhang, X., Sun, J.: Channel pruning for accelerating very deep neural networks. In: Proceedings of IEEE ICCV, vol. 2, pp. 1389–1397 (2017)
28. Hu, J., Shen, L., Sun, G.: Squeeze-and-excitation networks. In: CVPR (2018), pp. 7132–7141. arXiv:1709.01507v2
29. Hinton, G.E., Srivastava, N., Krizhevsky, A., Sutskever, I., Salakhutdinov, R.R.: Improving neural networks by preventing co-adaptation of feature detectors (2012). arXiv:1207.0580
30. Howard, A.G., Zhu, M., Chen, B., Kalenichenko, D., Wang, W., Weyand, T., Andreetto, M., Adam, H.: MobileNets: efficient convolutional neural networks for mobile vision applications. In: CoRR (2017). arXiv:1704.04861v1
31. Huang, G., Sun, Y., Liuy, Z., Sedra, D., Weinberger, K.Q.: Deep networks with stochastic depth. In: ECCV (4), vol. 9908 of Lecture Notes in Computer Science, pp. 646–661 (2016). arXiv:1603.09382v3
32. Huang, G., Liu, Z., van der Maaten, L., Weinberger, K.Q.: Densely connected convolutional networks. In: CVPR (2017). arXiv:1608.06993v5
33. Huang, J., Rathod, V., Sun, C., Zhu, M., Korattikara, A., Fathi, A., Fischer, I., Wojna, Z., Song, Y., Guadarrama, S., Murphy, K.: Speed/accuracy trade-offs for modern convolutional object detectors. In: CVPR (2017). arXiv:1611.10012v3
34. Hubara, I., Courbariaux, M., Soudry, D., El-Yaniv, R., Bengio, Y.: Quantized neural networks: training neural networks with low precision weights and activations. In: CoRR (2016). arXiv:1609.07061
35. Huh, M., Agrawal, P., Efros, A.: What makes ImageNet good for transfer learning? In: CoRR (2016). arXiv:1608.08614
36. Fu, C.-Y., Liu, W., Ranga, A., Tyagi, A., Berg, A.C.: DSSD: Deconvolutional single shot detector (2016). arXiv:1701.06659
37. Iandola, F.N., Moskewicz, M.W., Ashraf, K., Han, S., Dally, W.J., Keutzer, K.: SqueezeNet: AlexNet-level accuracy with 50x fewer parameters and <1 MB model size. In: CoRR (2016). arXiv:1602.07360v4
38. Ioffe, S., Szegedy, C.: Batch normalization: accelerating deep network training by reducing internal covariate shift. In: ICML, Lille, France, pp. 448–456 (2015)
39. Jia, Y., Shelhamer, E., Donahue, J., Karayev, S., Long, J., Girshick, R.B., Guadarrama, S., Darrell, T.: Caffe: convolutional architecture for fast feature embedding. In: Proceedings of ACM International Conference on Multimedia, pp. 675–678 (2014). arXiv:1408.5093
40. Krizhevsky, A.: Learning multiple layers of features from tiny images. Technical Report, University of Toronto (2009)
41. Krizhevsky, A., Sutskever, I., Hinton, G.E.: Imagenet classification with deep convolutional neural networks. Adv. Neural Inf. Process. Syst. 1097–1105 (2012)

42. Krizhevsky, A.: cuda-convnet: High-performance c++/cuda implementation of convolutional neural networks (2012)
43. LeCun, Y., Boser, B., Denker, J.S., Henderson, D., Howard, R.E., Hubbard, W., Jackel, L.D.: Backpropagation applied to handwritten zip code recognition. Neural Comput. 1(4), pp. 541–551 (1989)
44. LeCun, Y., Bottou, L., Bengio, Y., Haffner, P.: Gradient-based learning applied to document recognition. Proc. IEEE 86(11), pp. 2278–2324 (1998)
45. Lin, T.-Y., Maire, M., Belongie, S., Hays, J., Perona, P., Ramanan, D., Dollar, P., Zitnick, C.L.: Microsoft COCO: Common objects in context. In: ECCV (2014). arXiv:1405.0312v3
46. Lin, T.-Y., Dollar, P., Girshick, R., He, K., Hariharan, B., Belongie, S.: Feature pyramid networks for object detection. In: CVPR (2017). arXiv:1612.03144v2
47. Lin, T.-Y., Goyal, P., Girshick, R., He, K., Dollar, P.: Focal loss for dense object detection. In: ICCV (2017). arXiv:1708.02002v2
48. Liu, W., Rabinovich, A., Berg, A.C.: ParseNet: looking wider to see better. In: ICLR (2016). arXiv:1506.04579v2
49. Liu, W., Anguelov, D., Erhan, D., Szegedy, C., Reed, S.: SSD: single shot multibox detector. In: ECCV (2016)
50. Liu, S., Jia, J., Fidler, S., Urtasun, R.: SGN: sequential grouping networks for instance segmentation. In: ICCV, pp. 3516–3524 (2017)
51. Liu, S., Qi, L., Qin, H., Shi, J., Jia, J.: Path aggregation network for instance segmentation. In: CVPR, pp. 8759–8768 (2018). arXiv:1803.01534v1
52. Long, J., Shelhamer, E., Darrell, T.: 'Fully convolutional networks for semantic segmentation'. In: CVPR, pp. 3431–3440 (2015). arXiv:1411.4038v2
53. Luo, W., Li, Y., Urtasun, R., Zemel, R.L.: Understanding the effective receptive field in deep convolutional neural networks. In: CoRR (2017). arXiv:1701.04128v2
54. Maas, A.L., Hannun, A.Y., Ng, A.Y.: Rectifier nonlinearities improve neural network acoustic models. In: ICML (2013)
55. Mahajan, D., Girshick, R., Ramanathan, V., He, K., Paluri, M., Li, Y., Bharambe, A., Maaten, L.: Exploring the limits of weakly supervised pretraining. In: ECCV (2), vol. 11206 of Lecture Notes in Computer Science, pp. 185–201 (2018). arXiv:1805.00932v1
56. Markidis, S., Chien, S.W.D., Laure, E., Peng, I.B., Vetter, J.S.: NVIDIA tensor core programmability, performance & precision. In: IEEE IPDPS Workshops, pp. 522–531 (2018). arXiv:1803.04014v1
57. Mavromoustakis, C.X., Batalla, J.M., Mastorakis, G., Markakis, E., Pallis, E.: Socially oriented edge computing for energy awareness in IoT architectures. IEEE Commun. Mag. 56(7), pp. 139–145 (2018)
58. Miyashita, D., Lee, E.H., Murmann, B.: Convolutional neural networks using logarithmic data representation. In: CoRR (2016). arXiv:1603.01025
59. Nair, V., Hinton, G.E.: Rectified linear units improve restricted Boltzmann machines. In: Proceedings of the 27th ICML, pp. 807–814 (2010)
60. Neuhold, G., Ollmann, T., Bulo, S.R., Kontschieder, P.: The mapillary vistas dataset for semantic understanding of street scenes. In: ICCV (2017)
61. Pleiss, G., Chen, D., Huang, G., Li, T., Maaten, L., Weinberger, K.Q.: Memory-efficient implementation of DenseNets. In: CoRR (2017). arXiv:1707.06990v1
62. Peng, C., Zhang, X., Yu, G., Luo, G., Sun, J.: 'Large kernel matters—improve semantic segmentation by global convolutional network. In: CVPR, pp. 1743–1751 (2017). arXiv:1703.02719v1
63. Rastegari, M., Ordonez, V., Redmon, J., Farhadi, A.: XNOR-Net: imagenet classification using binary convolutional neural networks. In: Leibe, B., Matas, J., Sebe, N., Welling, M., (eds.) ECCV, vol. 4, pp. 525–542 (2016)
64. Redmon, J., Divvala, S., Girshick, R., Farhadi, A.: You only look once: unified, real-time object detection. In: CVPR, pp. 779–788 (2016). arXiv:1506.02640v5
65. Redmon, J., Farhadi, A.: YOLO9000: better, faster, stronger. In: CVPR (2017). arXiv:1612.08242v1

66. Ren, S., He, K., Girshick, R., Sun, J.: Faster R-CNN: towards real-time object detection with region proposal networks. In: NIPS, pp. 91–99 (2015)
67. Ronneberger, O., Fischer, P., Brox, T.: 'U-Net: Convolutional networks for biomedical image segmentation. In: MICCAI (2015). arXiv:1505.04597v1
68. Ruan, J., Hu, X., Huo, X., Shi, Y., Chan, F.T.S., Wang, X., Manogaran, G., Mastorakis, G., Mavromoustakis, C.X., Zhao, X.: An IoT-based E-business model of intelligent vegetable greenhouses and its key operations management issues. In: Neural Comput. Appl. **2019**, pp. 1–16 (2019)
69. Rumelhart, D.E., Hinton, G.E., Williams, R.J.: Learning representations by back-propagating errors. In: Nature **323**(6088), pp. 533–536 (1986)
70. Russakovsky, O., Deng, J., Su, H., Krause, J., Satheesh, S., Ma, S., Huang, Z., Karpathy, A., Khosla, A., Bernstein, M., Berg, A.C., Fei-Fei, L.: ImageNet large scale visual recognition challenge. In: IJCV **115**(3), pp. 211–252 (2015)
71. Sanchez, J., Perronnin, F., Mensink, T., Verbeek, J.: Image classification with the fisher vector: theory and practice. In: RR-8209, INRIA (2013)
72. Sandler, M., Howard, A.G., Zhu, M., Zhmoginov, A., Chen, L.C.: MobileNetV2: Inverted Residuals and Linear Bottlenecks. In: 'CVPR' IEEE Computer Society, pp. 4510–4520 (2018). arXiv:1801.04381v3
73. Simonyan, K., Zisserman, A.: Very deep convolutional networks for large-scale image recognition. In: ICLR (2015). arXiv:1409.1556v6
74. Srivastava, N., Hinton, G., Krizhevsky, A., Sutskever, I., Salakhutdinov, R.: Dropout: a simple way to prevent neural networks from overfitting. In: JMLR 15, pp. 1929–1958 (2014)
75. Srivastava, R.K., Greff, K., Schmidhuber, J.: Training very deep networks. In: NIPS (2015). arXiv:1507.06228v2
76. Sun, C., Shrivastava, A., Singh, S., Gupta, A.: Revisiting unreasonable effectiveness of data in deep learning era. In: Proceedings of ICCV (2017). arXiv:1707.02968v2
77. Szegedy, C., Liu, W., Jia, Y., Sermanet, P., Reed, S., Anguelov, D., Erhan, D., Vanhoucke, V., Rabinovich, A.: Going deeper with convolutions. In: CVPR, pp. 1–9 (2015). arXiv:1409.4842v1
78. Szegedy, C., Vanhoucke, V., Ioffe, S., Shlens, J., Wojna, Z.: Rethinking the inception architecture for computer vision. In: CVPR, pp. 2818–2826 (2016). arXiv:1512.00567v3
79. Szegedy, C., Ioffe, S., Vanhoucke, V.: Inception-v4, inception-resnet and the impact of residual connections on learning. In: ICLR Workshop, p. 12 (2016). arXiv:1602.07261v2
80. Taigman, Y., Yang, M., Ranzato, M., Wolf, L.: Deepface: closing the gap to human-level performance in face verification. In: CVPR (2014), pp. 1701–1708 (2014)
81. Uijlings, J.R., van de Sande, K.E., Gevers, T., Smeulders, A.W.: Selective search for object recognition. In: IJCV (2013)
82. Welinder, P., Branson, S., Mita, T., Wah, C., Schroff, F., Belongie, S., Perona, P.: Caltech-UCSD Birds 200. Technical Report, Caltech (2010)
83. Xie, S., Girshick, R., Dollar, P., Tu, Z., He, K.: Aggregated residual transformations for deep neural networks. In: CVPR, pp. 5987–5995 (2017). arXiv:1611.05431v2
84. Zeiler, M.D., Fergus, R.: Visualizing and understanding convolutional neural networks. In: ECCV (2014). arXiv:1311.2901v3
85. Zhang, D., Yang, J., Ye, D., Hua, G.: Lq-nets: learned quantization for highly accurate and compact deep neural networks. In: Proceedings of ECCV (2018). arXiv:1807.10029v1
86. Zhang, X., Li, Z., Loy, C., Lin, D.: Polynet: a pursuit of structural diversity in very deep networks. In: CVPR (2017). arXiv:1611.05725v2
87. Zhang, X., Zhou, X., Lin, M., Sun, J.: ShuffleNet: an extremely efficient convolutional neural network for mobile devices. In: CVPR (2018). arXiv:1707.01083v2
88. Zhao, H., Shi, J., Qi, X., Wang, X., Jia, J.: Pyramid scene parsing network. In: CVPR (2017). arXiv:1612.01105v2
89. Zhou, A., Yao, A., Guo, Y., Xu, L., Chen, Y.: Incremental network quantization: towards lossless cnns with low-precision weights. In: Proceedings of ICLR (2017)

90. Zhou, B., Lapedriza, A., Khosla, A., Oliva, A., Torralba, A.: Places: a 10 million image database for scene recognition. In: IEEE TPAMI (2017)
91. Zhuang, B., Shen, Ch., Tan, M., Liu, L., Reid, I.: Structured binary neural networks for accurate image classification and semantic segmentation. In: CoRR (2018). arXiv:1811.10413v2

Printed in the United States
by Baker & Taylor Publisher Services